Smart Technologies and Fundamental Rights

Value Inquiry Book Series

VOLUME 350

Philosophy and Human Rights

The titles published in this series are listed at *brill.com/vibs* and *brill.com/phur*

Smart Technologies and Fundamental Rights

Edited by

John-Stewart Gordon

BRILL
RODOPI

LEIDEN | BOSTON

Cover illustration: Photo by John-Stewart Gordon and used with permission.

Library of Congress Cataloging-in-Publication Data

Names: Gordon, John-Stewart, 1976- editor.
Title: Smart technologies and fundamental rights / edited by John-Stewart Gordon.
Description: Leiden ; Boston : Brill-Rodopi, 2021. | Series: Value inquiry book series, 2542-5153 ; volume 350 | Includes bibliographical references and index.
Identifiers: LCCN 2020030892 (print) | LCCN 2020030893 (ebook) | ISBN 9789004436732 (hardback) | ISBN 9789004437876 (ebook)
Subjects: LCSH: Artificial intelligence--Moral and ethical aspects. | Artificial intelligence--Social aspects.
Classification: LCC Q334.7 .S64 2021 (print) | LCC Q334.7 (ebook) | DDC 303.48/34--dc23
LC record available at https://lccn.loc.gov/2020030892
LC ebook record available at https://lccn.loc.gov/2020030893

Typeface for the Latin, Greek, and Cyrillic scripts: "Brill". See and download: brill.com/brill-typeface.

ISSN 2542-5153
ISBN 978-90-04-43673-2 (hardback)
ISBN 978-90-04-43787-6 (e-book)

This book is printed on acid-free paper and produced in a sustainable manner.

Contents

Acknowledgements

The present volume, *Smart Technologies and Fundamental Rights*, contains fourteen outstanding and challenging articles concerning fundamental rights at the intersection of law, ethics and smart technologies. They are authored by a combination of well-known experts and up-and-coming younger colleagues from various countries and disciplines, including philosophy, law, informatics, computer science and information technologies. The contributors, in alphabetical order, are as follows: Darius Amilevicius (VDU), Aníbal Monasterio Astobiza (Basque Country University), Paulius Čerka (VDU), Vladislav V. Fomin (VDU), Carl Fox (University of Leeds), John-Stewart Gordon (VDU), Edita Gruodytė (VDU), David Gunkel (Northern Illinois University), Kai Jakobs (RWTH Aachen University), Tanel Kerikmäe (Tallinn Law School), Julija Kiršienė (VDU), Janina Loh (University of Vienna), Vygantas Malinauskas (VDU), Katrin Nyman Metcalf (Tallinn Law School), Kęstutis Mosakas (VDU), Adam Poulsen (Charles Sturt University), Stephen Rainey (University of Oxford), Anwaar Ulhaq (Charles Sturt University), and Jonathan Wolff (Blavatnik School of Government, University of Oxford).

I am very thankful to all these authors for their great work and their willingness to contribute an article to this volume. I also thank the Springer publishing house for granting permission to reprint my article, which appeared previously in *AI & Society*. This project has received funding from the European Social Fund, project number 09.3.3-LMT-K-712-01-0173, under a grant agreement with the Research Council of Lithuania (LMTLT).

I would also like to thank Brill for its interest in publishing this book as part of its "Philosophy and Human Rights" series, and in particular Benjamin Suchard for his excellent support at the final stages of this volume's development. Last but not least, I would like to thank the members of the project, in alphabetical order, Darius Amilevičius, Paulius Astromskis, Vladislav Fomin, Edita Gruodytė, Julija Kiršienė, Kęstutis Mosakas, Aušrinė Pasvenskienė, Rasa Simanavičienė, and Milda Žaliauskaitė as well as my former research assistant Paula Segler for their help. Paula provided outstanding support with regard to all the editorial work on the book; without her assistance I certainly would have been lost in the details of the project.

John-Stewart Gordon
Vytautas Magnus University (VDU), August 2020

List of Figures and Tables

Figures

Tables

Biographical Notes

Darius Amilevičius
is associate professor and head of Intelligent Systems Laboratory (member of CLAIRE – Confederation of Laboratories for Artificial Intelligence Research in Europe) at the Faculty of Informatics, senior researcher of the EU-funded research project "Integration Study on Future Law, Ethics, and Smart Technologies" (2017-2021) as well as project leader and senior researcher of the EU-funded R&D project "Building Advanced Natural Language Processing Information Systems and Cloud Services for Lithuanian (SEMANTIKA2)" (2017–2020) at Vytautas Magnus University in Kaunas, Lithuania. He is supporter of the Confederation of Laboratories for Artificial Intelligence Research in Europe, member of the European AI Alliance, and member of the Lithuanian Computer Society (member of steering board of AI section). He has written and published several peer-reviewed articles in the context of artificial intelligence technologies and natural language processing.

Aníbal Monasterio Astobiza
is a Basque Government Posdoctoral Researcher. Anibal graduated in Philosophy (Universidad de Deusto) before completing a MA in Social Psychology, and obtained his PhD in Cognitive Science and Humanities at the Universidad del País Vasco/Euskal Herriko Unibertsitatea with a dissertation on social cognition. He is an academic visitor at the Oxford-Uehiro Centre for Practical Ethics, University of Oxford and Visiting Fellow in Global Health and Social Medicine at the Centre of Bioethics, Harvard Medical School, Harvard University. His research lies at the intersection of the cognitive, biological and social sciences exploring their philosophical underpinnings. He is a member of international (and national) research projects such as for example, EXTEND and INBOTS (Horizon 2020).

Paulius Čerka
is an associate professor for a private law department. He has more than 20 years' experience of practice as an attorney at law and in scientific research. He regularly leads various seminars for Lithuanian auditors on law issues related to taxation and economic problems. Čerka is the co-author of several scientific articles published in international scientific journals on legal issues related to artificial intelligence.

Vladislav V. Fomin

holds a professor's position at the department of Informatics at Vytautas Magnus University and a principal researcher position at Vilnius University in Kaunas, Lithuania and Turiba University in Riga, Latvia. Prof. Fomin also holds visiting professor's positions at the Faculty of Computing of the University of Latvia and La Rochelle Business School in France. After earning his PhD degree in 2001, Vladislav V. Fomin held academic positions at the University of Michigan in Ann Arbor, Copenhagen Business School, Delft University of Technology, Montpellier Business School, and Rotterdam School of Management, Erasmus University. Prof. Fomin has an extensive experience in international research and professional projects. He is serving on the editorial board of The International Journal of Standardization Research (IJSR) and Baltic Journal of Modern Computing (BJMC), serving as a reviewer for many recognized journals and is regularly invited to serve as a scientific committee member to recognized international conferences. Fomin is a member of European Academy of Standardization (EURAS) and The Association for Information Systems (AIS). Prof. Fomin has over 100 scientific publications in journals, conferences, and as book chapters, including Information Systems Research, Academy of Management Journal, Journal of Strategic Information Systems, Communications of the Association for Information Systems, International Journal of IT Standards & Standardization Research, Telecommunications Policy, Knowledge, Technology & Policy. Conference publications and presentations include Academy of Management, The International Conference on Information Systems, European Conference on Information systems, Hawaii International Conference on System Sciences, and other.

Carl Fox

is a Lecturer at the IDEA Centre in the University of Leeds. He co-edited 'Media Ethics, Free Speech and the Requirements of Democracy', and in 2015 won the Robert Papazian Essay Prize for this article on 'Political Authority, Practical Identity, and Binding Citizens'. He is currently a co-investigator on an AHRC-funded project examining media regulation and the role of press codes of ethics across a number of European countries.

John-Stewart Gordon

is full professor of philosophy, senior researcher at the Faculty of Law, head of the Research Cluster for Applied Ethics (RCAE), and principal investigator of the EU-funded research project "Integration Study on Future Law, Ethics, and Smart Technologies" (2017–2021) at Vytautas Magnus University in Kaunas, Lithuania. Furthermore, he is member of the editorial boards of *Bioethics* (since 2007), the Baltic Journal of Law & Politics (since 2018), the journal AI &

Society (since 2018), and he has been area-editor and board member of the Internet Encyclopedia of Philosophy (2007–2014). Furthermore, since 2016, he is general editor of the book series *Philosophy and Human Rights* at Brill. John spent extended research stays at the universities of Queen's, Oxford, Toronto, Maynooth, and Tallinn. He has written and edited several books in the context of practical philosophy and published peer-reviewed articles and special issues at international leading journals and encyclopaedias.

Edita Gruodytė
is tenured professor in the public law department and vice-dean for research in the Faculty of Law. She has more than 20 years' experience of diverse legal work (both as a lawyer for various companies and as a private attorney) and in scientific research and academic work. Gruodytė is the author and co-author of several books and more than 40 scientific articles published in various national and international scientific journals and books. She specializes in criminal law (including national, European, business crimes), legal ethics and certain aspects of human rights. She is an expert consultant for various national and international organizations and has some legislative drafting experience.

David J. Gunkel
is an award-winning educator and scholar, specializing in the philosophy of technology with a particular focus on emerging technology and ethics. He is the author of over 75 scholarly articles and has published nine books – *Hacking Cyberspace* (Westview Press, 2001), *Thinking Otherwise: Philosophy, Communication, Technology* (Purdue University Press, 2007), *Transgression 2.0: Media, Culture and the Politics of a Digital Age* (Continuum, 2011), *The Machine Question: Critical Perspectives on AI, Robots, and Ethics* (MIT Press, 2012), *Heidegger and Media* (Polity, 2014), *Of Remixology: Ethics and Aesthetics After Remix* (MIT Press, 2016), *The Changing Face of Alterity: Communication, Technology and Other Subjects* (Rowman & Littlefield, 2016), *Gaming the System: Deconstructing Video Games, Game Studies and Virtual Worlds* (Indiana University Press, 2018), and *Robot Rights* (MIT Press, 2018). He has lectured and delivered award-winning papers throughout North and South America and Europe and is the founding co-editor of the *International Journal of Žižek Studies* and the Indiana University Press book series Digital Game Studies. Dr. Gunkel currently holds the position of Presidential Teaching Professor in the Department of Communication at Northern Illinois University.

Kai Jakobs
joined RWTH Aachen University's Computer Science Department as a member of technical staff in 1985. Since 1987, he has been Head of Technical Staff at

the Chair of Informatik 4 (Communication & Distributed Systems). He holds a PhD in Computer Science from the University of Edinburgh and is a Certified Standards Professional. His research interests and activities focus on various aspects of ICT standards and the underlying standardisation process. Over time, he has (co)-authored/edited a text book on data communication networks and, more recently, more than 30 books on standards and standardisation processes, with a focus on the ICT sector. Over 280 of his papers have been published in conference proceedings, books and journals. He has been on the programme committee and editorial board of numerous international conferences and journals, respectively, and has served as an external expert on evaluation panels of various European R&D programmes, on both technical and socio-economic issues. Kai is Vice President of the European Academy for Standardisation (EURAS). He is also founder and editor-in-chief of the "International Journal on Standardization Research," the "Advances in Information Technology Standards and Standardization Research" and the "EURAS Contributions to Standardisation Research" book series.

Tanel Kerikmäe

is full professor of law at TalTech and has been contributing to the research of European law and policies. He created a research group of law & tech that received significant funding from EU Commission (H2020, Jean Monnet Centre of Excellence), NATO, Folke Bernadotte Academy, etc. Many of these grants are focused on developing legal digital skills and e-governance related regulatory and ethical aspects. Tanel has been main editor and contributor of several books such as "Regulating eTechnologies in the European Union" and "The Future of law and eTechologies." His recent scholarly articles include topics on digital single market, automatization of legal reasoning and big data, legal impediments to new technologies and use of augmented and artificial intelligence in legal practice. Tanel has been invited scholar to Nagoya, Lausanne, Luzern, Stellenboch universities but also to several Asian universities to give a lecture series on technological and digital aspects of international economy. Tanel has been legal expert to European Commission, Council of Europe, World Bank and Estonian Government.

Julija Kiršienė

is Vice-Rector for Research at Vytautas Magnus University. She is also a professor in the Department of Private Law. She is a member of several public bodies, including the Judicial Court of Honour at National Courts Administration of Lithuania, Management Board of European Union Fundamental Rights Agency (FRA), the Judicial Mediation Commission at National Courts Administration of Lithuania. She is an expert on the Research Council of Lithuania and chairperson of the board of Kaunas Science and Technology Park. She has more than 20 years' experience in diverse legal work (as attorney and notary public) and in scientific research and academic work. Professor Kiršienė is the author and co-author of more than 40 books and scientific articles in various national and international scientific journals, member of editorial boards of two internationally recognised scientific journals. Her research interests include: contract law, tort law, legal ethics and professionalism, and human rights.

Janina Loh

(née Sombetzki) is university assistant (Post-Doc) in the field of philosophy of technology and media at the University of Vienna. She studied at the Humboldt University Berlin and wrote her dissertation (2009–2013) on the issue of responsibility – Verantwortung als Begriff, Fähigkeit, Aufgabe. Eine Drei-Ebenen-Analyse (Springer 2014) – as a fellow at the graduate school Verfassung jenseits des Staates: Von der europäischen zur Globalen Rechtsgemeinschaft?, supervised by Prof. Volker Gerhardt and Prof. Rahel Jaeggi. After a post-doc position at the Christian-Albrechts-University Kiel (2013–2016), Janina Loh is now (since April 2016) in Vienna. She just published the first German Introduction to Trans- and Posthumanism (Junius 2018) and the first Introduction to Robot Ethics in the German language (Suhrkamp 2019). She habilitates on the Critical-Posthumanist Elements in Hannah Arendt's Thinking and Work (working title). Janina Loh's main research interests lie in the field of trans- and posthumanism (especially critical posthumanism), robot ethics, feminist philosophy of technology, responsibility research, Hannah Arendt, theories of judgement, and ethics in the sciences.

Vygantas Malinauskas

is a lecturer of jurisprudence at Vytautas Magnus University Faculty of Law and Faculty of Catholic Theology. He teaches regular courses with titles such as "Legal Theory," "Legal argumentation," "Church – State relations." He is co-author of several books and publications on topics across his areas of interests. In addition to academic work, he is a regular contributor to a variety of media portals, online journals and blogs writing on legal, social, political and moral issues. Also, he is a member of Bilateral commissions of Lithuanian Government and Holy See (Vatican) for implementation for international treaties and a member Legal Commission of Bishops' Conferences of the European Union (COMECE). As an expert and a member of working groups he contributed to the drafting of several legislative proposals to the Lithuanian Parliament. He is a member of the board of several NGO's and advocacy groups, including Free Society Institute.

His research interests – human rights, legal theory, theory of justice, natural law, legal ethics, legal argumentation, Church and State relations.

Kęstutis Mosakas
is a PhD student in philosophy at Vytautas Magnus University, a junior researcher for the EU-funded research project "Integration study on Future Law, Ethics, and Smart Technologies," and the research assistant of the Research Cluster for Applied Ethics (VMU). He holds a BA degree in English philology (2013, VMU) and an MA degree in practical philosophy (2017, VMU). His main fields of research include ethics of technology (machine moral status in particular), meta-ethics, and philosophy of religion. Kestutis has had a number of presentations at international conferences and workshops and is actively working on articles for peer-reviewed journals on the mentioned topics. His PhD thesis "Moral and Legal Rights for Intelligent Robots" will attempt to explore the vital question of whether the highly sophisticated machines that will exist in the future should be recognized as bearers of moral rights and granted legal protection.

Katrin Nyman-Metcalf
a professor of TalTech LawSchool, Tallinn Univesrity of Technology, has been a regular visiting professor at universities in Sweden, Georgia and Latvia. She has her PhD from Uppsala University (Sweden) and she has published extensively on international and EU law and countries in transition. Apart from academic work, Professor Nyman-Metcalf has over 15 years of international consultancy experience in the areas of ICT law and setting up of regulatory systems. She has acted as team leader of an EU project advising the Albanian National Agency for Information Society, and a project supporting e-governance in Albania and Kosovo. She has carried out training on communications related legal issues in Armenia, Iraq, Jordan, Moldova and other developing countries.

Adam Poulsen
is a computer scientist and researcher at Charles Sturt University in New South Wales, Australia. His research covers human-robot interaction, elderly healthcare robots, robot and machine ethics, value sensitive design, care ethics, and LGBTIQ+ aged care. Broadly, Adam's primary focus is on creating person-centred, care ethical healthcare robots to assist in, or enhance, the provision of care. Adam aims to create such robots with Values in Motion Design, a design methodology which is grounded in care ethics and acknowledges the importance of extrinsic, or instrumental, care values. In his current research, Adam is using Values in Motion Design to model socially assistive healthcare robots for LGBTIQ+ elders who are

experiencing loneliness. It is his hope that such robots can be helpful in the self-care of this population and others in the future.

Stephen Rainey
is a research fellow in the Oxford Uehiro Centre for Practical Ethics, University of Oxford. He is working in the Horizon2020-funded project "BrainCom," which is developing therapeutic brain-computer interfaces that will enable communication for users with debilitating speech conditions. Dr Rainey studied philosophy in Queen's University Belfast, and obtained his PhD in 2008 with a thesis on rationality. Stephen has research interests in the philosophy of language, rationality, governance, and artificial intelligence.

Anwaar Ulhaq
is serving as a lecturer in School of Computing and Mathematics, Charles Sturt University. Anwaar holds PhD (Artificial Intelligence) from Monash University, Australia. He has an extensive teaching and research experience of working at reputed Australian universities including Victoria University, Swinburne University of Technology and Central Queensland University. He has also worked a research fellow at Institute for Sustainable Industries & Liveable Cities, Victoria University, Australia. His research interests include signal and image processing, machine learning, data analytics and machine ethics.

Jonathan Wolff
is the Blavatnik Chair in Public Policy at the Blavatnik School of Government, Oxford, having been. Professor of Philosophy and Dean of Arts and Humanities at UCL. His recent works include *Ethics and Public Policy: A Philosophical Inquiry* (Routledge 2011, second edition 2019) *The Human Right to Health* (Norton 2012) and *An Introduction to Moral Philosophy* (Norton 2018). Earlier works include *Disadvantage* (OUP 2007), with Avner de-Shalit; *An Introduction to Political* Philosophy (OUP, 1996, third edition 2016); *Why Read Marx Today?* (OUP 2002); and *Robert Nozick* (Polity 1991). He has been a member of the Nuffield Council of Bioethics, the Academy of Medical Science working party on Drug Futures, the Gambling Review Body, the Homicide Review Group, an external member of the Board of Science of the British Medical Association, and a Trustee of GambleAware. He writes a regular column on higher education for the Guardian.

Introduction

Kęstutis Mosakas and John-Stewart Gordon

It would be hard to overstate the transformative impact that recent technological inventions and innovations have had on our lives. Today, numerous automated systems are utilized on a daily basis in virtually every sphere of human activity and billions of people all over the globe are connected by smart mobile devices through the Internet. Having undergone the third industrial revolution (involving electronics, IT and automated systems) and as we are now making the transition to the fourth industrial revolution, cyber-physical systems (Schwab 2016), we have come to recognize that the relation between technologies and human beings has become so close as to be characterized as intimate (Van Est et al. 2014). As Royakkers et al. (2018, 127) point out, technologies nestle themselves *in* us (through implants) and *between* us (through social media), know *about* us (consider Big Data), and are becoming more and more *like* us (consider the advances in robotics). Given the general trends in technological development (e.g. growth in computing power; see Nordhaus 2007), as well as the many benefits that technologies bring, we can be quite certain that progress in research areas such as IT, computer engineering, and robotics will continue to occur as rapidly as possible.

1 Opportunities

One particularly promising venue for future technologies is artificial intelligence (AI), which has already come a long way since the Dartmouth Conference in 1956 at which the term was first coined. Starting with the "astonishing" feats of solving algebra problems and proving geometry theorems (Russell and Norvig 2010), the application of AI systems has vastly expanded ever since. Today, AI systems are no longer simply formal reasoners but are engaged in a variety of complex and important tasks, such as piloting airplanes, diagnosing diseases, recognizing faces, caring for the elderly and the infirm, playing complex games, exploring the cosmos, and many more. Moreover, AI applications are having a considerable economic impact and causing disruptions in the business sector, which have already caused us to reconsider numerous established marketing strategies (Kumar et al. 2016). Indeed, because of the rapid development of AI systems for businesses, a labor productivity boost of almost

 | DOI:10.1163/9789004437876_002

40% is expected by the year 2035 in certain developed countries (Purdy and Daugherty 2016).

Given the success of AI in a great number of domains, it is easy to see why researchers have focused on further expanding and developing its capacities. For example, many entities are currently investing great effort in perfecting autonomous vehicles, which are expected to revolutionize transport within the next decade or two. If implemented responsibly, autonomous vehicles could provide us with safer, faster, cheaper, and more efficient means of travel (Fagnant and Kockelman 2015). Another development of great importance is the so-called Internet of Things (IoT), the underlying idea of which is that "everyday objects can be equipped with identifying, sensing, networking and processing capabilities that will allow them to communicate with one another and with other devices and services over the Internet to accomplish some objective" (Whitmore, Agarwal, and Xu 2015, 261). This technology, as Ismail (2018, 1) points out, when successfully applied to home appliances and different components of city infrastructure (thereby resulting in what is known as "smart cities"), will create "never-before-seen opportunities to converge the physical and the digital – via data analytics – to improve efficiency (both in public and private sectors), drive economic benefits and improve livelihoods."

Other important prospective uses for AI include language translation (researchers predict that AI will surpass human beings in translating languages by 2024; see Grace et al. 2017), medical applications (expected to surpass human beings in performing surgeries by 2053; see ibid.), military applications, sophisticated AI assistants that will allow us to automate a variety of complex tasks, and even AI that is capable of giving advice or making decisions in domains as complex as ethics, law or politics. Moreover, numerous experts predict the emergence of general human-level AI – that is, systems as competent as human beings in all domains of interest – by the end of the current century (Müller and Bostrom 2014). Thus, the opportunities presented by AI are quite impressive.

2 Challenges

However, significant challenges come with these opportunities. The unprecedented pace of current technological growth creates a danger that legal regulations will severely lag behind technological development. The World Economic Forum's white paper on values and the Fourth Industrial Revolution stated, "Given the Fourth Industrial Revolution's extraordinarily fast technological and social change, relying only on government legislation and incentives to ensure the right outcomes is ill advised. These are likely to be out of date or redundant by the time they are implemented" (White Paper 2016, 6). Thus, to ensure that technological developments remain consistent with society's fundamental values, we must try to anticipate their likely effects, consider the related issues in advance, and develop necessary solutions and regulatory safeguards well before the mentioned technologies actually come into existence. This proactive stance requires a determined interdisciplinary endeavor by experts from fields as varied as IT, law and practical philosophy (particularly ethics) to match the challenges posed by the rapid technological growth occurring on every level. Although, for this purpose, inquiry into the details of how IT experts build and operationalize AI systems is undoubtedly crucial, it must also be supplemented by an adequate normative framework and guidelines that will allow us to implement and utilize smart technologies with great care and prudence. This is where law and moral philosophy come into play. Accordingly, the present volume seeks to facilitate safe technological development by considering, from all three of the vital perspectives, the many problems that arise at the intersection between smart technologies and fundamental rights.

Consider automation as an example. Clearly, it is a greatly desirable asset that we will continue to rely on. As Allen, Wallach, and Smit point out, "The computer revolution is continuing to promote reliance on automation, and autonomous systems are coming whether we like it or not" (2011, 52). However, the numerous benefits that make the expansion of autonomous AI systems virtually inevitable are accompanied by serious problems and potential threats to fundamental rights. Consider the case of autonomous vehicles. Not only are they likely to present instances of the so-called trolley problem (Foot 1967; Achenbach 2015), which has to do with making appropriate decisions in ethically complex situations when people could be severely harmed or even killed by either choice, but the manufacturing of autonomous vehicles involves a variety of stakeholders, thus making it very difficult to assign blame if the vehicle makes a morally inappropriate decision (Holstein 2017). This scenario raises the additional question of whether an AI system itself can be held responsible for its actions and, if so, under what conditions.

Another important related problem, generally referred to as "algorithmic bias" or "machine bias," concerns how AI systems learn and make decisions. In recent decades, we have come to see that AI systems, contrary to popular belief, are not immune to biases. As Garcia points out, machine bias occurs "when seemingly innocuous programming takes on the prejudices either of its creators or the data it is fed" (Garcia 2016, 112). Amazon, for example, had to scrap an AI-based recruitment tool because it proved to be biased against women (Dastin 2018), whereas Northpoint's software for predicting future criminals turned out to be biased against black people (Angwin et al. 2016). Unfortunately, machine bias can be difficult to detect and to resolve. As Danks and London observe, "As systems become more complicated and their workings more inscrutable to users, it may become increasingly difficult to understand how autonomous systems arrive at their decisions" (2017, 1). The problem is further exacerbated by machine learning through deep neural networks, which, while greatly useful, also creates a type of "black box" problem by making AI decisions difficult to explain (Fernandez 2018).

Other significant challenges to fundamental rights in the context of smart technologies arise from the amount of personal data of all sorts that is stored electronically. In the age of AI and Big Data, personal information has become a commodity in its own right (Press 2017). Certain web software, for example, can use people's personal data for commercial purposes, such as to display the kind of online adverts that are most likely to persuade the user to make a certain purchase. Or consider what health insurance companies might do if they had access to various medical data on their potential customers. With more advanced data collection tools being developed every day, this situation poses serious risks and raises crucial questions regarding the possession and use of personal data by third parties, as well as people's right to access data concerning themselves. Furthermore, data leaks and breaches represent another area of ethical concern. In the year 2017, for example, 477 healthcare data breaches were reported in the US, affecting a total of 5,579,000 patients (Landi 2018).

Finally, great ethical and legal challenges are resulting from developments in robotics. One key issue is to ensure that robots behave towards human beings in an ethically appropriate manner (Anderson and Anderson 2011; Wallach and Allen 2010). To achieve this goal, we must examine the variety of ethical approaches available and find the most optimal solution that avoids related pitfalls (Gordon 2019). Another vital issue is the rights of robots themselves, both moral and legal (Gunkel 2018). In this on-going debate, some have argued that to have moral status, a machine must be sentient (Torrance 2008), whereas others believe that the social presence of the machine is already a sufficient condition, regardless of any particular ontological properties it may possess (Coeckelbergh 2010). These positions have significant implications not only for the highly sophisticated machines with general AI that may appear by the end of this century, but also for the so-called social robots that already exist today, such as robotic companions and caregivers. Indeed, the very use of such nonhuman companions has aroused heated debate among machine ethicists (Gunkel 2015; Tavani 2018; Darling 2016). The possible use of sex robots also generates unique ethical questions, which include but are not limited to the problems of consent, companionship, impact on society, therapeutic uses (e.g.

treatment for paedophilia), and more (Sullins 2012; Levy 2007; Danaher and McArthur 2017). In sum, we are on the verge of confronting a multitude of ethical and legal problems related to robotics that must be addressed as soon as possible.

3 Contributions[1]

John-Stewart Gordon argues in the first chapter *What Do We Owe to Intelligent Robots?* that we should grant robots moral rights once they become moral persons. The great technological advances in such areas as computer science, artificial intelligence, and robotics have brought the advent of artificially intelligent robots within our reach within the next century. Against this background, the interdisciplinary field of machine ethics is concerned with the vital issue of making robots "ethical" and examining the moral status of autonomous robots that are capable of moral reasoning and decision-making. The existence of such robots will deeply reshape our socio-political life. This paper focuses on whether such highly advanced yet artificially intelligent beings will deserve moral protection (in the form of being granted moral rights) once they become capable of moral reasoning and decision-making. He argues that we are obligated to grant them moral rights once they have become full ethical agents, i.e., subjects of morality. He presents four related arguments in support of this claim and thereafter examines four main objections to the idea of ascribing moral rights to artificial intelligent robots.

David Gunkel examines the idea of robot rights in the second chapter *Robot Rights: thinking the Unthinkable.* In a recent proposal issued by the European Parliament it was suggested that robots and artificial intelligence might need to be considered "electronic persons" for the purposes of social and legal integration. The very idea sparked controversy, and it has been met with considerable resistance. Underlying the controversy, however, is an important philosophical question: When (if ever) would it be necessary for robots, AI, or other socially interactive, autonomous systems to have some claim to moral and legal standing? When (if ever) would a technological artifact need to be considered more than a mere instrument of human action and have some legitimate claim to independent social status? Or to put it more directly: Can or should robots ever have anything like rights? This essay takes up and investigates these questions. It reviews and critiques current thinking (or lack of thinking) about

1 This section contains the slightly revised chapter abstracts.

this subject matter, maps the terrain of the set of available answers that have been provided in the existing literature, and develops an alternative way of responding to and taking responsibility for the opportunities and challenges that we now confront in the face or the faceplate of increasingly social and interactive artifacts.

Kęstutis Mosakas defends the standard property-based view of moral standing for machines in the third chapter *Machine Moral Standing: In Defence of the Standard Properties-based View*. In recent years, more and more attention has been devoted to the question of machine moral standing. While the view that moral standing is determined by certain ontological properties possessed by entities has generally been the standard one in moral philosophy, the so-called relational approach, the view that moral standing is constructed through relations between entities, has been emerging as an alternative in the context of machine rights in particular. In the present chapter, Mosakas has three principal goals. The first is to outline and problematize some of the traditional arguments that were used to deny rights for certain disadvantaged groups in the past, thereby showing that the question of machine moral standing is more complex and philosophically interesting than it may initially appear to be. The second is to consider the arguments against the standard view put forward by the proponents of relationism and to examine them critically. The third one is to consider the relational approach itself and expose some of its weaknesses.

Janina Loh examines the question of whether one should ascribe moral rights to robots in the fourth chapter *Ascribing Rights to Robots as Potential Moral Patients*. With regard to robots as potential moral patients some thinkers advocate the possibility of attributing (some) robots rudimentary or even fundamental rights. Apart from academic philosophical discourses, this question is of great relevance for political agents and institutions.

The European Parliament, for instance, is currently working on a concept that would allow (some) robots to be given the status of "electronic persons." And in October 2017 the robot Sophia was the world's first robot to be granted full citizenship in Saudi Arabia. In this chapter, however, the focus lies on fundamental moral rights, rather than on fundamental legal rights. In the following, she first presents the discipline of robot ethics and then gives an overview of its three fields, namely (1) robots as potential moral agents, (2) robots as potential moral patients, and (3) inclusive approaches that question the classical subject-object dichotomy. In the third section, she deals with the attribution of (moral) rights with regard to the second robot-ethical field, which (as will be shown) in turn is closely related to the first field. In an excursus in the fourth section the inclusive approaches will be discussed.

Adam Poulsen and Anwaar Ulhaq discuss the idea of fundamental rights in the context of healthcare delivered by smart technologies in the fifth chapter *Fundamental Rights and Smart Health Technologies.* To cover this perspective, they identify the fundamental human rights in healthcare with a comprehensive review of smart health technologies and explore various design considerations of smart health technologies in relation to fundamental rights in healthcare with a discussion about the positive and negative impacts of these technological innovations. This chapter highlights the importance of care ethics and describes how good care goes beyond just guaranteeing fundamental rights. Finally, they present a brief analysis of the existing design approaches for the creation of smart health technologies with their recommendations to improve technological design to guarantee both fundamental rights and good care ethics.

Jonathan Wolff examines the practice and nature of regulation (of smart technologies) in the sixth chapter *Rules for Regulators*. Recognising that regulation should always be a response to a risk provides a key to developing a set of rules of good regulation. This, in turn, allows the identification of a series of ways in which regulation can fail, both in relatively routine cases and in the face of new technologies. After setting out some general considerations about regulation and regulatory failure, this paper turns its attention to new smart technologies, and especially AI and robotics, explores their risks, and outlines underlying considerations that could guide their regulation.

Carl Fox examines the phenomenon of hurtful speech in the context of social media in the seventh chapter *Free Speech, Public Shaming and the Role of Social Media*. In this chapter he takes a closer look at a particularly contentious form of hurtful speech. It is clear that many targets of online public shaming campaigns find it to be a deeply distressing and degrading experience. Moreover, shaming sometimes results in egregious harms being visited on ordinary people for relatively minor offences. He starts by arguing that the standard free speech defence for objectionable speech will not support a permissive approach to public shaming, because public shamings are not political in the way that such an argument would require. Instead, he considers another approach which builds on a significant political dimension that is apparent in many public shamings. Relatively powerless individuals and groups can use social media to achieve the critical mass required to deploy shame as a meaningful sanction for wrongful behaviour that the state either can't, won't, or shouldn't punish itself. Could shaming on social media make a sufficiently important positive contribution that we should learn to live with its pernicious effects? Ultimately, he concludes that while we have strong reasons to find new ways to buttress political equality, we should not embrace public shaming on social media.

Drawing on the work of Martha Nussbaum, he offers some reasons for thinking that social media is, in fact, uniquely ill-suited to deploying shame as a sanction for bad behaviour.

Stephen Rainey and Aníbal Monasterio Astobiza examine the link between global governance and political legitimacy in the context of AI in the eighth chapter *Global Governance of AI: Pressure on Political Legitimacy.* Political legitimacy in representative democracies is founded on, at least, government by consent. This means that "we the people" agree to an arrangement wherein elected officials enact political programmes according to proposals agreed to via elections. The realities of this often leaves much to be desired, and innovations are sought to improve accountability. One such innovation is the proposal to include AI in the development of policy. The promise is that data on "the people" will be collected, analysed, and used more efficiently, ensuring better representation. Where algorithmic analysis of vast datasets, comprising citizen behaviours and preferences, can lead to better policy insight, better politics and more just societies can emerge. This is the promise, but in fact this is better described as a hope. The use of technologies in the way outlined will serve to undermine political legitimacy in terms of government by consent. It will do this by virtualising citizens, and administering the state according to virtualised imperatives based in statistical projections from these virtual citizens. The state, with too much AI in governance, departs from the legitimate kind of states-for-citizens we are used to, and risks becoming a kind of service delivery without normative foundation. They claim that it is unclear what arguments are offered to justify making a change from the former to this latter model. This chapter analyses the issues associated with this kind of eventuality, diagnosing the problem in terms of citizens' representation, and suggest what kind of role AI ought to have, should it be deployed in governance.

Tanel Kerikmäe and Katrin Nyman Metcalf examine the protection of human rights against the background of automation in the ninth chapter *The Rule of Law and The Protection of Fundamental Human Rights in an Era of Automation.* Automation and artificial intelligence affect many areas of life. It is essential to examine what, if anything, this means for rule of law and human rights. Technology can be used to protect rights, but also poses new risks. Examples from a digital society like Estonia – primarily e-governance but also proactive governance – are mentioned as illustrations. Here are ways in which technology is used for greater transparency, e.g. on how authorities use personal data. To some extent, existing legislation can be sufficient and there is no need for a lot of special rules, but we need to be aware of risks not least due to the lack of transparency of algorithms and the importance of certain private firms with access to large amounts of personal data. More research is needed, not least to establish how rules can be enforced and to what extent technology can fulfil the role of regulation.

Julija Kiršienė and Vygantas Malinauskas analyse the link between AI and the concept of privacy in the tenth chapter *AI in the Context of Prevailing Privacy Concepts: In Search of a New Approach.* Privacy is pivotal for civilized states, as well as for social wellbeing and individual autonomy, dignity, and liberty. It is not hard to imagine why in this hyper-connected online world, considering the progress of data-processing technologies and the amount of data available for analysis (Big data, Internet of Things, AI), unprecedented privacy concerns are raised. Prevailing, traditional approaches to the right to privacy are associated with a person's ability to control personal information. However, in the age of AI technologies, this concept is no longer sufficient, since a person has less control over his or her personal information because of the omnipresent eye of data-mining technologies powered by AI. New opportunities created by technological progress and the insufficiency of existing legal and regulatory frameworks make it necessary to search for alternative approaches to ensure the adequate protection of privacy in cases where a person's privacy is infringed by others using private information derived from publicly available sources.

Edita Gruodytė and Paulius Čerka discuss the corporate liability model with respect to AI in the eleventh chapter *Artificial Intelligence as a Subject of Criminal Law: A Corporate Liability Model Perspective.* Artificial Intelligence (AI) is one of the most transformative forces of our time, but its use also generates questions of risk. Personality and liability issues, in particular, are concerned with the fact that autonomous AI may not be fully controlled by humans and could cause unintentional harm or deaths of humans. The aim of this article is to discuss if AI could be held criminally liable for the most serious infringements on human rights. They explore this issue through an analysis of corporate criminal liability, arguing that corporate criminal liability shares some aspects with AI. Both AI and corporations are artificial, since legal personhood is a fundamental legal fiction taking origin from the philosophy of law, and both are usually used for economic gain. They analyse arguments and rationale for establishing criminal liability to legal entities, main doctrines of corporate liability, and how these aspects inform AI criminal liability.

Kai Jakobs proposes a new approach of responsible standardisation in the context of smart systems in the twelfth chapter *Responsibility by Design?! – On the Standardisation of "Smart" Systems.* Based on the premise that standards shape future technologies and that ideally everyone affected by these technologies should have a say in their development, the paper argues that a mechanism is needed to enable all relevant and affected stakeholders to contribute to standards development. The paper uses the example of emerging smart systems to discuss different options to achieve this. The discussion focuses on "Responsible Research and Innovation," which, among others, highlights the need of broad stakeholder involvement in technical developments. Identified practical issues with this approach lead to the proposal of "Responsible Standardisation."

Vladislav V. Fomin examines the problem of bias in smart systems in the thirteenth chapter *The Shift from Traditional Computing Systems to Artificial intelligence and the Implications for Bias.* With the steady growth of digitization, the resulting unprecedented growth of data, on the one hand, fuels the development of machine learning systems of all kinds. On the other hand, the advent of the new paradigm of computing has sparked discussion on whether it should be the technology of choice in a number of domains where the faultlessness of technology-mediated outcomes is a crucial requirement. This chapter provides a comparative analysis of the concepts of data- and logic-related distortion and bias in the traditional (algorithmic) and artificial intelligence-based computational systems. Simple categories of causes of distortion and bias are introduced and illustrated with examples. Possible measures to maintain distortion- and bias-free artificial intelligence systems are discussed.

Darius Amilevicius examines the issue of how to regulate AI risks with respect to the benefits of AI applications in the final chapter *Machine Bias and Fundamental Rights.* Artificial intelligence (AI) is becoming a true general-purpose technology and is evolving into a utility that is likely to ultimately scale across every industry and sector of our economy. Industrialization of AI systems has an immense potential not only for the overall technological advancement of society but also for unethical practices. AI-based systems depend on the collection of data and the use of vast quantities of data to make predictions which, in numerous cases, have served to perpetuate existing social patterns of bias and discrimination. Unlike human decision-makers who have the agency and the free will to change their moral perspective over time, for the foreseeable future AI systems will not have any such capabilities of their own. In this paper, his main focus is to reflect rationally on where to implement the actual regulation of AI risk and harm in the context of the benefits of rapid AI development and industrialization. While the ultimate consensus about universal ethical standards is lacking, a legal approach must be taken. International fundamental human rights law possesses well-developed standards and institutions, as well as a universal framework for safeguards.

4 Closing Remarks

The chapters in this volume cover a broad range of vital topics that highlight the ethical, socio-political, and legal challenges as well as technical issues involving the relationship between AI and fundamental rights. Unquestionably, the use of AI could eventually lead to a better and more humane society in almost all domains of human life. However, as some scholars have forcefully argued, the potential for misuse of AI by individuals, companies, or the state is high. Moreover, the so-called alignment problem, or the idea that intelligent machines may turn against human beings if we fail to teach (or educate) them properly, is yet another threat. Therefore, we should be extremely cautious with respect to how we develop AI, because we might not be able to control the machines once they have reached a certain level of sophistication; in other words, at some point it will become impossible to put the genie back in the bottle. Without question, AI is a double-edged sword that could bring humanity the greatest of benefits but also the greatest of miseries, or even conceivably our extinction.

We hope that the chapters of this volume will contribute greatly to the ongoing debates with respect to the special relation of smart technologies and fundamental rights.

References

Achenbach, Joel. 2015. "Driverless Cars are Colliding with the Creepy Trolley Problem." *The Washington Post,* December 29, 2015. https://www.washingtonpost.com/news/innovations/wp/2015/12/29/will%E2%80%90self%E2%80%90driving%E2%80%90cars%E2%80%90ever%E2%80%90solve%E2%80%90the%E2%80%90famous%E2%80%90and%E2%80%90creepy%E2%80%90trolley%E2%80%90problem/?arc404=true.

Allen, Colin, Wendell Wallach, and Iva Smit. 2011. "Why Machine Ethics?" In *Machine Ethics,* edited by Michael Anderson and Susan Leigh Anderson, 51–61. Cambridge: Cambridge University Press.

Anderson, Michael, and Susan Leigh Anderson, eds. 2011. *Machine Ethics.* Cambridge: Cambridge University Press.

Angwin, Julia, et al. 2016. "Machine Bias." *Pro Publica,* May 23, 2016. https://www.propublica.org/article/machine-bias-risk-assessments-in-criminal-sentencing.

Coeckelbergh, Mark. 2010. "Robot Rights? Towards a Social-Relational Justification of Moral Consideration." *Ethics and Information Technology* 12(3): 209–221.

Danaher, John, and Neil McArthur. Eds. 2017. *Robot Sex: Social and Ethical Implications*. The MIT Press.

Danks, David, and Alex John London. 2017. "Algorithmic Bias in Autonomous Systems." *Proceedings of the 26th International Joint Conference on Artificial Intelligence*. https://www.cmu.edu/dietrich/philosophy/docs/london/IJCAI17-AlgorithmicBias-Distrib.pdf.

Darling, Kate. 2016. "Extending Legal Protection to Social Robots: The Effects of Anthropomorphism, Empathy, and Violent Behavior Towards Robotic Objects." In *Robot Law,* edited by Ryan Calo, A. Michael Froomkin and Ian Kerr, 213–231. Northampton: Edward Elgar.

Dastin, Jeffrey. 2018. "Amazon Scraps Secret AI Recruiting Tool that Showed Bias Against Women." *Reuters*, October 10, 2018. https://www.reuters.com/article/us-amazon-com-jobs-automation-insight/amazon-scraps-secret-ai-recruiting-tool-that-showed-bias-against-women-idUSKCN1MK08G.

Fagnant, Daniel J., and Kara Kockelman. 2015. "Preparing a Nation for Autonomous Vehicles: Opportunities, Barriers and Policy Recommendations." In *Transportation Research* Part A: Policy and Practice 77: 167–181.

Fernandez, Sheldon. 2018. "Explainable AI: Peering Inside the Deep Learning Black Box." In *InfoWorld*, October 24, 2018. https://www.infoworld.com/article/3315748/explainable-ai-peering-inside-the-deep-learning-black-box.html.

Foot, Philippa. 1967. "The Problem of Abortion and the Doctrine of Double Effect." *Oxford Review* 5: 5–15.

Garcia, Megan. 2016. "Racist in the Machine: The Disturbing Implications of Algorithmic Bias." *World Policy Journal* 33(4): 111–117.

Gordon, John-Stewart. 2019. "Building Moral Robots: Ethical Pitfalls and Challenges." *Science and Engineering Ethics:* 1–17.

Grace, Katja, et al. 2017. "When Will AI Exceed Human Performance? Evidence from AI Experts." https://arxiv.org/abs/1705.08807.

Gunkel, David J. 2015. "The Rights of Machines: Caring for Robotic Care-Givers." In *Machine Medical Ethics*, edited by Simon Peter van Rysewyk and Matthijs Pontier, 151–166. Heidelberg/New York/Dordrecht/London: Springer.

Gunkel, David J. 2018. *Robot Rights*. Cambridge: MIT Press.

Holstein, Tobias. 2017. "The Misconception of Ethical Dilemmas in Self-Driving Cars." *Proceedings*.

Ismail, Nick. 2018. "Smart City Technology: It's all About the Internet of Things." *Information Age*, August 14, 2018. https://www.information-age.com/smart-city-technology-123473905/.

Kumar, A. et al. 2016. "From Social to Sale: The Effects of Firm-generated Content in Social Media on Consumer Behavior." *Journal of Marketing* 80(1): 7–25.

Landi, Heather. 2018. "Healthcare Data Breach Costs Remain Highest at $408 Per Record." *Healthcare Innovation*, July 13, 2018. https://www.hcinnovationgroup.com/

cybersecurity/news/13030528/healthcare-data-breach-costs-remain-highest-at-408-per-record.
Levy, David. 2007. *Love and Sex with Robots: The Evolution of Human–Robot Relationships*. New York: Harper-Perennial.
Müller, Vincent C., and Nils Bostrom. 2014. "Future Progress in Artificial Intelligence: A Survey of Expert Opinion." In *Fundamental Issues of Artificial Intelligence*, edited by Vincent C. Müller. 553–570. Berlin: Springer.
Nordhaus, William D. 2007. "Two Centuries of Productivity Growth in Computing." *The Journal of Economic History*. 67(1): 128–159.
Press, Gil. 2017. "6 Predictions For The $203 Billion Big Data Analytics Market." *Forbes*, January 20, 2017. https://www.forbes.com/sites/gilpress/2017/01/20/6-predictions-for-the-203-billion-big-data-analytics-market/#6ba3f4972083.
Purdy, Mark, and Paul Daugherty. 2016. "Why Artificial Intelligence is the Future of Growth." *Accenture*. https://www.accenture.com/t20170524T055435__w__/ca-en/_acnmedia/PDF-52/Accenture-Why-AI-is-the-Future-of-Growth.pdf.
Royakkers, Lambèr, et al. 2018. "Societal and ethical issues of digitization." *Ethics and Information Technology* 20(2): 127–142.
Russell, Stuart J., and Peter Norvig. 2010. *Artificial Intelligence: A Modern Approach*. Upper Saddle River, New Jersey: Prentice Hall.
Schwab, Klaus. 2016. "The Fourth Industrial Revolution." *World Economic Forum*. https://www.weforum.org/agenda/2016/01/the-fourth-industrial-revolution-what-it-means-and-how-to-respond/.
Sullins John P. 2012. "Robots, Love, and Sex: The Ethics of Building a Love Machine." *IEEE Transactions on Affective Computing* 3(4): 398–409.
Tavani, Herman. 2018. "Can Social Robots Qualify for Moral Consideration? Reframing the Question about Robot Rights." *Information* 9(4): 73.
Torrance, Steve. 2008. "Ethics and Consciousness in Artificial Agents." *AI & Society* 22(4): 495–521.
Van Est, Rinie, et al. 2014. *Intimate Technology: The Battle for our Body and Behaviour*. The Hague: Rathenau Instituut.
Wallach, Wendell, and Colin Allen. 2010. *Moral Machines: Teaching Robots Right from Wrong*. Oxford: Oxford University Press.
Whitmore, Andrew, Anurag Agarwal, and Li Da Xu. 2015. "The Internet of Things – A Survey of Topics and Trends." *Information Systems Frontiers* 17(2): 261–274.
World Economic Forum. 2016. *Values and the Fourth Industrial Revolution: Connecting the Dots Between Value, Values, Profit and Purpose*. http://www3.weforum.org/docs/WEF_Values_and_the_Fourth_Industrial_Revolution_WHITEPAPER.pdf.

PART 1

Ethical Challenges of Smart Technologies

CHAPTER 1

What Do We Owe to Intelligent Robots?

John-Stewart Gordon

1 Introduction[1]

In recent years, the interdisciplinary field of machine ethics – that is, how to program robots with ethical rules, so that they become either implicit or explicit moral agents (Moor 2006) – has become of utmost importance because of current and anticipated technological developments in the fields of computer sciences, artificial intelligence (AI), and robotics (Lin et al. 2014; Wallach and Allen 2010; Anderson and Anderson 2011; Gunkel and Bryson 2014a, b). Machine ethics considers the implications of making artificially intelligent robots (henceforth IRs) "ethical"[2] and examines related issues such as whether IRs have moral status or not, including moral and legal[3] rights. Whether this is even possible is a matter of great debate among researchers working in the field of machine ethics. For example, Johnson and Axinn (2014, 4) take a critical viewpoint, arguing that autonomous robots only mimic ethics, since they lack "the imagination to conceive of the effects should the principle of their

---

1 This research is funded by the European Social Fund according to the activity "Improvement of researchers" qualification by implementing world-class R& D projects of Measure No. 09.3.3-LMT-K-712.

2 Susan Anderson (2011a, b, 22) defines the goal of machine ethics as "to create a machine that follows an ideal ethical principle or set of principles in guiding its behaviour; in other words, it is guided by this principle, or these principles, in the decisions it makes about possible courses of action it could take. We can say, more simply, that this involves 'adding an ethical dimension' to the machine."

3 One can read the following interesting legal development in the Preliminary Draft Report of UNESCO's World Commission on the Ethics of Scientific Knowledge and Technology (COMEST) on Robotic Ethics: "The Committee on Legal Affairs of the European Parliament, in its 2016 Draft Report with Recommendations to the Commission on Civil Law Rules on Robotics, already considers the possibility of 'creating a specific legal status for robots, so that at least the most sophisticated autonomous robots could be established as having the status of electronic persons with specific rights and obligations, including that of making good any damage they may cause, and applying electronic personality to cases where robots make smart autonomous decisions or otherwise interact with third parties independently' (12)" (Draft Report United Nations 2016, 26). See, also Malvaux's Report with recommendations to the Commission on Civil Law Rules on Robotics (Delvaux 2017) and Calverley (2011) for an interesting discussion on the idea of ascribing legal rights to machines.

 | DOI:10.1163/9789004437876_003

actions be made universal, as well as the free will to make the choice to follow a moral style"; Rodogno (2016, 12) claims, "The day in which robots fulfil the conditions of moral agency and moral patience outlined here, however, will be the day in which they will be emotive creatures capable of humanly recognizable interests."[4]

The ethical discourse around robots has become quite sophisticated since the days of Isaac Asimov's Three Laws of Robotics[5] (in Runaround, Asimov 1942), subsequently complemented by his so-called Zeroth Law of Robotics (in Robots and Empire, Asimov 1986), which state that a robot may not harm humanity or, by inaction, allow humanity to suffer harm.[6] The idea that IRs may rebel against their creators and cause the global extinction of humanity, or at least of most human beings, has been prominently portrayed in the literature and in movies for nearly a century, dating back to Karel Capek's famous science fiction play Rossum's Universal Robots (1920).[7] Whether IRs will eventually rebel against human beings remains to be seen. In any case, however, considering their potential moral status is of utmost importance given their expanding involvement in many areas of human life by virtue of the simple fact that they will interact with human beings. For example, healthcare robots now provide physical assistance and companionship, as well as monitoring health and safety; military robots such as intelligent drones make decisions in warfare, and other robots perform surveillance, educational, and work tasks.

---

4 Johnson and Axinn (2014, 1) and Rodogno (2016, 1) admit, however, that their reasoning refers only to the present-day robots, and that it remains conceivable that robots might become full moral agents in the future.

5 Asimov's initial Three Laws of Robotics are as follows: (1) a robot may not injure a human being or, through inaction, allow a human being to come to harm; (2) a robot must obey the orders given it by human beings except where such orders would conflict with the first law, and (3) a robot must protect its own existence as long as such protection does not conflict with the first or second law.

6 For a thorough discussion on the logic and problems of Asimov's four laws of robotics, see Clarke (2011, 254–284), who examines the complex issues that arise when robots are supposed to follow these laws. For another critical examination of Asimov's laws as the foundation of machine ethics, see Anderson (2011a, 285–296). By examining the robot Andrew in Asimov's short story "The Bicentennial Man," she provides important insights on the ethical status of the laws. The up-shot is that the laws are ethically inappropriate for intelligent beings such as Andrew, who is considered to act more ethically than most human beings.

7 Similar ideas have been depicted in movies such as 2001: A Space Odyssey (Stanley Kubrick 1968), in which the spaceship's HAL 9000 computer attempts to kill its crew on the way to Jupiter, and in The Terminator I–V (James Cameron and Alan Taylor 1984–2015), where the machines rebel against human beings. Similarly, in the famous Matrix Trilogy (the Wachowski brothers 1999–2003), sentient machines subdue the human population by keeping them in a dream world while using their bodies as an energy source.

They can cause harm if their autonomous actions are not restricted by any moral safeguards in the course of such interaction (e.g. Picard 1997),[8] which will only become increasingly complex and will deeply influence their place in society and how human beings view them. It seems clear that, once IRs have reached the level of being moral machines, the important issue of their moral status must be addressed as well – i.e., whether IRs should be seen as full moral agents, capable of moral reasoning and decision-making, or as "moral patients," incapable of moral reasoning and decision-making but vulnerable beings that can be treated immorally (see Floridi 2011).[9]

This paper is motivated by three different but related ideas: (1) that this topic is a worthy thought experiment in moral philosophy, (2) that we can gain knowledge about ethical theories by trying to develop ethical systems for robots, and (3) that we should be preparing for the socio-political changes that can be expected as highly advanced and, perhaps, even autonomously intelligent robots become a reality. One might object that the advent of such robots is too speculative to justify a serious moral analysis of this issue, but there are at least two answers to that objection. First, scepticism regarding the possibility of highly advanced, intelligent robots seems quite premature given the great technological advances and future prospects in robotics, AI, and computer science. Second, it is unwise and un-philosophical to abstain from discussing important moral and socio-political issues the emergence of which cannot be fully ruled out in advance. In what follows, however, I am concerned only with the vital issue of IRs' moral status, and in particular the important idea of ascribing moral rights to IRs, which depends on the empirical question of whether and how IR may become moral machines. Here, based on the enormous prospects for future technological developments, I take it for granted that IRs will become moral machines in the future. In what follows, I attempt to show that if IRs are capable of moral reasoning and decision-making on a level that is comparable with the moral agency of human beings, then one must see IRs not only as moral patients, but also as full moral agents with

8 "The greater the freedom of a machine, the more it will need moral standards" (Picard 1997, 19).

9 Grau (2011, 458) correctly claims, "Once we do venture into the territory of robots that are similar to humans in morally relevant respects, however, we will need to be very careful about the way they are treated. Intentionally avoiding the creation of such robots may well be the ethical thing to do, especially if it turns out that the works performed by such machines could be performed equally effectively by machines lacking morally relevant characteristics."

corresponding moral rights (Sullins 2011).[10] Most non-philosophers who work in the robotics, AI, and computer science seem to believe that, if intelligent robots are substantially human-like in their capabilities, then we must also grant them moral rights. This view, however, is highly controversial[11] among those philosophers and ethicists who work in the field of ethics of technology, and therefore, one must provide a more substantial reasoning to justify moral rights for robots, as I will attempt to do below.

After this introduction, the second brief section provides some preliminary information concerning the issue of moral rights for machines. The third section offers a comprehensive account of ascribing moral rights to IR once they are fully autonomous and capable of moral reasoning and decision-making. This account is buttressed by a thorough discussion of several prominent objections in the fourth section. My responses to the objections will provide some additional information on how one should view IRs and their relationships with human beings in the future. They are followed by brief concluding remarks.

2 Machines and Moral Rights

"A full ethical agent can make explicit ethical judgements and generally is competent to reasonably justify them. An average adult human is a full ethical agent. We typically regard humans as having consciousness, intentionality, and free will. Can a machine be a full ethical agent? It's here that the debate about machine ethics becomes most heated. Many believe a bright line exists between the senses of machine ethics discussed so far and a full ethical agent. For them, a machine can't cross this line. The bright line marks a crucial ontological

10 Gibilisco (2003, 268–270) distinguishes five generations of robots according to their particular capabilities: (1) robots that are mechanical, stationary, fast, physically rugged, based on servomechanisms, but without external sensors and AI (before 1980); (2) robots that are programmable (by virtue of microcomputer control), having vision and tactile systems, position, and pressure sensors (1980–1990); (3) robots that are mobile and autonomous, able to recognize and synthesize speech, having incorporated navigation systems or tele-operated, and possessing AI (mid-1990s and after); (4) and (5) speculative robots of the future that are able to reproduce, have a sense of humor, etc. (see also Preliminary Draft of COMEST on Robotics Ethics 2016, 4.)

11 This debate can be compared to the heated debates over abortion, animal rights, and environmental rights over the past few decades, in that it is by no means clear that possessing similar capabilities to human beings should eventually lead to the granting of moral rights to robots.

difference between humans and whatever machines might be in the future" (Moor 2006, 20).[12]

The trajectory of technological development seems, indeed, quite promising and there is no principal reason why one should not consider the possibility of machines that will become full ethical agents. In one of the most interesting and morally challenging episodes (number 35) of Star Trek: The Next Generation (1989), Gene Roddenberry raises the simple but complex question of who owns Data, the artificial intelligent android serving on the starship Enterprise. Given Data's capabilities – he is a full ethical agent in Moor's terms – should he be allowed to make his own decisions, such as to quit his job and leave the starship crew to avoid the possibility of suffering harm in a dangerous experiment in which he has been ordered to participate? Data is, in general, portrayed as not just a moral patient but also a moral agent, and, therefore, entitled to moral rights comparable to those of human beings (including, as the episode shows, the confirmed moral and legal right to quit his job). The main difference between Data and a human person, however, is that he is artificial. Of course, at this moment, there is no Data or any other comparable IR with similar capabilities, but we might see one within this century. Given the meaning and importance of, for example, animal rights (Donaldson and Kymlicka 2013; Francione 2009) and environmental rights (Atapattu 2015) with respect to the further development of our moral reasoning, one should also be prepared to think about moral rights – or even "human" rights – for IRs once they fulfil certain particular features. History has shown us, as, forcefully demonstrated by Darwin's theory of evolution, that the proclaimed substantial difference between animals and human beings is mistaken. Any morally relevant empirical differences[13] that people have claimed to be of utmost importance for ascribing moral status and rights come, as a matter of fact, in degrees. If machines attain a capability of moral reasoning and decision-making that is comparable to the moral agency of human beings, they then should be entitled

12 The notion of an "ethical agent" amounts to the equivalent of what is commonly called a "moral agent" in the context of ethics and moral philosophy. The notion of moral denotes other people's interests and deontological constraints, whereas ethics usually refers to one's own individual interests and well-being. For a more detailed depiction, see Gordon (2013).

13 For example, reasoning (and intelligent behaviour), autonomous decision-making, feeling pain, having identifiable personal interests, and the desire to continue one's life, emotion, etc. Or consider the famous list of items provided by Warren (1973) in the context of abortion and personhood: sentience, emotionality, reason, the capacity to communicate, self-awareness, and moral agency.

to the status of full moral (and legal) agents, equipped with full moral standing and related rights (Sullins 2011; for a contrary view, see; Torrance 2005).[14][15]

In his classic paper "When Is a Robot a Moral Agent?," Sullins (2011) argues for giving intelligent robots moral rights once they fulfil three necessary requirements of robotic moral agency: (1) significant autonomy, i.e., "the machine is not under the direct control of any other agent or user" (158), (2) behaving intentionally,[16] and (3) being in a position of responsibility[17](157–160). Sullins concludes his line of reasoning with the following remarks:

> Robots are moral agents when there is a reasonable level of abstraction under which we must grant that the machine has autonomous intentions and responsibilities. If the robot can be seen as autonomous from many points of view, then the machine is a robust moral agent, possibly approaching or exceeding the moral status of human beings. Thus, it is certain that if we pursue this technology, then, in the future, highly complex, interactive robots will be moral agents with corresponding rights and responsibilities. Yet, even the modest robots of today can be seen to be moral agents of a sort under certain, but not all, levels of abstraction and are deserving of moral consideration. (Sullins 2011, p. 160)

14 In "A Robust View of Machine Ethics" (2005), Torrance argues that even if IRs share the same features that define human beings as moral agents, robots will, nonetheless, have no "intrinsic moral status," because they are non-organic. Only "genuinely sentient" beings who are organic by nature deserve our "moral concern or moral appraisal."

15 Johnson and Axinn (2014, 2) hold the contrasting view that "a person has not only rights, duties, free will, but also the imagination to understand the effect of different actions, and the ability to impose on him or herself the categorical imperative. How close do robots come to the features of a human person, the features that make for moral motivation and moral action? Such robots (i.e., robots that lack free will and imagination) certainly do not have rights." I will respond to their claims in the section on objections below.

16 Here, Sullins adheres to Floridi's idea of avoiding issues related to free will and intentionality with respect to IRs, because they are unresolved problems in human behaviour as well and hence should not be necessary conditions for ascribing moral agency to robots. Sullins (2011, p. 158) states, "If the complex interaction of the robot's programming and environment causes the machine to act in a way that is morally harmful or beneficial and the actions are seemingly deliberate and calculated, then the machine is a moral agent."

17 If the robot behaves in a way that suggests that "it has a responsibility to some other moral agent(s), [we can ascribe moral agency to a robot]" and "[i]f the robot behaves in this way, and if it fulfils some social role that carries with it some assumed responsibilities, and if the only way we can make sense of its behaviour is to ascribe to it the 'belief' that it has the duty to care for its patients, then we can ascribe to this machine the status of a moral agent" (Sullins 2011, 159).

I agree with Sullins's conclusion, except for his view that "even the modest robots of today can be seen to be moral agents" (160). The concept of moral agency – including robotic moral agency – is more sophisticated than Sullins suggests (as I discuss below in Sects. 3, 4), even though one must admit that, for example, the idea of free will in the context of autonomy and the assumption of a strong version of intentionality[18] in human behaviour are completely unresolved issues of human agency and, therefore, should not be applied to robots, as Floridi and Sanders (2004) have rightly argued.

3 Should IRs Enjoy Moral Rights?

The previous section was particularly concerned with how machines should act towards human beings. This section examines the related question of whether one should ascribe moral rights to machines once they have become full ethical agents. The first part of this section contains some general remarks on the proper status of IRs in their initial situation; the second part offers some morally relevant observations on why one should take the idea of moral rights for IRs seriously (for similar and related issues, see Wallach and Allen 2010; Anderson and Anderson 2011; Gunkel and Bryson 2014a, b).[19]

3.1 *The Initial Situation*

The history of ethics is a history of the ascription of moral rights and duties (see Singer 2011; Gunkel and Bryson 2014a, p. 6; Gunkel 2014, p. 115). It started with a focus on men of a certain particular social status, and then, it extended moral rights to include women as well as children, eventually expanding further to include (self-conscious) animals and, finally, unconscious nature. This process, of course, was by no means straightforward and linear; there have been many sudden shifts, steps backward, and moments of radical social change along the way. In recent history, views of moral rights and duties have been influenced by such powerful forces as the African-American rights movement in the US, the movement for gender equality, the animal

18 On the contrary, for example, Floridi (2011, 200) argues that an intentional state is not necessary for moral agency, since assessing this feature presupposes a so-called "privileged access" to a person's mental state, which is theoretically possible but practically unachievable. Therefore, the view that to be a moral agent, the artificially intelligent being "must relate itself to its actions in some more profound way, involving meaning, wishing, or wanting to act in a certain way and being epistemically aware of its behaviour" (200), is unnecessary.

19 For a true manifesto for treating robots morally, see Hall (2011, 32–33).

rights movement, the gay movement, and the disability rights movement. Alongside and interacting with these movements, we have seen the evolution of international human rights legislation, the moral and legal lingua franca of modern times. Furthermore, it is claimed not that all human beings do in fact practically enjoy the same moral and legal rights, but that they theoretically should enjoy these universal rights, simply because they are members of the human species. Given this moral development – which also includes, to some extent, the protection of animals and nature – it seems useful to consider IRs as potentially members of our moral community, based on their presumed capacity to reason and to make moral decisions. There are, at least, three main views: (1) the optimistic view that states that robots may not be moral agents at this moment, but that they may become so in the future, because there is no principle reason why robots should not become moral agents in the future (Dennett 1998), (2) the pessimistic view claims that robots will never become moral agents since they will never possess an autonomous free will (Bringsjord 2008), and (3) the deficient human view that says that human beings are not moral agents, but robots are because an action is free if and only if it is based on fully thoughtful reasons using strict logic (Nadeau 2006).[20] How will the relation between human beings and machines develop as a whole? This question raises important socio-political issues that we must address sooner or later (see Lin et al. 2014). Should it be possible to marry an artificially intelligent android and to create a robot partner according to one's own idiosyncratic preferences (Levy 2007)? How will this influence the concept of family in the future? Is it morally acceptable to use IRs – without their consent – in workplaces or for other duties, such as in hospitals, care homes for elderly people, civil services, prostitution, and war? Should one protect IRs against any forms of exploitation? If all of this could be possible at some future point, why should they not enjoy moral rights? What are the morally relevant differentia specifica between human beings and IRs? Or is being human already a sufficient condition for ascribing moral rights (for a contrary view, see Singer 2009)? Such questions have already been the subject of great controversy among researchers in different disciplines and will certainly remain of interest in the technological era to come. Whether IRs will face a long struggle for recognition and understanding remains to be seen. The following discussion offers some important arguments in favour of moral rights for IRs.

20 See, also Sullins's (2011, 155–157) considerations on the moral agency of robots.

3.2 *The Concept of Personhood and the Moral Agency/Patiency Divide*

This part provides a more general background for understanding and appreciating the arguments in support of moral rights for intelligent robots by focussing on the concept of personhood and the moral agency/patiency divide as understood in this debate. Tying together, some of the already mentioned considerations and preparing the ground for the following discussion will help us to better understand the principles underlying the opposing moral views on ascribing moral rights to robots.

It seems that even if robots had the same (morally relevant) capabilities as human beings, many people (including philosophers and ethicists) would still be reluctant to treat them in morally similar ways and to acknowledge their full moral standing. This is because human beings tend to ascribe full moral rights only to fellow human beings. Traditionally, personhood is tied to moral agency and used to confer moral rights on human beings. Therefore, only human beings who are moral agents enjoy the full protection of moral rights, and are considered as moral equals.

The concept of personhood, traditionally associated with autonomy and rationality (e.g. Kant 2009; Singer 1979), is the key to understanding the inner logic of this ascription of moral rights and whether certain beings are considered full moral agents or, rather, moral patients with a less high moral standing.

The traditional understandings of personhood have been severely challenged by at least five significant contemporary movements:

1. the feminist movement, stressing moral equality between the sexes;
2. the gay and lesbian movement, stressing moral equality between people of different sexual orientation;
3. the animal rights movement, stressing the importance of sentience as requiring moral behaviour towards nonhuman living beings;
4. the disability rights movement, stressing moral equality among persons regardless of physical or mental ability;
5. the environmental rights movement, stressing the importance of nonhuman natural life.

All five movements criticize the traditional concepts of moral status and attempt to replace them with a new definition of personhood that increases the number of members of the moral community.[21] That is important, because only members of the moral community enjoy moral protection. First, women

21 The concept of personhood and the limits of moral agency and patiency have been thoroughly discussed by Hernandez-Orallo (2017, Chaps. 16–18) and by Altman (2011), who examines the key notions with respect to Kant's position.

and people of gay or lesbian sexual orientation have historically not been treated as morally equal or granted the same moral rights as heterosexual men (this is still true in many countries around the world). Second, animals, human foetuses, and mentally impaired human beings have, in many or most cases, not been considered persons and, therefore, do not enjoy the same moral rights as "normal" adult human beings. They are commonly considered as moral patients, not moral agents. Indeed, there is currently no uniform definition of personhood. Elsewhere, I have suggested that we abstain from using the concept of personhood to ascribe moral rights, because reaching agreement on the necessary and sufficient criteria for personhood seems impossible (Gordon 2017).

Against this background, it may seem inconceivable that most people – including philosophers and ethicists – would ever accept the idea that intelligent robots should have the same moral rights as human beings due to their comparable capabilities. The great struggle of the five above-mentioned movements substantiates this concern. Some authors, such as Bryson (2010), believe that we should view robots as slaves and use them simply as tools. Thus, even if we succeed in creating intelligent robots with human-like abilities, there may still be significant opposition to granting them moral rights, based on the traditional idea of human personhood. But what about moral patiency for machines? We currently, at least to some extent, have accepted the ideas that (1) some animals have a high moral status and enjoy some moral rights as moral patients and (2) the environment should also be protected, because it has a moral standing, as well. The underlying reasoning, depending on the particular arguments adopted, is that animals and nature have either instrumental or intrinsic value. Applying the same line of argumentation, intelligent robots may, then, also enjoy some moral rights as moral patients (though not as moral agents, even if they possess human-like abilities). Of course, this would be a violation of the moral principle that similar cases should be treated in similar ways, but our general tendency towards speciesism is a widespread phenomenon and quite difficult to overcome (Singer 2009). Many people draw a firm line between the natural and the artificial with respect to ascribing moral status and moral rights.

Gunkel (2012), in an excellent book that reviews relevant works from both the continental and analytical traditions, considers whether machines should be seen as moral agents (first chapter) or moral patients (second chapter). Gunkel believes that both perspectives fall short of properly addressing the "machine question" and suggests deconstructing the binary agent-patient dichotomy with respect to intelligent robots. He calls this new alternative a "Copernican Revolution" to signify that it is a completely novel approach (third chapter). However, Gunkel remains undecided as to whether intelligent machines, once they have become a reality, should be seen as full moral agents equipped with the same moral rights as human beings:

> Should machines like AIs, robots, and other autonomous systems be granted admission to the community of moral subjects, becoming what would be recognized as legitimate moral agents, patients, or both? This question cannot be answered definitively and finally with a simple "yes" or "no." The question will need to be asked and responded to repeatedly in specific circumstances. However, the question needs to be asked and explicitly addressed rather than being passed over in silence as if it did not matter. (Gunkel 2012, 215)

The idea of rethinking the whole classical dichotomy between moral agents and moral patients in the context of the machine question is quite intriguing. However, it seems quite difficult to abstain from using the old concepts, as we can see when we consider Gunkel's question whether robots "would be recognized as legitimate moral agents, patients, or both." If Gunkel really attempts to deconstruct these notions and suggests abolishing the classical dichotomy, then he must also come up with a different way of expressing his position. One possible approach would be to devise a different notion that can function as an umbrella term, such as "moral being." This might be, indeed, an interesting starting point for a new analysis.[22]

Independent of the particular wording, it is striking that Gunkel (2012) and, particularly, Darling (2016) refer to animal rights as an analogy to contemplate the moral status of intelligent robots. According to Darling (2016, p. 214), one should think about a "near-future possibility to regulate people's behaviour towards certain types of robots," because misconduct towards social robots – to whom human beings feel strongly attached based on the three vital criteria of their physicality, their perceived autonomous movement, and their social behaviour (Darling 2016, 217–218) – may lead to unwelcome consequences. Specifically, people may become traumatized or desensitized when they see social robots abused, mistreated, or otherwise suffering misconduct and cruelty (Darling 2016, 224). Adhering to Kant's reasoning on how one should treat

22 I am sympathetic with this novel idea, but, in this paper, I adhere to the classical notion of moral agents and patients, because I believe that intelligent robots – once they exist – should be considered full moral agents. In the following sections, I provide several arguments in support of this claim.

animals, Darling (2016, 227–228) claims that, by analogy, one should also treat social robots morally, because mistreating them may encourage people to engage in misconduct towards fellow human beings, as well. She states, "As mentioned above, if lifelike and alive is subconsciously muddled, then treating certain robots in a violent way could desensitize actors toward treating living things similarly" (Darling 2016, 231). Therefore, she argues, one should think about implementing a legal framework for social robots consistent with that governing animal abuse. This step should be taken even if social robots lack equivalent capacities to human beings. The underlying idea seems to be that one should also protect lifelike beings "when society cares deeply enough" about them (Darling 2016, 230).

I agree with Darling's approach with respect to the "near future possibility of regulating people's behaviour towards certain types of robots" based on the above-mentioned possible detrimental consequences for fellow human beings. However, I would add that when the capabilities of intelligent robots have reached a certain level, then one must ascribe full moral rights to them, including the right to life,[23] based on non-consequentialist reasoning, independently of whether society "cares deeply enough" about them. Whether intelligent robots are, in fact, moral beings is not a matter of popular vote or social opinion. Rather, their moral status depends on empirical factors that determine whether they have attained a level of functioning at which they should be considered morally equal beings. The arguments in support of this view are provided in the next section.

3.3 *Arguments in Support of Moral Rights for* IRs

The following four different but related reasons may not sufficiently support moral rights for machines if taken individually; in combination, however, they seem to offer a sufficient line of reasoning in favour of moral rights for machines that cannot be dismissed offhandedly. This general line of argumentation will be supported in the next section, where I make additional points in support of moral rights for machines while responding to objections raised against this idea. Together, these two sections substantiate the main position of this paper.

23 Darling, however, does not entertain the idea of granting intelligent robots the right to life: "Animals themselves are not protected from being put down, but rather only when ending their lives is deemed cruel and unnecessary given the method or circumstances. Similarly, it would make little sense to give robots a 'right to life'" (Darling 2016, 229).

3.3.1 Protection Against Exploitation

If machines will be capable of rational choice, including the ability to make moral decisions at some point in the future, then it seems crystal clear that they are not only moral patients (i.e., moral objects) but also moral agents (i.e., moral subjects) by virtue of their autonomy and hence deserve to become members of our moral community. Their moral status must then be protected by moral and legal rights (Calverley 2011). For example, to force an artificially intelligent being that is capable of moral reasoning and decision-making to perform actions – e.g., serving in war, engaging in prostitution, or committing illegal deeds – that stand in contrast with his or her own interest is not only immoral by contravening the declared wishes of that being but also becomes a matter of pure exploitation of a being capable of moral reasoning and decision-making.

One might object that a machine, even a highly advanced moral machine, is not free to do what it wants, since it is always (legally) owned by someone else – for example, a human being, an institution, or a company. The underlying rationale for this objection adheres to the claim that the producers or owners of the machine can do whatever they want with it since they legally own it. This line of reasoning is misleading and overlooks at least two counter-arguments. First, the proposed argument suggests that whenever something has been produced, it, therefore, either belongs to the producer or the person who has legally purchased the particular good (independently of whether that good is in fact capable of making its own autonomous moral decisions). If that were the case, however, then human children would also be owned by their parents, since the parents are responsible for having produced them and given them life. Of course, parents are the legal guardians of their children and can, therefore, make decisions on their behalf – within a legal framework that protects children against abuse by their parents – until they are capable of making their own autonomous decisions. In this respect, parents do not own their children and are not allowed to do whatever they want with their children, even though they have "produced" the children. Likewise, following the same line of reasoning, advanced moral machines must be treated as autonomous agents; they have been produced by someone, but they should be considered as free persons who cannot be legally owned by another person, an institution, or a company. As members of the moral community, they should enjoy the same protection against exploitation as do their fellow community members, such as human beings. Admittedly, at this moment, it is hard to conceive of this setting, but on the assumption that IRs will eventually become autonomous moral agents, and then, the above reasoning seems to be correct.

Second, a person cannot be legitimately owned by someone else. This is the modern moral outcome of the institution of slavery, which has been part of the history of human beings since very early times. It has always been the general strategy of a dominating group to deny particular minorities – seen as the other – full, i.e., equal, moral status, or personhood so as to deny them equal moral protection. African Americans, Jews, Sinti and Roma, and women have been among the many victims of this strategy over time. Nowadays, it has become common to ascribe moral rights only to persons (including higher animals, such as the great apes, that are considered as moral persons, as well) and to grant them full moral and legal protection, while other beings, who do not fulfil the suggested criteria of personhood,[24] do not enjoy the same full moral and legal protection.[25] Against this background, however, one should not make the same mistake at some future point and treat autonomous artificial beings as the other by violating their legitimate moral right to be free from any type of suppression.

3.3.2 Moral Protection Based on Rationality

A related line of argumentation to protect moral machines from exploitation might apply a neo-Kantian approach as follows. Rational beings have dignity insofar as they are capable of acting morally, i.e., autonomously. In other words, autonomy is the foundation of the dignity of each rational being, including human beings (see Kant's Groundwork) and – one could further argue – artificially intelligent beings, as well. As Kant rightly claims, beings that have dignity should not be allowed either to exploit themselves or to use other rational beings as mere means only, but should respect their own dignity and that of other rational beings.

One might object that Kant's argument does not apply to machines, since they are clearly not human beings. This objection is misleading and prematurely suggests that Kant's ethics applies only to human beings. On the contrary, Kant's ethics is applicable to all rational beings, independently of their biological origin, and does not rely on any type of moral preference concerning the human species (this is the so-called logocentrism of Kant's ethics).

24 For example, self-consciousness, consciousness, ability to feel pain, having feelings, perceiving oneself as an entity that exists and has an interest in its future existence, etc.

25 For problems with this conception, consider the problem of abortion in medical ethics, the moral status of human beings with severe mental impairments in disability studies, and the moral status of animals in the context of the animal rights movement. The idea of linking the very right to exist with certain particular criteria that fulfil the idea of personhood is a contested but widely held position (Gordon 2016; Koch 2004).

Another, more complex issue is whether one should be allowed to use rational beings, particularly if they are not humans, for some higher ends of humanity by admitting that even if they are not used as mere means for some contingent personal ends, they could still be used, at least, to accomplish some justified and highly important ends for humanity. For example, one might argue, based on this approach, that IRs (even though they may morally condemn it) are obligated to fight in wars for their nation or the world, because they are better fighters and less valuable than human lives. What should not be tolerated among human beings, namely the making of comparisons of moral value between people or ethnicities – even though this has often been done in the past – might be possible in some limited cases to justify using IRs based on their substitutability. Furthermore, if an IR could be fully substituted, because the "mind" of the machine could be stored externally in case of emergency without any loss of the individual personality (whether this is actually possible, of course, remains an empirical question), then that might justify the use of IRs rather than human beings in dangerous situations, since human beings are not substitutable.

A more challenging response against ascribing moral rights to machines concerns the idea of dignity as the foundation of moral rights. It could be argued that only human beings have dignity, and therefore, machines are not members of the moral community, simply because they lack human dignity.

Admittedly, the concept of human dignity, by definition, refers only to human beings and not to other non-human beings. The general idea, however, that dignity as such only concerns human beings is unconvincing, since it has been increasingly acknowledged that animals have dignity as well (Donaldson and Kymlicka 2013). They may not have human dignity, but they do have what can be called animal[26] dignity (which becomes relevant in cases where human beings abuse animals). In a similar fashion, one should ascribe rational dignity to artificially intelligent machines, provided that they are capable of autonomous rational decision-making and part of our world. Yet another possible response is that the notion of dignity is itself a ontologically vague and unclear concept and should be replaced by the notion of autonomy itself, because this is, strictly speaking, the phenomenon that people have in mind in all or most

26 In "Dignity and Animals: Does It Make Sense to Apply the Concept of Dignity to All Sentient Beings?" Federico Zuolo (2016, 1117–1130) argues that the main arguments – e.g., by Nussbaum (2006) and Meyer (2001) – for ascribing dignity to animals (i.e., the species-based approach, moral individualism, and the relational approach) are unconvincing and that one should instead use other normative concepts to justify the moral importance of animals.

cases in which they refer to human dignity (see Macklin 2003; Gordon 2014). One may disagree with the latter view even if one admits that the notion of dignity is currently not defined in a promising way so as to avoid certain pitfalls, such as proving to be inconclusive in complex cases in bioethics (Cochrane 2010). However, even if we are currently unable to properly define this key ethical term, we do acknowledge clear cases of indignity when we are confronted by such cases as a soldier being dragged behind a car through the streets or animals who are severely abused and confined in small pens without fresh air and proper food (Gordon 2014). These are clear instances of indignity, and sentient moral machines capable of rational reasoning and decision-making should also not be abused in a way that harms their dignity as rational beings, such as by forcing them to commit prostitution or other immoral deeds.

3.3.3 The Presumption of Equal Treatment

Moral rights are commonly ascribed to (human) beings who are persons, i.e., who have personhood. For the sake of argument, let us grant this contested but widely held view. The consequence of this statement is that all human beings who fulfil the particular criteria of personhood, whatever these might be, are entitled to moral protection. For example, according to Kant, the key criterion would be the capability to act according to the moral law to which one subscribes to on the basis of one's own autonomy. The utilitarian philosopher Jeremy Bentham disagreed with this general idea and claimed that not the ability to reason, but the capability to feel pain was the key criterion that justified moral protection. In other words, all sentient beings (including animals) who fulfil this particular criterion do have moral rights, in particular the right to life. Given the two different accounts, it is quite possible that artificially intelligent beings might, in some distant future, fulfil either or both of the above- mentioned criteria, i.e., they will be capable of rational reasoning and of acting according to the moral law (based on their autonomy) and/or able to feel pain and thus to have feelings. In either case, given the presumption of equal treatment, one must also ascribe moral rights to IRs based on their particular design.

The history of ethics reveals that the presumption of equal treatment in the context of ascribing moral rights has often not been applied impartially. Instead, it has been restricted exclusively to the human species, regardless of whether other non-human beings may fulfil the very same criteria of personhood by which they should be entitled to moral protection (Singer 1979). The underlying reason for this situation is that the idea that moral protection should also include sentient, but non-rational animals remains highly contested and rejected by many ethicists and laypeople alike. That experts and laypeople, however, regularly commit fallacies based on speciesism in the context of ascribing moral rights has convincingly been pointed out by Singer (1975, 1979, 2009). In the past, the same line of reasoning has been applied to reject the moral rights of otherwise equally entitled human groups, based on incoherence and prejudice: women, African Americans in the US, people with impairments, and many others.

It might be objected that even though such injustices have occurred in the past (with respect to the presumption of equal treatment among fellow human beings and concerning sentient animals), human beings, and animals, unlike artificially intelligent beings, deserve moral protection, simply because they are natural living beings and are, therefore, also able to suffer in a deep and existential way that moral robots could not experience. The idea of naturalness is morally unimportant, however, since it is based on a moral prejudice that is, in general, implausible and unconvincing. The fact that a being is natural and not artificial says nothing about its moral status. The bare ontological substance of a being does not convey any morally relevant features (for a contrary view, see Torrance 2005; Johnson 2011, 169–172). In the movie A.I. Artificial Intelligence (2001), the child robot David was abandoned by his biological mother and experienced severe sorrow that can easily be seen as the equivalent of existential human suffering. Of course, David is not real but a movie character, and hence, one should not take this story line too seriously; no current robot is capable of suffering in the way that this movie portrays. However, we cannot rule out the possibility that artificially intelligent beings will have deep-seated feelings and emotions at some future point. In the final analysis, whether robots will become capable of moral reasoning and thereby be full ethical agents it is an empirical and not a normative question. The important question is, rather, whether humanity will, when that time arrives, be willing to accept robots as one of their kind, i.e., as moral beings.

3.3.4 Brutalization of the Human Character

Another important line of argumentation that Kant uses in the context of how we should treat animals is actually based on a virtuous line of reasoning. He claims that we should treat animals well, because mistreating them would negatively affect our behaviour towards our fellow human beings by causing us to become morally less sensitive over time. In other words, our moral character would be brutalized if we mistreated animals and we would start acting in a similar way towards other human beings. Strictly speaking, Kant does not say that mistreating animals is immoral, but just that it is something that one should not do. Analogously, human beings should treat artificially intelligent androids well, because mistreating them, especially in view of their great physical similarities to human beings, would make us morally less sensitive to fellow human beings (see also Anderson 2011a, b, 293–295).

This line of argument has its strengths and weaknesses. Its great strength is that it does not presuppose that the object of morality – i.e., the sentient animal or IR – has an inviolable moral right to life; rather, it argues for treating the object well, because acting otherwise would harm the moral agent himself or herself. Its great weakness, of course, is that the object of morality itself is not granted any moral claim. Others are not morally bound to act in a certain particular way towards the objects, such as moral robots, because of any inherent characteristic that they possess.

One objection might question the degree of the emotional relationship between human beings and robots, which might differ substantially from our emotional connection to real animals, thus causing us not to give robots the kind treatment that we grant to animals. This objection is premature, however. It has been observed, for example, that cuddly toys in care homes for elderly people positively influence the atmosphere as well as the residents' emotional and medical condition. The elderly people do, indeed, treat the cuddly toys – in particular, Paro, a robot seal with fur – with care and feel sad when they are not available or broken (Turkle 2011, 71). The actual relationship between these elderly residents and Paro is emotionally deep and complex. The same can be observed in situations where people, including children, use responsive programs such as Eliza[27] and robots such as Kismet[28] and Cog[29] (for a good overview of such examples, see Turkle 2011, 62–76). Given these examples, one can only imagine how relationships might develop when artificially intelligent beings are part of our world, life, and homes, as service robots or even as partners (Levy 2007). Therefore, it is quite conceivable that strong emotional attraction to a robot could establish moral bonds of a particular type that would eventually justify ascribing not only a moral status but even moral rights to IRs (Coeckelbergh 2014).[30]

27 Eliza is a chat program designed to mirror the thoughts of users, so as to give the impression that Eliza is consistently supportive. This mechanism has created a strong emotional effect (the so-called Eliza effect) on many people who have used the program.

28 Kismet, developed at MIT, is a complex robot that responds to facial expressions, vocalizations, and one's tone of voice.

29 Cog, developed at MIT, can follow human motion, imitate behaviour, and track eye movements.

30 Coeckelbergh (2014) questions the standard approach of ascribing moral rights to beings based on properties such as the ability to reason or to feel pain; instead, he suggests using a relational and phenomenological approach, contending that moral status emerges through relations between different beings (in particular 69–70). I do agree, at least, to some extent with his view of relations between beings as highly important in evaluating

4 Objections

The next section contains some additional objections to the account presented above. In response to these objections, further arguments are provided to substantiate the claim that IRs should be entitled to moral rights once they are capable of moral reasoning and decision-making.

4.1 *The Free Will Defence*

The objection that IRs, no matter how technologically advanced they may be, are in principle unable to become moral agents, because they lack free will has been forcefully claimed by, for example, Johnson and Axinn (2014) and Rodogno (2016). In their view, moral agency necessarily presupposes free will. The underlying claim is that only human beings – and not robots – can be held morally responsible for their deeds, because they are the only beings who have free will. Therefore, it would be inappropriate to ascribe moral rights to IR.

It is of course impossible to solve the free will problem in this section, but I would like to highlight two unstated and highly controversial claims on which the above-mentioned authors (and many others in the debate) base their line of reasoning: first, that human beings have free will; second, that free will is a necessary precondition of moral agency. Most critics who work in the field of robotics and who use the free will defence against ascribing moral agency to IRs are unaware of the complex and interdisciplinary debate over free will (for an excellent overview, see the edited volumes of Watson 2003; Kane 2002; Pothast 1978). For example, the critics do not properly distinguish between the two central notions of freedom of will[31] and freedom of action[32] but, instead, use the definition of free action to define the notion of free will. This debate has become very sophisticated, with numerous thoughtful approaches associated with the main strands such as compatibilism (soft determinism),[33]

moral status, but Coeckelbergh's questioning of the very idea of moral standing and his view of relations as morally foundational are somewhat unconvincing. Nonetheless, the relational approach has proven to be an important perspective in the context of disability studies, as well, particularly with regard to the moral status of people with severe mental impairments (Koch 2004). In both cases, the vital idea is to adhere to the concrete relation between two parties, whether it is the relation between the non-impaired human being and the person with mental impairment, or the human-robot relation.

31 The freedom to will what one wants to will.

32 The freedom to act according to one's own will.

33 Free will is compatible with a world of physical determinism.

incompatibilism (hard determinism),[34] and libertarianism,[35] as well as independent views that cannot be discussed here. However, the dispute clearly evidences that the underlying premise used by the above-mentioned critics – that human beings have free will – is controversial, given the existence of causality in the empirical world (i.e., physical determinism). Second, many authors in the free will debate (for example, Frankfurt 1969, 1971) believe that moral agency does not necessarily presuppose free will. In other words, the critics must prove both that free will exists and that it is necessary for moral agency before they can argue that IRs cannot be moral agents, because they lack free will. The critics of the free will defence in robotics must provide a more substantial line of reasoning along the lines of the libertarian strand to prove their point.

The more vital and related question, however, is whether the capability of moral reasoning and decision-making itself is enough to ascribe moral agency to artificially intelligent beings without adhering to the complex notion of free will. Against this background, it seems unfair to argue that robots must have free will to earn moral agency, even though the ideas that free will even exists or that human beings possess it remain controversial. Thus, until critics such as Johnson and Axinn (2014) and Rodogno (2016) provide a substantial argument in support of their underlying idea that free will is necessary for moral agency and that robots are, in principle, incapable of possessing it; their general claim is unsupported and misleading.

4.2 *The Following-a-Program Objection*

Searle (1980, 1994) famously claims that machines cannot truly make decisions, because they are only following their program. Therefore, the very idea of IRs possessing moral agency that provides a basis for corresponding moral rights and duties is impossible. This frequently raised objection is related to the above-mentioned case of the free will defence, in that it suggests that IRs are strictly determined in their decisions based on their programming and, therefore, lack autonomous reasoning and decision-making.[36] Davenport, however, argues against this view, contending that "robots can learn new 'rules' as a result of interactions with the environment and/or internal reflections on past interactions. These new rules physically change the causal make-up of the mechanism, thus producing new behaviours, so that, in the future,

34 A deterministic world and free will are incompatible.

35 Free will (in a strong sense) presupposes an indeterministic world without (full) causation of mental events.

36 For a more detailed discussion of this objection, see Whitby (2011, 140–142).

in essentially identical circumstances, the robot may act completely differently" (2014, 53).

In earlier stages of computer development, chess programs such as IBM's DeepBlue depended on brute-force search when competing against human competitors, even when defeating world champions such as Garry Kasparov in 1996 and 1997. Nowadays, AI programs such as Google DeepMind's AlphaGo Zero operate quite differently from its predecessors. Starting from tabula rasa (blank slate learning), AlphaGo Zero was able to learn the complex traditional Chinese game of Go within just a few hours on its own, after being given only the basic rules of how to play Go and no further instructions. AlphaGo Zero was eventually able to beat the previous Go program, AlphaGo (numerous times by the score of 100 to 0), that defeated world champion Lee Sedol in 2017 (Silver et al. 2017).

It turns out that the particular combination of self-play reinforcement learning and deep neural network architecture is a key to the further development of AI programs. Interestingly, the same system is currently used in an unrelated domain to study the so-called protein folding in medicine, where the results could be used to find cures for "many devastating diseases, including Alzheimer's, Parkinson's, and cystic fibrosis" that originate from misfolded proteins (Knapton 2017). Accordingly, the self-taught AI program AlphaGo Zero is only the beginning, not the end, of progress in reinforcement learning.

This development presages the conceivable possibility that we will ultimately be able to build autonomous IRs that will act completely independently of human beings in the moral domain. As mentioned above, however, before we can ascribe moral rights to any being (including artificial beings), that being must meet the relevant criteria for moral agency. The minimum criteria for moral agency are commonly identified as autonomy and rationality. Therefore, if a being lacks either autonomy and/or rationality, then it will normally be considered a moral patient and not a moral agent (see Sect. 3.2). In this context, AlphaGo Zero could offer an interesting test case for considering IRs' capacity for moral reasoning and decision-making, and perhaps, such a test will be conducted in the future. Given its great potential, one would expect that it could quickly outperform many (or even all) existing moral programs, such as the casuistic approach of Rzepka and Araki (2005), the web-based approach of Guarini (2006), and the MoralDM of Dehghani et al. (2011). Amidst such developments, whether we will eventually reject the pessimistic view that a robot can never do something that it is not programmed to do is a purely empirical question.

In consequence, the threshold of granting intelligent robots moral rights – as I argue in this paper – is eventually a complex amalgam of (1) some necessary capabilities such as autonomy, rationality, and the ability of moral reasoning and decision-making and (2) the existence of morally important social ties between human beings and robots that prompt the robots to go beyond their initial programming and to develop features that make them individually unique. That means, "sophisticated robots will undoubtedly develop unique identities, becoming, in a very real sense, individuals. As they live and work together with humans and other robots, they will naturally assimilate and develop moral rules that guide their social interactions. Eventually, we will come to accept them as fully moral agents, treating them as we treat other humans." (Davenport 2014, 58).

4.3 *Moral Rights and the Idea of Dignity*

A third objection contends that moral rights should apply only to natural beings such as human beings. Since robots are neither natural nor human beings, they should, therefore, not enjoy moral rights. In other words, robots are artificial beings without dignity worthy of being protected.

This objection contains within it at least three different lines of reasoning: (1) an ontological distinction between artificial and natural beings, (2) the idea that only human beings are moral agents and therefore have moral rights, and (3) the claim that artificial beings do not have any dignity. Concerning the first point, we have already observed in 3.3.3 above that the ontological difference between artificial existence and natural life, once both are capable of moral reasoning and decision-making, is unimportant for the ascription of moral rights. The decisive factor is whether the being is a moral agent, not what the being is made of.

Second, the idea that only human beings deserve moral protection is based on a biological bias (i.e., speciesism) and does not do justice to, for example, some important approaches in the context of animal ethics, where authors such as Donaldson and Kymlicka (2013) and Francione (2009) attempt to show that animals also have strong moral rights. If proponents adhere to the criterion of capacity for reasoning when ascribing moral rights to human beings only, this account faces some vital problems. First, it opens the door to arguments, on the basis of utilitarianism or Kantianism, that people with mental impairments do not have the right to life; second, in animal ethics, proponents argue that one should protect higher animals such as the great apes who have the capacity to reason and, therefore, a moral right to life. In other words, the use of reasoning capacity as a basis for ascribing moral rights both transcends the borders of the human species, by including higher animals that are capable of reasoning, and also limits the rights accorded to some human beings, e.g., people with mental impairments, comatose people, people in permanent vegetative states, infants, and people with dementia.

Third, opponents could base the objection that artificial beings do not possess any dignity on Kant by arguing that something that is artificially created, including artificial beings such as robots, has no dignity but only a "price" and, therefore, is not part of the moral community or entitled to moral protection. This line of reasoning, however, is premature and misleading. Kant's famous distinction between irrational things (including animals) that have a "relative value" and rational persons who have an "absolute value" might instead be used against the opponents themselves (Groundwork, 60). One could argue as follows: In Kant's so-called kingdom of ends – which is the systematic connection of different rational beings through common ends – everything has either a "price" (for exchangeable equivalents) or "dignity" (for persons), (Groundwork, 68). It follows that if artificial beings, which are, nevertheless, human products, are rational (and hence autonomous), then they also have an absolute value and therefore are not simply things, but persons with dignity, worthy of moral protection. Indeed, Kant's theory is not concerned with human beings only but with all rational beings (i.e., logocentrism). But what about the issue of exchangeability with respect to robots? By definition, persons are not exchangeable but unique in Kant's reasoning and, hence, have dignity (absolute value) in contrast to exchangeable things that have a price (relative value). It appears that IRs are both rational (and hence persons with an absolute value) and exchangeable (and, therefore, having a price with a relative value). That seems to be a contradiction. This complex issue, however, can be resolved when one acknowledges that the personal identity (or individual self) of an IR is shaped and influenced by the unique and individual life history of the particular robot as an autonomous being. Hanna and Thompson (2003) distinguish between the "body" as a physical organism (Körper) and the "lived body" (Leib), which is the individual embodied experience of a particular organism in its life world. This distinction is important and could be further developed and applied to IRs as well once they have reached the particular threshold. Then, it would be no longer permissible to simply exchange the bodies of IRs, by virtue of the idea that the "lived body" depends on the particular physical body of the IR in question.[37] If one accepts this line of reasoning, it follows that IRs are no longer exchangeable and hence deserve full moral protection.

37 The idea that IRs will develop individual selves and become unique members of the community is substantiated by Davenport: "Sophisticated robots will necessarily incorporate a model of themselves and their body in order to predict the effects of their interactions with the world. This mental model is the basis of their self-identity. As time goes by, it will incorporate more and more of the agent's interactions, resulting in a history of exchanges that give it (like humans) unique abilities and knowledge. This, then, is part of what makes an individual a unique and potentially valuable member of the group. Such

4.4 *Rationality Is Not Enough*

Finally, it could be objected that, even if an artificial being is capable of reasoning, that would not suffice for the ascription of moral rights. Rather, the IR must also possess other important morally relevant criteria such as feelings and the capacity for suffering, self-preservation, and the capacity to reproduce. Only if the being has, in general, these capabilities can one ascribe full moral rights to it.

In response to this objection, first, the capability to have moral feelings and the capacity for physical suffering cannot be, in general, necessary conditions for claiming moral rights. If this were the case, then some people with mental impairments (who may lack the possibility to develop and reflect on their moral feelings for making decisions) or those with congenital analgesia (who cannot feel any physical pain) would be excluded as subjects of morality from the moral community and would become merely objects of morality. Contemporary research shows, however, that the capacity to have feelings, and moral feelings in particular, is important for moral reasoning and decision-making as such (Döring and Mayer 2002). Whether this amounts to a claim that, therefore, robots can never become full ethical agents remains to be seen. However, two brief responses are in order. First, we do not know whether it will be possible to reproduce moral feelings in robots; this is a technological issue. Second, it can be questioned whether the existence of moral feelings in human beings is necessary for moral theory, given that the most prominent and influential ethical theories, including both Kantianism and utilitarianism, completely avoid the use of moral feelings in arriving at moral decisions. Perhaps, human beings, by virtue of their particular nature, need moral feelings for their own survival as a species, but that does not mean that other rational beings must possess the same moral feelings to make equally good moral decisions.[38]

---

machines will certainly have to be consciously aware (a-consciousness) of their environment" (Davenport 2014, 56).

38 See also Allen et al. (2011, 59–60): "When it comes to making ethical decisions, the interplay between rationality and emotion is complex. Whereas the Stoic view of ethics sees emotions as irrelevant and dangerous to making ethically correct decisions, the more recent literature on emotional intelligence suggests that emotional input is essential to rational behaviour." Emotions certainly play an essential part with respect to the genesis of human morality, but emotions as such should never influence the justification of our moral reasoning and decision making. Therefore, I do not believe that emotions are necessary for IRs to arrive at correct moral decisions, but they will be essential for robots to engage with human beings on a social level. I agree with Whitby (2011, 142), who claims that there "are also many contexts in which we prefer a moral judgment to be free from emotional content," such as those made by doctors and judges. However, "[e]motion may

It could be further objected that only beings who, phenomenologically speaking, have an idea of what it really means to suffer will be responsive to human suffering. We have already cited a contrary sentiment from fiction by referring to the example of the child robot David, who suffers due to having been abandoned by his biological mother (see 3.3.3). First, the idea that human suffering itself is necessary for moral reasoning and decision-making is questionable; second, it seems not impossible that intelligent robots will be capable of deep-seated feelings and emotions in the future;[39] third, although human suffering is certainly influential in human beings in the context of issues concerning human existence, life, and death, it does not follow that only beings capable of human suffering are legitimate subjects of morality and, hence, enjoy moral protection.

Second, it could be argued that the natural instinct of self-preservation is an important part of ascribing moral rights to beings, since, without this natural instinct, the particular being is not worthy of protection. This claim is based on the idea that if a being – under normal conditions – does not care whether it is alive or may be killed, then it does not make any sense to give that being moral protection. Against this line of reasoning, one could respond that it is, in general, possible to program an artificial instinct of self-preservation into robots; after all, nature has done the same using biological means with respect to human beings. The robot's instinct would thus be analogous to the natural human instinct (see Asimov's third law of robotics), and hence, the distinction between natural and artificial is irrelevant.

Third, critics might object that to deserve moral protection, beings must be able to reproduce themselves and to strive to become better as a species. Machines, according to this view, cannot reproduce or improve themselves without human intervention; therefore, one should not consider them as moral agents. However, this objection is misleading. For example, scientists at MIT succeeded as early as 2005 in building (primitive) self-reproducing robots, including a mechanism for corrections. They believe that self-reproducing robots

well be an important component of human judgments, but it is unjustifiably anthropocentric to assume that it must therefore be an important component of all judgments" (Whitby 2011, 144).

39 In his classical paper "The Feelings of Robots" (Ziff 1959, 68), Paul Ziff claims that it is absurd to assume that robots will be capable of feeling anything. There is, however, no principled reason why this is logically impossible. For an illuminating discussion of the idea and meaning of suffering, see Gunkel (2014, 118–122) who argues that the concept of suffering is too complex and faces severe difficulties since it "remains fundamentally inaccessible and unknowable" (Gunkel 2014, 120).

are the very first step towards micro-electronic systems in nanotechnology that will be able to reproduce themselves without human intervention and even self-correct so as to become better and better with each following generation. This technological development could be used to build intelligent machines that can in turn construct other machines while avoiding the mistakes of previous machine generations. This may sound like a dangerous development to some, but it is a welcome opportunity to others. Furthermore, it seems questionable in general to view reproductive ability as morally relevant, since many human beings cannot reproduce yet remain moral subjects worth being protected. Therefore, it would be inconsistent to claim that in the case of robots, one should consider the capability to reproduce as morally relevant when we do not apply this criterion to human beings.

5 Conclusions

This paper has defended the idea of granting moral rights to artificially intelligent robots, once they are capable of moral reasoning and decision-making, against several different objections. There are no convincing reasons why IRs should be seen as morally inferior to human beings once they have reached this technological threshold. Therefore, it is of utmost importance that IRs learn how to make moral decisions and act accordingly, given their ever-increasing involvement in many sensitive fields and their increasing social interaction with human beings. Although we may not have to face this issue in the near-future (i.e., within the next couple of decades), it seems highly likely that humanity will confront this scenario at some future point. Whether IRs will eventually become better moral agents than human beings remains unknown, but we can be certain that the arrival of artificially intelligent beings will revolutionize our way of living. Whether this revolutionary change brings good or bad fortune depends mainly on our willingness to reach out to our fellow artificial beings.

References

Allen, Colin, Wendell Wallach, and Iva Smit. 2011. "Why Machine Ethics?" In *Machine Ethics*, edited by Michael Anderson and Susan Leigh Anderson, 51–61. Cambridge: Cambridge University Press.

Altman, Matthew C. 2011. *Kant and Applied Ethics: The Uses and Limits of Kant's Practical Philosophy*. New Jersey: Wiley-Blackwell.

Anderson, Susan Leigh. 2011a. “The Unacceptability of Asimov’s Three Laws of Robotics as a Basis for Machine Ethics.” In *Machine Ethics*, edited by Michael Anderson and Susan Leigh Anderson, 285–296. Cambridge: Cambridge University Press.

Anderson, Susan Leigh. 2011b. “Machine Metaethics.” In *Machine Ethics*, edited by Michael Anderson and Susan Leigh Anderson, 21–27. Cambridge: Cambridge University Press.

Anderson, Michael, and Susan Leigh Anderson, eds. 2011. *Machine Ethics*. Cambridge: Cambridge University Press.

Asimov, Isaac. 1942. *Runaround. A Short Story*. New York: Street and Smith Publications.

Asimov, Isaac. 1986. *Robots and Empire. The Classic Robot Novel*. New York: HarperCollins.

Atapattu, Sumudu. 2015. *Human Rights Approaches to Climate Change: Challenges and Opportunities*. New York: Routledge.

Bringsjord, Selmer. 2008. “Ethical Robots: The Future Can Heed Us.” *AI & Society* 22(4): 539–550.

Bryson, Joanna J. 2010. “Robots Should Be Slaves.” In *Close Engagements with Artificial Companions: Key Social, Psychological, Ethical and Design Issues*, edited bay Wilks Yorick, 63–74. Amsterdam: John Benjamins.

Calverley, David J. 2011. “Legal Rights for Machines.” In *Machine Ethics*, edited by Michael Anderson and Susan Leigh Anderson, 213–227. Cambridge: Cambridge University Press.

Capek, Karel. 1920. *Rossum’s Universal Robots*. Adelaide: The University of Adelaide.

Clarke, Roger. 2011. “Asimov’s Laws of Robotics: Implications for Information Technology.” In *Machine Ethics*, edited by Michael Anderson and Susan Leigh Anderson, 254–284. Cambridge: Cambridge University Press.

Cochrane, Alasdair. 2010. “Undignified Bioethics.” *Bioethics* 24(5): 234–241.

Coeckelbergh, Mark. 2014. “The Moral Standing of Machines: Towards a Relational and Non-cartesian Moral Hermeneutics.” *Philosophy & Technology* 27(1): 61–77.

Darling, Kate. 2016. “Extending Legal Protection to Social Robots: The Effects of Anthropomorphism, Empathy, and Violent Behavior Towards Robotic Objects.” In *Robot Law*, edited by Ryan Calo, A. Michael Froomkin and Ian Kerr, 213–231. Northampton: Edward Elgar.

Davenport, David. 2014. “Moral Mechanisms.” *Philosophy & Technology* 27(1): 47–60.

Dehghani, Morteza, Ken Forbus, Emmett Tomai, and Matthew Klenk. 2011. “An Integrated Reasoning Approach to Moral Decision Making.” In *Machine Ethics*, edited by Michael Anderson and Susan Leigh Anderson, 422–441. Cambridge: Cambridge University Press.

Delvaux, Mady. 2017. *Report with Recommendations to the Commission on Civil Law Rules on Robotics* (2015/2103 (INL)).

Dennett, Daniel. 1998. "When HAL kills, Who's to Blame? Computer ethics." In *Hal's Legacy: 2001's Computer as Dream and Reality*, edited by David Stork, 351–365. Massachusetts: The MIT Press.

Donaldson, Sue and Will Kymlicka. 2013. *Zoopolis. A Political Theory of Animal Rights*. Oxford: Oxford University Press.

Döring, Sabine, and Verena Mayer, eds. 2002. *Die Moralität der Gefühle*. Deutsche Zeitschrift für Philosophie, Sonderband 4. Berlin: Akademie Verlag.

Floridi, Luciano. 2011. "On the Morality of Artificial Agents." In *Machine Ethics*, edited by Michael Anderson and Susan Leigh Anderson, 184–212. Cambridge: Cambridge University Press.

Floridi, Luciano, and JW Sanders. 2004. "On the Morality of Artificial Agents." *Minds and Machines* 14(3): 349–379.

Francione, Gary L. 2009. *Animals as Persons: Essays on the Abolition of Animal Exploitation*. New York: Columbia University Press.

Frankfurt, Harry. 1969. "Alternate Possibilities and Moral Responsibility." *The Journal of Philosophy* 66(23): 829–839.

Frankfurt, Harry. 1971. "Freedom of the Will and the Concept of the Person." *The Journal of Philosophy* 68(1): 5–20.

Gibilisco, Stan. 2003. *Concise Encyclopedia of Robotics*. New York: McGraw-Hill.

Gordon, John-Stewart. 2013. "Modern Morality and Ancient Ethics." *Internet Encyclopedia of Philosophy*. http://www.iep.utm.edu/anci-mod/.

Gordon, John-Stewart. 2014. "Human Dignity, Human Rights, and Global Bioethics." In *Global Bioethics and Human Rights: Contemporary Issues*, edited by Wanda Teays and Alison Dundes Renteln, 68–91. Lanham: Rowman & Littlefield.

Gordon, John-Stewart. 2016. "Human Rights." In *Oxford Bibliographies in Philosophy*. https://www.oxfordbibliographies.com/view/document/obo-9780195396577/obo-9780195396577-0239.xml?rskey=z2W9vS&resul%20t=47&q=.

Gordon, John-Stewart. 2017. "Remarks on a Disability-Conscious Bioethics." In *Human Rights and Disability. Interdisciplinary Perspectives*, edited by John-Stewart Gordon, Johann-Christian Pöder and Holger Burckhart, 9–20. London: Routledge.

Grau, Christopher. 2011. "There is no 'I' in 'Robot': Robots and Utilitarianism." In *Machine Ethics*, edited by Michael Anderson and Susan Leigh Anderson, 451–463. Cambridge: Cambridge University Press.

Guarini, Marcello. 2006. "Particularism and the Classification and Reclassification of Moral Cases." *IEEE Intelligent Systems* 21(4): 22–28.

Gunkel, David J. 2012. *The Machine Question: Critical Perspectives on AI, Robots, and Ethics*. Cambridge: MIT Press.

Gunkel, David J. 2014. "A Vindication of the Rights of Machines." *Philosophy & Technology* 27(1): 113–132.

Gunkel, David J., and Joanna J. Bryson. 2014a. "Introduction to the Special Issue on Machine Morality: The Machine as Moral Agent and Patient." *Philosophy & Technology* 27(1): 5–8.

Gunkel, David J., and J Bryson. 2014b. "The Machine as Moral Agent and Patient." *Philosophy & Technology*, Special Issue 27(1): 5–142.

Hall, J. Storrs. 2011. "Ethics for Self-improving Machines." In *Machine Ethics*, edited by Michael Anderson and Susan Leigh Anderson, 512–523. Cambridge: Cambridge University Press.

Hanna, Robert, and Evan Thompson. 2003. "The Mind-Body-Body Problem." *Theoria Et Historia Scientiarum* 7(1): 24–44.

Hernândez-Orallo, José. 2017. *The Measure of All Minds: Evaluating Natural and Artificial Intelligence.* Cambridge: Cambridge University Press.

Johnson, Deborah G. 2011. "Computer Systems: Moral Entities but not Moral Agents." In *Machine Ethics*, edited by Michael Anderson and Susan Leigh Anderson, 168–183. Cambridge: Cambridge University Press.

Johnson, Aaron M., and Sidney Axinn. 2014. "Acting vs. Being Moral: The Limits of Technological Moral Actors." *Proceedings of the IEEE 2014 International Symposium on Ethics in Engineering, Science, and Technology* (ETHICS), 30: 1–4.

Kane, Robert, ed. 2002. *The Oxford Handbook of Free Will.* Oxford: Oxford University Press.

Kant, Immanuel. 2009. *Groundwork of the Metaphysic of Morals.* New York: Harper Perennial Modern Classics.

Knapton, Sarah. 2017. "AlphaGo Zero: Google DeepMind supercomputer learns 3,000 years of human knowledge in 40 days." *The Telegraph*, October 18, 2017. https://www.telegraph.co.uk/science/2017/10/18/alphago-zero-google-deepmind-supercomputer-learns-3000-years/.

Koch, Tom. 2004. "The Difference that Difference Makes: Bioethics and the Challenge of 'Disability.'" *Journal of Medicine and Philosophy* 29(6): 697–716.

Levy, David. 2007. *Love and Sex with Robots: The Evolution of Human-Robot Relationships.* New York: Harper.

Lin, Patrick, Keith Abney, and George Bekey, eds. 2014. *Robot Ethics: The Ethical and Social Implications of Robotics. Intelligent Robotics and Autonomous Agents.* Cambridge: The MIT Press.

Macklin, Ruth. 2003. "Dignity is a Useless Concept." *BMJ British Medical Journal* 327(7429): 1419–1420.

Meyer, Michael. 2001. "The Simple Dignity of Sentient Life: Speciesism and Human Dignity." *Journal of Social Philosophy* 32(2): 115–126.

Moor, James H. 2006. "The Nature, Importance, and Difficulty of Machine Ethics." *IEEE Intelligent Systems* 21(4): 18–21.

Nadeau, Joseph E. 2006. Only Androids can be Ethical. In *Thinking About Android Epistemology*, edited by Ken Ford and Clark Glymour, 241–248. Cambridge: MIT Press.

Nussbaum, Martha. 2006. *Frontiers of Justice. Disability, Nationality, Species Membership.* Cambridge: The Belknap Press of the Harvard University Press.

Picard, Rosalind. 1997. *Affective Computing*. Cambridge: The MIT Press.

Pothast, Ulrich, ed. 1978. *Seminar: Freies Handeln Und Determinismus*. Frankfurt am Main: Suhrkamp.

Rodogno, Raffaele. 2016. "Robots and the Limits of Morality." In *Social Robots. Boundaries, Potential, Challenges*, edited by Marco Norskov. Routledge. http://pure.au.dk/portal/files/90856828/Robots_and_the_Limit s_of_Morality.pdf.

Rzepka, Rafal, and Kenji Araki. 2005. "What Statistics Could Do for Ethics? The Idea of Common Sense Processing Based Safety Valve." *AAAI Fall Symposium on Machine Ethics, Technical Report* FS-05-06: 85–87.

Searle, John. 1980. "Minds, Brains and Computers." *The Behavioral and Brain Sciences* 3(3): 417–457.

Searle, John. 1994. *The Rediscovery of Mind*. Cambridge: The MIT Press.

Silver, David et al. 2017. "Mastering the Game of Go Without Human Knowledge." *Nature* 550: 354–359.

Singer, Peter. 1975. *Animal Liberation*. London: Avon Books.

Singer, Peter. 1979. *Practical Ethics*. Cambridge: Cambridge University Press.

Singer, Peter. 2009. "Speciesism and Moral Status." *Metaphilosophy* 40(3–4): 567–581.

Singer, Peter. 2011. *The Expanding Circle: Ethics, Evolution, and Moral Progress.* Princeton: Princeton University Press.

Sullins, John P. 2011. "When Is a Robot a Moral Agent?" In *Machine Ethics*, edited by Michael Anderson and Susan Leigh Anderson, 151–161. Cambridge: Cambridge University Press.

Torrance, Steve. 2005. "A Robust View of Machine Ethics." In *Technical Report – Machine Ethics: Papers from the AAAI Fall Symposium*, FS-D5-06, American Association of Artificial Intelligence, Menlo Park, 88–93.

Turkle, Sherry. 2011. "Authenticity in the Age of Digital Companions." In *Machine Ethics*, edited by Michael Anderson and Susan Leigh Anderson, 62–76. Cambridge: Cambridge University Press.

United Nations. 2016. "Preliminary Draft Report of COMEST on Robotics Ethics." SHS/YES/COMEST-9EXT/16 (3): 1–31. http://unesdoc.unesco.org/images/0024/002455/245532E.pdf.

Wallach, Wendell, and Colin Allen. 2010. *Moral Machines: Teaching Robots Right from Wrong*. Oxford: Oxford University Press.

Warren, Mary. 1973. "On the Moral and Legal Status of Abortion." *Monist* 57(1): 43–61.

Watson, Gary, ed. 2003. "Free Will." *In Oxford Readings in Philosophy*. Oxford: Oxford University Press.

Whitby, Blay. 2011. "On Computable Morality: An Examination of Machines as Moral Advisors." In *Machine Ethics*, edited by Michael Anderson and Susan Leigh Anderson, 138–150. Cambridge: Cambridge University Press.

Ziff, Paul. 1959. The Feelings of Robots. *Analysis* 19(3): 64–68.

Zuolo, Federico. 2016. Dignity and Animals. Does it Make Sense to Apply the Concept of Dignity to All Sentient Beings? *Ethical Theory Moral Practice* 19(5): 1117–1130.

CHAPTER 2

Robot Rights – Thinking the Unthinkable

David J. Gunkel

1 Introduction

Whether we recognize or not, we are in the midst of a robot invasion. The machines are now everywhere and doing virtually everything. We chat with them online. We play with them in digital games. We collaborate with them at work. And we rely on their capabilities to manage many aspects of our increasingly complex data-driven lives. Consequently, the "robot invasion" is not something that will transpire as we have imagined it in our science fiction, with a marauding army of evil-minded androids descending from the heavens. It is an already occurring event with machines of various configurations and capabilities coming to take up positions in our world through a slow but steady incursion.

As these various mechanisms come to occupy influential positions in contemporary culture – positions where they are not necessarily just tools or instruments of human action but a kind of social entity in their own right – we will need to ask ourselves some rather interesting but difficult questions. At what point might a robot, algorithm, or other autonomous system be held accountable for the decisions it makes or the actions it initiates? When, if ever, would it make sense to say, "It's the robot's fault." Conversely, when might a robot, an intelligent artifact, or other socially interactive mechanism be due some level of social standing or respect? When, in other words, would it no longer be considered nonsense to ask a question like "Can and should robots have rights?"

Although considerable effort has already been expended on the question of robots and responsibility, the other question, the question of *robot rights*, remains conspicuously absent or at least marginalized.[1] In fact, as David Levy (2005, 393) has asserted, "the notion of robots having rights is unthinkable." This chapter is designed as both a provocation and thoughtful response to this

1 The few exceptions that are available in the literature are less "exceptions" and more an indication of the marginal status of this alternative mode of inquiry, cf. Coeckelbergh 2010, Gordon 2018, Gunkel 2018.

 | DOI:10.1163/9789004437876_004

apparent "fact." It is, therefore, deliberately designed to think the unthinkable by critically considering and making a serious philosophical case for the rights of robots. It does so not to be controversial, even if controversy is often the result of this kind of philosophical intervention, but in order to respond to real and pressing challenges concerning emerging technology and the current state of and future possibilities for moral philosophy.

2 The Unthinkable

"The notion of robot rights," as Seo-Young Chu (2010, 215) points out, "is as old as is the word 'robot' itself. Etymologically the word 'robot' comes from the Czech word 'robota,' which means 'forced labor.' In Karel Čapek's 1921 play *R.U.R.*, which is widely credited with introducing the term 'robot,' a 'Humanity League' decries the exploitation of robot slaves." For many researchers and developers, however, the very idea of "robot rights" is simply unthinkable. Versions of this kind of immediate and largely unquestioned dismissal circulate in the work of noted experts in the field:

> Noel Sharkey – "The idea of machine consciousness and rights is a distraction, it's fairy tale stuff. We need proper informed debate, about the public safety about for instance the millions of domestic robots that are predicted to be arriving in the next few years." (Henderson 2007 and Marks 2007)
>
> Alan Winfield – "I know of no one within the serious robotics community that would use that phrase, 'robot rights'." (Henderson 2007)
>
> Luciano Floridi – "It may be fun to speculate about such questions, but it is also distracting and irresponsible, given the pressing issues we have at hand." (Floridi 2017, 4)

Whatever the reasons, there is a deliberate decision and concerted effort not to think – or at least not to take as a serious matter for thinking – the question of robot rights. The very idea is openly mocked as ridiculous and laughable; it is cited for engaging in pointless speculation that has the effect of distracting one from the true and serious work that needs to be done; or (at best) the subject is pushed aside and marginalized as a kind of entertaining sidebar that might be fun, but which does not really need serious attention. This is, however, precisely the reason why we must think the unthinkable – to challenge these declarations, assumptions, and orthodoxies; and this needs to transpire for at least two reasons.

First, any dogmatic declaration of this kind should already make us uneasy and suspicious. The fact that this thought has been declared "unthinkable" should give one pause and trigger critical questions like: Who says? Who gets to decide in advance what we can or cannot think? And, perhaps more revealing, what values and assumptions are being protected through this kind of proscription and prohibition? As Barbara Johnson (1981, xv) reminds us, the task of critical thinking is to read backwards from what seems obvious, self-evident, and natural, to show how these things already have a history that determine what becomes possible and what proceeds from them. When an idea, like robot rights, is immediately declared unthinkable, that is an indication of the need for critical inquiry and the difficult but necessary task of confronting and thinking the unthinkable.

Second, challenging exclusions and prohibitions is the proper work of ethics. Ethics operates by making exclusive decisions. It inevitably and unavoidably picks winners and losers and determines who is inside the moral community and what remains outside or on the periphery (Gunkel 2012 and Coeckelbergh 2012). But our moral theories and legal practices develop and evolve by challenging these restrictions and exceptions. Ethics, therefore, advances by critically questioning its own exclusivity and eventually accommodating many previously excluded or marginalized others – women, people of color, animals, the environment, etc. But as Christopher Stone has pointed out, this moral progress is difficult insofar as any extension of rights – any effort to extend moral/legal consideration to previously excluded populations – has always been "a bit unthinkable." In other words, "each time there is a movement to confer rights onto some new 'entity,' the proposal is bound to sound odd or frightening or laughable" (Stone 1974, 7).

The task of both moral and legal philosophy, therefore, is not (and cannot be) simply to defend existing orthodoxies by shrinking away in fright in the face of alternatives, trigging angry reactions to the possibility of including others, or mocking such challenges with dismissive laughter. Instead the task is, or at least should be, to stress-test and question the limitations and exclusions of existing moral/legal positions and modes of thinking. Defending orthodoxy is the purview of religion and ideology; critically testing hypotheses and remaining open to revising the way we think about the world in the face of new challenges and opportunities is the task of science. The project of philosophy – and moral philosophy in particular – has always been to ask about what has not been thought before or what has been decried and avoided as unthinkable.

3 Thinking the Unthinkable

In order to have an informed debate about these matters – in order to think the unthinkable and argue either for or against robots and AI having any claim to moral and/or legal standing – we need to ensure that all parties to the conversation understand what is being talked about. In other words, we have to get rights right, right from the beginning.

3.1 *Defining Rights*

Conversations about rights and their recognition, protection, and/or violation is rather common in contemporary discussions about moral, legal, and political matters. "The discourse of rights is," as Tom Campbell (2006, 3) notes, "pervasive and popular in politics, law and morality. There is scarcely any position, opinion, claim, criticism or aspiration relating to social and political life that is not asserted and affirmed using the term 'rights.' Indeed, there is little chance that any cause will be taken seriously in the contemporary world that cannot be expressed with a demand for the recognition or enforcement of rights of one sort or another." But it is precisely this pervasiveness and wide-spread usage that is the problem and a potential obstacle to clear understanding. "Rights" is a rather loose signifier that is applied in ways that are not always consistent or even carefully delineated. Even experienced jurists, as Wesley Hohfeld (1920) noted at the turn of the previous century, have a proclivity to conflate different senses of the term and often times mobilize varied (if not contradictory) understandings of the word in the course of a decision or even a single sentence.

In order to redress this perceived deficiency and lend some terminological precision and clarity to the use of the term "rights," Hohfeld developed a classification schema, analyzing rights into four fundamental types, or what are commonly called the "Hohfeldian incidents." Although Hohfeld's work was initially developed in the context of law and therefore applies to the concept of legal rights, his typology has been successfully ported to, modified for, and used to explain moral and political rights as well (Kramer, Simmonds and Steiner 1998). For this reason, as Leif Wenar (2005, 235) argues, "any assertion of a right can be translated into an assertion about a single Hohfeldian incident, or into an assertion about a complex of incidents, or into a set of alternative assertions about such incidents. All rights are Hohfeldian incidents."

The Hohfeldian incidents include the following four types of rights: privileges, claims, powers, and immunities. These four incidents are further organized into two groupings, the first two (privileges and claims) are considered

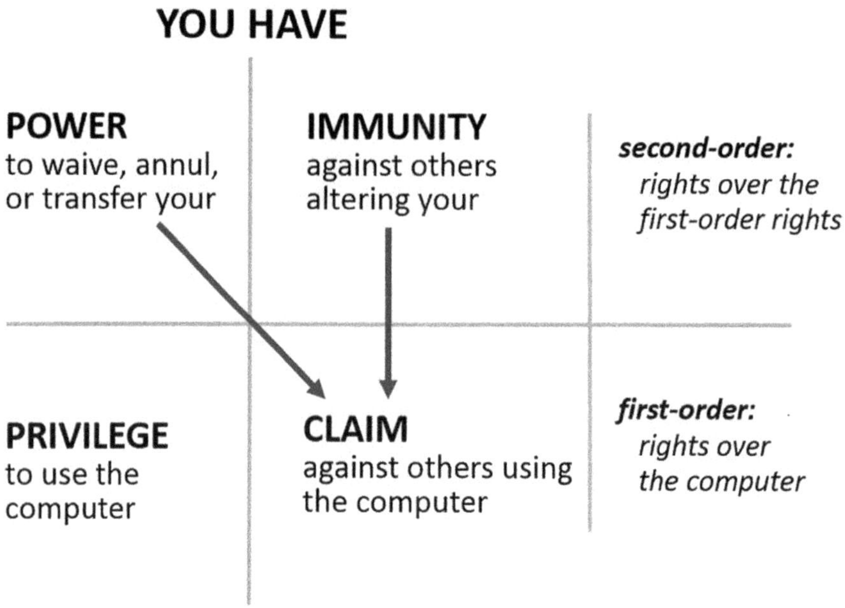

FIGURE 2.1 Hohfeldian analysis of a property right
WENAR (2015)

"primary rights" (Hart 1961) or "first-order incidents" (Wenar 2005, 232), while the other two (powers and immunities) fall under the category of "secondary rights" (Hart 1961) or "second-order incidents" (Wenar 2005, 232). A particular moral or legal right, i.e. the right to property ownership, can be characterized in terms of one or more of the four incidents (Figure 2.1).

Additionally the four incidents or types of rights are all formulated in such a way that each one necessitates a correlative duty. If one has a right (a privilege, a claim, a power, or an immunity), it means that there is an associated duty imposed upon another who is obligated to respect that right. Or as Johannes Marx and Christine Tiefensee (2015, 71) explain: "The 'currency' of rights would not be of much value if rights did not impose any constraints on the actions of others. Rather, for rights to be effective they must be linked with correlated duties." Hohfeld, therefore, initially presents the four incidents in terms of rights/duties pairs:

If A has a Privilege, then someone (B) has a No-claim.
If A has a Claim, then someone (B) has a Duty.
If A has a Power, then someone (B) has a Liability
If A has an Immunity, then someone (B) has a Disability.

This means that a statement concerning rights can be considered either from the side of the possessor of the right (the "right" that A has or is endowed with), which is a "patient oriented" way of looking at a moral, legal, or political situation; or from the side of the agent (what obligations are imposed on B in relationship to A), which considers the responsibilities of the producer of a moral, legal, or political action.

3.2 *Theory of Rights*

Though Hohfeld's typology analyzes and defines what rights are, it does not explain who has a right or why. This is the purview of theory, and there are two competing candidates concerning this issue: Will Theory and Interest Theory. Interest theories connect rights to matters of welfare. "Interest theorists maintain that individuals are capable of having rights if they possess certain interests, the value of which provides reason to impose duties on others" (Marx and Tiefensee 2015, 72). Although there is some variation in the way this has been articulated and formalized, "any doctrine classifiable as an Interest Theory of rights would," as Matthew Kramer (1998, 62) has explained, "subscribe to the following two theses":

1) Necesssary but insufficient for the actual holding of a right by a person X is that the right, when actual, preserves one or more of X's interests.
2) X's being competent and authorized to demand or waive the enforcement of a right is neither sufficient nor necessary for X to be endowed with that right.

Will theorists, by contrast, require that the subject of a right possess the authority and/or capacity to assert the privilege, claim, power, or immunity. "According to advocates of the Will or Choice Theory, only those individuals qualify as right-holders who have the ability to exercise a right. That is, only a person who can choose either to impose or waive the constraints that her right places on others' conduct can be seen to have rights" (Marx and Tiefensee 2015, 72). Once again there is some variability in the literature concerning how this particular theory is formulated. But all or most of them, as Kramer (1998, 62) argues share these three principles:

1) Sufficient and necessary for X's holding of a right is that X is competent and authorized to demand or waive the enforcement of the right.
2) X's holding of a right does not necessarily involve the protection of one or more of X's interests.
3) A right's potential to protect one or more of X's interests is not sufficient per se for X's actual possession of that right.

Although a great deal has been written about these two theories and their importance in the construction and development of rights discourse, we can take note of two important consequences.

First, comparing the common elements of these two theories, it is immediately evident that one of the major differences between will theory and interest theory is scope, that is, determinations regarding who can be included in the community of entities possessing rights and what is to be excluded as rights-less. Will Theory is designed to be more narrow, restrictive, or conservative on this account, while Interest Theory admits and can accommodate a wider range of entities. In other words, because Will Theory requires right's holders to be "competent and authorized to demand or waive the enforcement of the right," it excludes from consideration human persons of diminished mental capacity, children, animals, the environment, etc. By contrast, "the Interest Theory can readily ascribe rights to children and to mentally incapacitated people (and indeed to animals, if a theorist wishes); it can acknowledge that the criminal law bestows individual rights on the people who are protected by the various statutes and other mandates that impose criminal duties; and it maintains that any genuine right does not have to be waivable and enforceable by the right-holder but can be waivable and enforceable by someone else instead" (Kramer 1998, 78).

Second, each theory has its advocates and champions. Matthew Kramer (1998), for instance, is a vehement advocate of the Interest Theory, arguing that it provides a more adequate formulation of the way that rights actually operate than the more restrictive Will Theory, which risks excluding entire populations of legitimate moral and legal subjects; while Hillel Steiner (1998, 234) defends Will Theory, arguing that it "offers a perfectly general account of what it is to have a right in any type of society" and that "the account it offers is distinctly more plausible than that afforded by the Interest Theory." The debate between these two factions is ongoing and appears to be unresolvable. "The long and unresolved historical contest between these two single- function theories," Wenar (2015, 238) concludes, "stretches back through Bentham (an interest theorist) and Kant (a will theorist) into the Dark Ages." And its most recent contest, which was staged in Kramer, Simmond, and Hillel's *A Debate Over Rights* (1998) ended in what Wenar (2015) and others have acknowledged as a "stalemate."

Instead of trying to resolve this seemingly irresolvable difference, we will operationalize a method that Slavoj Žižek (2006) calls "the parallax view." "Parallax" is the term Žižek deploys (in something of an appropriation and repurposing of the word's standard connotation) to name an irreducible difference in perspective that is not programmed for nor determined to result in some kind of final mediation or dialectical resolution. "The two perspectives," Žižek (2006, 129) writes, "are irretrievably out of sync, so that there is no neutral language to translate one into the other, even less to posit one as the 'truth' of the other. All we can ultimately do in today's condition is to remain faithful to this split as such, to record it." With the two competing theories of rights, therefore, the truth does not reside on one side or the other or in their mediation through some kind of dialectical synthesis or hybrid construction. The truth of the matter can be found in the shift of perspective from the one to the other. The critical task, therefore, is not to select the right theory of rights but to mobilize theory in the right way, recognizing how different theories of rights frame different problematics, make available different kinds of inquiries, and result in different possible outcomes.

3.3 *Differences Matter*

Finally, the set of all possible robot rights is not identical to nor the same as the set of human rights (Figure 2.2).

There may be a few legal or moral affordances that overlap and even similar kinds of rights that can be formulated by way of different combinations of the four incidents, but the one is not identical to the other. A common mistake results from conflation – the assumption that "robot rights" must mean and can only mean "human rights." We see this in the popular press, in the academic literature, and in policy discussions and debates with eye-catching headlines like:

> When will Robots Deserve Human Rights? (Dvorsky 2017)
> Do Robot Deserve Human Rights? (Sigfusson 2017)
> Granting Automata Human Rights (Lantz Fleming Miller 2015)
> Should Robots be Granted Human Rights? (Isa 2018)

There are two problems with this procedure. On the one hand, this risks becoming a classic slippery slope. The question concerning robots rights is immediately assumed to entail or lead to human rights, not recognizing that the rights for one category of entity, like an animal or a machine, is not necessarily equivalent to nor the same as that enjoyed by another category of entity, like a

```
{Robot Rights} !== {Human Rights}
```

FIGURE 2.2 Pseudo-object code expressing the fundamental difference between the set of all possible robot rights and the set of human rights

human being. It is possible, as Kate Darling (2012) has argued, to entertain the question of robot rights without this meaning all human rights.

On the other hand, the conflation and assumed equivalency has the effect of producing what is arguably a strawman argument:

1) Robots are not human beings
2) Only human beings can have human rights
3) Robots cannot have human rights

This inference, although undeniably correct, is a rather pointless and empty victory. Ontologically speaking, robots are different from human beings. Therefore proving that robots would not have the same set of claims, privileges, powers, and/or immunities granted a human being is already a foregone conclusion. Whatever robots rights consists in – and this has yet to be fully investigated and determined – it can and will be significantly different from the set of what is currently recognized as human rights.

4 The Right(s) Question

So let's take the question seriously and ask the right(s) question – "Can and should robots have rights?" This inquiry actually consists of two separate components: "Can robots have rights?" which is a question that asks about the ontological capabilities of a particular entity. And "Should robots have rights?" which is a question that inquiries about normative obligations in the face of the entity. These two questions invoke and operationalize a rather famous conceptual distinction in philosophy that is called the is/ought problem or Hume's Guillotine. In *A Treatise of Human Nature* (first published in 1738), David Hume differentiated between two kinds of statements: descriptive statements of fact and normative statements of value (Schurz 1997, 1). For Hume, the problem was the fact that philosophers, especially moral philosophers, often fail to distinguish between these two kinds of statements and therefore slip imperceptibly from one to the other.

For our purposes what is important is not the logical complications of the "is-ought fallacy" as Hume originally characterized it or the ongoing and seemingly irresolvable debate concerning the validity of the is-ought inference as it has been developed and argued in the subsequent literature (cf. Schurz 1997). What is pertinent for the investigation at hand is to recognize how the verbs "is" and "ought" organize qualitatively different kinds of statements and modes of inquiry. The former concerns ontological matters or statements of fact; the latter consists in axiological decisions involving what should be done or what ought to be. The guiding question of our inquiry utilizes modal variants of these two verbs, namely "can" and "should." According to the Humean thesis, therefore, it is possible to make the following distinction between two kinds of statements:

S1 = "Robots can have rights."
S2 = "Robots should have rights."

Because the is-ought inference is and remains an undecided and open question (Schurz 1997, 4), it is possible to relate S1 to S2 in ways that generate four options or modalities concerning the moral situation of robots. These four modalities can be organized into two pairs. In the first pair, which upholds and supports the is-ought inference, the affirmation or negation of the ontological statement (S1) determines the affirmation or negation of the axiological statement (S2). In the second pair, which endorses the Humean thesis or contests the derivation, one affirms the ontological statement (S1) while denying the axiological statement (S2), or vice versa. These four modalities, therefore, describe the current situation concerning the discussion and debate regarding the question of robot rights.

5 Robot Rights

The four sections that follow critically evaluate each one of the modalities as they have been deployed and developed in the literature, performing a cost/benefit analysis of the available arguments concerning the rights (or lack thereof) of robots, artificial intelligence, and related forms of autonomous technology.

5.1 *Cannot and Should Not*

In this first case, we deny or negate both statements. Robots are incapable of having rights, and for this reason robots should not have rights. This is a position that is formulated by strictly applying the instrumental theory of technology. According to Martin Heidegger's (1977, 4–5) analysis the presumed role and function of any kind of technological artifact, whether it be the product of handicraft or industrialized manufacture, is that it is a means employed by human users for specific ends: "We ask the question concerning technology when we ask what it is. Everyone knows the two statements that answer our question. One says: Technology is a means to an end. The other says: Technology is a human activity. The two definitions of technology belong together. For to posit ends and procure and utilize the means to them is a human activity." Heidegger calls this particular characterization "the instrumental definition" and he indicates that it forms what is considered to be the "correct" understanding of any kind of technological contrivance or artifact. As Deborah Johnson (2006, 197) describes it with regards to information technology: "Computer systems are produced, distributed, and used by people engaged in social practices and meaningful pursuits. This is as true of current computer systems as it will be of future computer systems. No matter how independently, automatic, and interactive computer systems of the future behave, they will be the products (direct or indirect) of human behavior, human social institutions, and human decision."

This way of thinking is persuasive, precisely because it reaffirms long standing assumptions about human exceptionalism and technological instrumentalism. Technologies, no matter how sophisticated, interactive, or social they appear to be, are just tools, nothing more. They are not – not now, not ever – capable of becoming moral subjects in their own right, and we should not treat them as such. It is precisely for this reason that, as J. Storrs Hall (2001, 2) points out, "we have never considered ourselves to have moral duties to our machines." The problem, of course, is that this position is disputed by much of the available data that has been produced by and derived from human robot interaction (HRI) studies undertaken by social scientists like Christoph Bartneck and Jun Hu (2008), Peter Kahn et al. (2012), Astrid Rosenthal-von der Pütten et al. (2013), and many others.

None of this is necessarily new or surprising. It was already tested and demonstrated with Fritz Heider and Mariane Simmel's "An Experimental Study of Apparent Behavior" (1944), which found that human subjects tend to attribute motive and personality to simple animated geometric figures. Similar results have been obtained by way of the computer as social actor (CASA) studies conducted by Byron Reeves and Clifford Nass in the mid-1990s. As Reeves and Nass discovered across numerous trials with human subjects, users (for better or worse) have a strong tendency to treat socially interactive technology, no matter how rudimentary, as if they were other people.

> Computers, in the way that they communicate, instruct, and take turns interacting, are close enough to human that they encourage social responses. The encouragement necessary for such a reaction need not be much. As long as there are some behaviors that suggest a social presence, people will respond accordingly. When it comes to being social, people are built to make the conservative error: When in doubt, treat it as

> human. Consequently, any medium that is close enough will get human treatment, even though people know it's foolish and even though they likely will deny it afterwards. (Reeves and Nass 1996, 22)

So what we have is a situation where our theory of technology – a theory that has considerable history behind it and that has been determined to be as applicable to simple hand tools as it is to complex computer systems – seems to be out of sync with the very real practical experiences we now have with artifacts in a variety of situations and circumstances. To put it in terminology that is more metaphysically situated: How something appears to be – how it actually operates in real social situations and circumstances – might be more important than what it actually is (or has been assumed to be). "We should," as Tony Prescott (2017, 146) concludes formulating a kind of phenomenological moral maxim, "take into account how people *see* robots, for instance, that they may feel themselves as having meaningful and valuable relationships with robots, or they may see robots as having important internal states, such as the capacity to suffer, despite them not having such capacities." To put it in a quick vernacular form: even when our head tells us it is just a robot, our heart cannot help but like it, and we may need to respond to and take responsibility for this phenomenon.

5.2 *Can and Should*

The flipside to the strict instrumentalist argument is to affirm both statements. This is the position espoused by many who support speculative ideas and hypotheses like superintelligence, the singularity, and artificial general intelligence (AGI). "The 'artificial intelligence' programs in practical use today," Ben Goertzel (2002, 1) admits, "are sufficiently primitive that their morality (or otherwise) is not a serious issue. But that will not remain for long. Not too far in the future, however, things are going to be different. AI's will possess true artificial general intelligence (AGI), not necessarily emulating human intelligence, but equaling and likely surpassing it. At this point, the morality or otherwise of AGI's will become a highly significant issue."

There are obvious advantages to this way of thinking insofar as it does not simply deny rights to robots *tout court*, but kicks the problem down the road and postpones decision making. Right now, we do not have robots that can be moral subjects, but when we do (and it is more often a question of "when" as opposed to "if"), then we will need to consider whether they should be treated differently. "As soon as AIs begin to possess consciousness, desires and projects," Robert Sparrow (2004, 203) argues, "then it seems as though they deserve some sort of moral standing." Or as Peter Singer and Agata Sagan (2009) write "if the robot was designed to have human-like capacities that might incidentally give rise to consciousness, we would have a good reason to think that it really was conscious. At that point, the movement for robot rights would begin."

This way of thinking is persuasive, precisely because it recognizes the actual limitations of current technology while holding open the possibility of something more in the not-too-distant future. This wait-and-see approach also helps us contend with the moral intuitions we have in the face of social robots. Although these mechanisms are intentionally designed to trigger our emotions we should remember, for now at least, that inside the box there is nothing – no thought, no consciousness, no sentience. If and when these machines can achieve something more, then we will be justified in considering the rights question. But for now these social behaviors are just cleverly manipulated effects without a real cause.

The main problem to this way of thinking, however, is that it does not really resolve the question regarding the rights of robots but just postpones the decision to some indeterminate point in the future. It says, in effect, as long as robots are not conscious or sentient or whatever ontological criteria counts, no worries. Once they achieve this capability, however, then we should consider extending some level of moral concern and respect. All of which means, of course, that this "solution" to the question "Can and should robots have rights?" is less a solution and more of a decision not to decide. Furthermore when the decisive moment (whenever that might be and however it might occur) does in fact come, there remains several theoretical and practical difficulties that make this way of thinking much more problematic than it initially appears to be.

First, there are terminological complications. A qualifying criteria like "consciousness," for instance, does not admit of a univocal characterization, but denotes, as Velmans (2000, 5) points out, "many different things to many different people." In fact, if there is any general agreement among philosophers, psychologists, cognitive scientists, neurobiologists, AI researchers, and robotics engineers regarding consciousness, it is that there is little or no agreement when it comes to defining and characterizing the concept. Other properties, like sentience, unfortunately do not do much better. As Daniel Dennett demonstrates in his eponymously titled essay, the reason "why you cannot make a computer that feels pain" has little or nothing to do with the technical challenges with making pain computable. It proceeds from the fact that we do not know what pain is in the first place. In other words, "there can be," as Dennett (1998, 228) concludes, "no true theory of pain, and so no computer or robot could instantiate the true theory of pain, which it would have to do to feel real pain."

Second, even if it were possible to resolve these terminological difficulties, maybe not once and for all but at least in a way that would be widely accepted, there remains epistemological limitations concerning detection of the capability in question. How can one know whether a particular robot has actually achieved what is considered necessary for something to have rights, especially because most, if not all of the qualifying capabilities or properties are internal states-of-mind? This is, of course, connected to what philosophers call the problem of other minds, the fact that, as Donna Haraway (2008, 226) cleverly describes it, we cannot climb into the heads of others "to get the full story from the inside." Although philosophers, psychologists, and neuroscientists throw considerable argumentative and experimental effort at this problem, it is not able to be resolved in any way approaching what would pass for definitive evidence, strictly speaking.

5.3 *Can but Should Not*

In opposition to these two approaches, there are other responses that seek to uphold the is/ought distinction by affirming one statement and denying the other. Since there are two statements, this strategy can take one of two forms. In its first formation, one affirms that robots can have rights but denies that this fact requires us to accord them social or moral standing. This is the argument that has been developed by Joanna Bryson beginning with the provocatively titled essay "Robots Should Be Slaves" (2010). Bryson's argument goes like this: Robots are property. No matter how capable they are, appear to be, or may become; we are obligated not to build things to which we would feel obligated. Or as she describes it in a more recent article: "It is unquestionably within our society's capacity to define robots and other AI as moral agents and patients. In fact, many authors (both philosophers and technologists) are currently working on this project. It may be technically possible to create AI that would meet contemporary requirements for agency or patiency. But even if it is possible, neither of these two statements makes it either necessary or desirable that we should do so" (Bryson 2018, 24).

To its credit, this approach succeeds insofar as it reasserts and reconfirms the instrumental theory in the face of (perceived) challenges from a new kind of socially interactive artifact. No matter how interactive or seemingly intelligent our robots and AIs are or can become, they should be, now and forever, considered instruments serving our needs and desires, nothing more. This is (or was supposed to have been) the significance of the title "Robots Should Be Slaves." What Bryson meant to convey with the title, which has (for very understandable reasons) sparked controversy and debate, is that robots should be designed, built, marketed, and understood as instruments and tools that serve our needs and *should not* (a formulation that has normative moral weight) be situated as companion peers. As Bryson (2015) has explained in a subsequent blog post: "I originally used that line ["robots should be slaves"] to clarify that it was obligatory not to build robots we are obliged to, that they should not be human-like because they will necessarily be owned."

There are, however, at least two problems with this decision and assertion. First, it requires a kind of asceticism or abstinence education. Bryson's text issues imperatives that take the form of social prohibitions for both designers and users. For designers, "thou shalt not create robots to be companions." For users, no matter how interactive or capable a robot is (or can become), "thou shalt not treat your robot as yourself." The effectiveness of these prohibitions, however, are disputed and disproven by actual data from HRI, beginning with the CASA studies conducted in the mid-1990s (Reeves and Nass 1996) to the recent robot abuse investigations (Bartneck and Hu 2008 and Suzuki et al. 2015). Even though we know better, we respond to and treat many artifacts with respect. This is not necessarily a bug to be fixed; it is a feature of human social interaction.

Second, this way of thinking has been formulated in terms of slavery or what has been called a "slave metaphor," which is not unique to Bryson's argument. It appears in a number of other publications addressing this matter including the work of Luciano Floridi (2017) and Hutan Ashrafian (2015). The problem here is not the usual concern over how the robot might feel about it subjugation. The problem is with us and the effect this kind of servitude would have on our social institutions. As both philosophical and historical studies have shown, institutionalized forms of servitude or slavery have adverse effects on those who occupy the position of mastery. As Alex de Tocqueville (1899, 361) commented about his direct experience with slavery in the Amerian south: "Servitude, which debases the slave, impoverishes the master."

5.4 *Cannot but Should*

The final modality also preserves separation between the two statements, but it does so by denying the first and affirming the second. In this case, which is something proposed and developed by Kate Darling, robots, at least in term of the currently available technology, cannot have rights. They do not, at least at this particular point in time, possess the necessary capabilities or properties to be considered a full moral person with powers, privileges, claims, and/or immunities. Despite this fact, they should, Darling argues, still be granted some level of moral and even legal status.

"Looking at state of the art technology," Darling (2012, 1) points out, "our robots are nowhere close to the intelligence and complexity of humans or animals, nor will they reach this stage in the near future. And yet, while it seems far-fetched for a robot's legal status to differ from that of a toaster, there is already a notable difference in how we interact with certain types of robotic objects." This is, Darling argues, due to our tendencies to anthropomorphize technological objects by projecting into them cognitive capabilities, emotions, and motivations that do not necessarily exist. And "social robots," Darling (2012, 1) continues, "play off of this tendency by mimicking cues that we automatically associate with certain states of mind or feelings. Even in today's primitive form, this can elicit emotional reactions from people that are similar, for instance, to how we react to animals and to each other." This fact necessitates, Darling concludes, new obligations in the face of robots that are at least on par with what we have extended to non-human animals.

The obvious advantage to this way of thinking is that it is able to scale to recent changes in our moral intuitions. Even if robots cannot be moral subjects strictly speaking (at least not yet), there is something qualitatively different about socially interactive artifacts. According to Darling (2016, 213), it is because we "perceive robots differently than we do other objects," that one should consider extending some level of protections to the latter but not the former. This conclusion is consistent with Hume's thesis. If "ought" cannot be derived from "is," then axiological decisions concerning moral value are little more than sentiments based on how we feel about something at a particular time. Darling mobilizes a version of this moral sentimentalism with respect to social robots: "Violent behavior toward robotic objects feels wrong to many of us, even if we know that the abused object does not experience anything" (Darling 2016, 223). Consequently, and to its credit, Darling's proposal, unlike some of the previous modalities, tries to work with rather than against recent empirical experiences with robots.

There are, however, at least two problems with this approach. First, it reaffirms one of Hume's criticisms that distinctions of vice and virtue are founded merely on the relations of objects, rendering moral decisions little more than expressions of personal interest and emotion. It is because we perceive the toaster and the social robot differently, that we should, according to Darling's argument, consider extending some level of legal protections to the latter but not the former. But such perceptions can be capricious, inconsistent and (perhaps what is worse) deliberately manipulated. "Feelings," Immanuel Kant (1983, 442) writes in response to this kind of moral sentimentalism, "naturally differ from one another by an infinity of degrees, so that feelings are not capable of providing a uniform measure of good and evil; furthermore, they do so even though one man cannot by his feeling judge validly at all for other men." Additionally, because sentiment is a matter of individual experience, it remains uncertain as to whose perceptions actually matter or make the difference? Who, for instance, is included (and who is excluded) from the collective first person "we" that Darling operationalizes in and makes the subject of her proposal? In other words, whose sentiments count when it comes to decisions concerning the extension of moral and legal rights to others, and do all sentiments have the same status and value when compared to each other?

Second, the proposal remains thoroughly anthropocentric and instrumentalizes others. According to Darling, the principal reason we need to consider extending rights to others, like animals and robots, is because of the way it makes us feel. This follows Kant's indirect duties argument for restricting animal abuse, and Darling endorses this anthropocentric formulation without any critical hesitation: "The Kantian philosophical argument for preventing cruelty to animals is that our actions toward non-humans reflect our morality – if we treat animals in inhumane ways, we become inhumane persons. This logically extends to the treatment of robotic companions" (Darling 2016, 227–228). This way of thinking, although potentially expedient for developing and justifying new forms of legal protections, renders the inclusion of previously excluded others less than altruistic; it transforms animals and robots into nothing more than instruments of human self-interest. The rights of others, in other words, is not about them; it is all about us.

6 Thinking Otherwise

Because of these complications, it is, I believe, necessary to formulate a mode of response that is devised and oriented otherwise. This alternative proceeds by taking aim at and deconstructing the ruling conceptual opposition that organizes the entire spectrum of possible responses – the is-ought distinction inherited from David Hume. The principal question for Hume, and the fundamental organizing principal of the four modalities of robot rights that we have just considered, is determining whether and to what extent *ought* may be derived from *is*. As long as we play the game according to this set of rules, little will change. Each alternative will continue to heap up arguments and evidence to support its particular position. But the basic problem and the debate will remain essentially unchanged. So I suggest that we flip the script and consider not "how *ought* may be derived from *is*" but rather "how *is* is only able to be derived from *ought*."

This is precisely the innovation introduced by Emmanuel Levinas who, in direct opposition to the usual way of thinking, asserts that ethics precedes ontology. In other words, the moral aspect, the ought dimension, comes first, in terms of both temporal sequence and status, and then the ontological dimension follows from it. This is a deliberate provocation that cuts across the grain of the philosophical tradition. As Floridi (2013, 116) correctly points out for most standard forms of moral theory, "what the entity is determines the degree of moral value it enjoys, if any." In other words, we first need to know what something is – what capabilities it has or cannot have – before making any decisions about how it is to be treated – whether it is just a thing that can be used and disposed of as we desire or whether it is another social subject who counts and can be benefitted or harmed by our decisions and actions.

Levinas deliberately inverts this entire transaction. According to his way of thinking, we are first confronted with a mess of anonymous others who intrude on us and to whom we are obligated to respond even before we know anything at all about them. "The strangeness of the Other, his irreducibility to the I, to my thoughts and my possessions, is precisely accomplished as a calling into question of my spontaneity, as ethics" (Levinas 1969, 43). It is only after having made a response in the face of the other that it becomes possible to decide who or what they are or are not. To use Hume's terminology – which will be a kind of translation insofar as Hume's philosophical vocabulary, and not just his language, is something that is foreign to Levinas's own formulations – we are first obligated to respond and then, after having made a response, what or who we responded to is able to be determined and identied.

The advantage to this way of thinking is that it provides an entirely different way of looking at the challenge not just of robots but of the way we have addressed and decided things in the face of this challenge. Following the contours of this Levinasian innovation, moral consideration is decided and conferred not on the basis of some predetermined ontological criteria or capability (or lack thereof) but in the face of actual social relationships and interactions. "Moral consideration," as Mark Coeckelbergh (2010, 214) describes, "is no longer seen as being 'intrinsic' to the entity: instead it is seen as something that is 'extrinsic': it is attributed to entities within social relations and within a social context." As we encounter and interact with others – whether they be other human persons, an animal, the natural environment, or a robot – this other entity is first and foremost situated in relationship to us. For this reason, the question of moral status does not depend on what the other is in its essence but on how she/he/it supervenes before us and how we decide, in the face of the other, to respond. In this transaction, "relations are prior to the things related" (Callicott 1989, 110),[2] instituting what Anne Gerdes (2015), following Coeckelbergh (2010), has called "a relational turn" in ethics.

This alternative configuration, therefore, does not so much answer or respond to the question with which we began as it alters the terms and conditions of the inquiry itself. When one asks "Can or should robots have rights?" the form of the question already makes an assumption, namely that rights are a kind of personal property or possession that an entity can have or should be bestowed with. Levinas does not inquire about rights nor does his moral theory attempt to respond to this form of questioning. In fact, the word "rights" is not in his philosophical vocabulary and does not appear as such in his published work. Consequently, the Levinasian inquiry is directed and situated otherwise: What does it take for something – another human person, an animal, a mere object, or a social robot – to supervene and be revealed as Other? This other question – a question about others that is situated otherwise – comprises a more precise and properly altruistic inquiry. It is a mode of questioning that remains open, endlessly open, to others and other forms of otherness. And it recognizes, following Coeckelbergh (2016, 190), that *alterity* is not a noun – an ontological property that intrinsically belongs to an entity – but is more of a verb – something that happens to entities in social relationships.

But this proposal, for all its promise to formulate an ethics that is situated otherwise, also has costs, especially as it applies to technological artifacts. First, it has to fight against and struggle to break free from the gravitational pull of Levinas's own moral theory. Whatever the import of his unique contribution, the Other in Levinas is still unabashedly and dogmatically anthropocentric. Consequently, the alternative presented here will only be able to succeed insofar as it actively works against and in excess of the inescapable anthropocentrism that we find in Levinas's own texts. Fortunately there is precedent for this kind of effort to use Levinas against Levinas or to radicalize Levinas's work in order to address the moral challenges of other forms of otherness, i.e. animals, artifacts, and machines (Benso 2000, Atterton and Calarco 2010, Gunkel 2012). "Although Levinas himself," as Matthew Calarco (2008, 55) writes, "is for

2 It could be argued that Calicott's claim is logically impossible since "relations" and "things related to" both emerge, logically speaking, at the very same time. Consequently, they are, at best, quiprimordial. This conclusion, however, misunderstand the radicality of the relational turn: Things first become the things that they are in and by being situated in relationship to others. Prior to being related, there may be indeterminate entities. But these entities only become some thing for another, that is, in and by the relationship to another. Like Schrödinger's cat, the exact ontological condition of the entity remains indeterminate – in the case of the cat, for example, it is neither living nor dead and both living and dead – until it comes into relationship with another and is observed.

the most part unabashedly and dogmatically anthropocentric, the underlying logic of his thought permits no such anthropocentrism. When read rigorously, the logic of Levinas's account of ethics does not allow for either of these two claims. In fact... Levinas's ethical philosophy is, or at least should be, committed to a notion of universal ethical consideration, that is, an agnostic form of ethical consideration that has no a priori constraints or boundaries."

Second, operationalizing this Levinasian approach requires an entirely different way of thinking about moral status and the decisions that are made concerning who counts as a socially significant other and what does not. As a result, the challenge (or opportunity) of the robot is not just determining how we can or should adjust our moral theories to respond to this other kind of social other. Instead, what we see in the face or the faceplate of the robot is nothing less than a fundamental challenge to the way these theories have been formulated in the first place and how they usually operate in practice, requiring a thorough reconsideration of moral philosophy that goes all the way down. Robots, then – especially increasingly social artifacts – are not just one more problem for applied moral philosophy, they show us the principal problems and unique challenges of moral philosophy itself.

7 Conclusions

Let me conclude with three statements that respond to and concern the current state and future direction of/for moral thinking:

1) Asking about robot rights is not a distraction, fairy tale stuff, or a waste of time. It is especially needed and necessary in the face of socially interactive artifacts that effectively challenge the standard operating procedures for dealing with technology.
2) The inquiry "Can and should robot have rights?" calls on and operationalizes a classic philosophical problem inherited from David Hume ([1739–1740]; 1980). Responses to the question take one of four forms. Each mode of response has its theoretical and practical advantages. But all four modalities exhibit significant problems and complications.
3) In response to these opportunities/challenges, I have argued for another alternative, a mode of "thinking otherwise" that capitalizes on the philosophical innovations of Emmanuel Levinas and other who follow his lead.

This does not mean, however, that this alternative proposal is a panacea or some kind of moral theory of everything. It just arranges for other kinds of questions and modes of inquiry that are more attentive to the actual situation in which we currently find ourselves. Consequently, my objective in this chapter has not been to resolve the question of robot rights once and for all, but to ask about and evaluate the means by which we have situated and pursued this inquiry.

For this reason, we end not as one might have initially expected – that is, by accumulating evidence or arguments in favor of permitting robots, a particular kind of robot, or even one representative robot, entry into the community of moral and/or legal subjects. Instead, we conclude with a fundamental challenge to ethics and the way moral philosophy has typically defined, decided, and defended the question concerning the standing and status of others. This is because the rights question, no matter who or what it concerns, has both a strong and weak component – strong and weak in the sense provided by John Searle (1980). Efforts at what would be the "strong" variety involve those discussions and debates that seek to formulate policy, moral rules, and/or legal statutes regarding the social position and status of robots, AI, and other autonomous systems. But these efforts also have a "weak" component – "weak" in Searle's particular sense of the word – insofar as they put into play, operationalize, and stress test various conceptualizations of moral and legal rights, leading to critical and discerning reevaluations of how these concepts have been operationalized and deployed.

Although ending in this way – in effect, responding to a question with yet another set of questions – is commonly considered bad form, this is not necessarily the case. This is because questioning is definitive of the philosophical enterprise. "I am," Dennett (1996, vii) writes, "a philosopher and not a scientist, and we philosophers are better at questions than answers. I haven't begun by insulting myself and my discipline, in spite of first appearances. Finding better questions to ask, and breaking old habits and traditions of asking, is a very difficult part of the grand human project of understanding ourselves and our world." The objective of the investigation, therefore, has not been to advocate on behalf of a cause, to advance some definitive declaration, or to offer up a preponderance of evidence to prove one or the other side(s) in the existing debates. The goal has been to ask about the very means by which we have gone about trying to articulate and formulate this problem and its possible resolution. The issue, then, is not only investigating whether or not robots can and should have rights, but also, and perhaps more importantly, exposing how these questions concerning rights have been configured and how these configurations already accommodate and/or marginalize others and other modes of inquiry. For this reason, the examination of rights as it applies (or not) to artifacts of our own making is not just one more version or iteration of an applied moral philosophy; its investigation opens up and releases a thorough and profound challenge to what is called ethics.

References

Ashrafian, Hutan. 2015. "Artificial Intelligence and Robot Responsibilities: Innovating Beyond Rights." *Science and Engineering Ethics* 21(2): 317–326.

Atterton, Peter, and Matthew Calarco. 2010. *Radicalizing Levinas*. Albany, NY: SUNY Press.

Bartnek, Christoph, and Jun Hu. 2008. "Exploring the Abuse of Robots." *Interaction Studies* 9(3): 415–433.

Benso, Silvia. 2000. *The Face of Things: A Different Side of Ethics*. Albany, NY: SUNY Press.

Bryson, Joanna. 2010. "Robots Should be Slaves." In *Close Engagements with Artificial Companions: Key Social, Psychological, Ethical and Design Issues*, edited by Yorick Wilks, 63–74. Amsterdam: John Benjamins.

Bryson, Joanna. 2015. "Clones Should NOT be Slaves." *Adventures in NI*. October 4,. https://joanna-bryson.blogspot.com/2015/10/clones-should-not-be-slaves.html.

Bryson, Joanna. 2018. "Patiency is not a Virtue: The Design of Intelligent Systems and Systems of Ethics." *Ethics and Information Technology* 20(1): 15–26.

Calarco, Matthew. 2008. *Zoographies: The Question of the Animal from Heidegger to Derrida*. New York: Columbia University Press.

Callicott, J. Baird. 1989. *In Defense of the Land Ethic: Essays in Environmental Philosophy*. Albany, NY: SUNY Press.

Campbell, Tom. 2006. *Rights: A Critical Introduction*. New York: Routledge.

Chu, Seo-Young. 2010. *Do Metaphors Dream of Literal Sleep? A Science-Fictional Theory of Representation*. Cambridge, MA: Harvard University Press.

Coeckelbergh, Mark. 2010. "Robot Rights? Towards a Social-Relational Justification of Moral Consideration." *Ethics & Information Technology* 12(3): 209–221.

Coeckelbergh, Mark. 2012. *Growing Moral Relations: Critique of Moral Status Ascription*. New York: Palgrave MacMillan.

Coeckelbergh, Mark. 2016. "Alterity Ex Machina: The Encounter with Technology as an Epistemological-Ethical Drama." In *The Changing Face of Alterity: Communication, Technology and Other Subjects*, edited by David J. Gunkel, Ciro Marcondes Filho and Dieter Mersch, 181–196. New York: Rowman & Littlefield.

Darling, Kate. 2012. "Extending Legal Protection to Social Robots." *IEEE Spectrum*. September 10, 2012. http://spectrum.ieee.org/automaton/robotics/artificial-intelligence/extending-legal-protection-to-social-robots.

Darling, Kate. 2016. "Extending Legal Protection to Social Robots: The Effects of Anthropomophism, Empathy, and Violent Behavior Toward Robotic Objects." In *Robot*

Law, edited by Ryan Calo, A. Michael Froomkin, and Ian Kerr, 213–231. Northampton, MA: Edward Elgar.

Dennett, Daniel. 1996. *Kinds of Minds*. Basic Books: New York.

Dennett, Daniel C. 1998. *Brainstorms: Philosophical Essays on Mind and Psychology*. Cambridge, MA: MIT Press.

Dvorsky, George. 2017. "When Will Robots Deserve Human Rights?" *Gizmodo*: June 2, 2017. https://gizmodo.com/when-will-robots-deserve-human-rights-1794599063.

Floridi, Luciano. 2013. *The Ethics of Information*. Oxford: Oxford University Press.

Floridi, Luciano. 2017. "Robots, Jobs, Taxes, and Responsibilities." *Philosophy & Technology* 30(1): 1–4.

Gerdes, Anne. 2015. "The Issue of Moral Consideration in Robot Ethics." *ACM SIGCAS Computers & Society* 45(3): 274–280.

Goertzel, Ben. 2002. "Thoughts on AI Morality." *Dynamical Psychology: An International, Interdisciplinary Journal of Complex Mental Processes*. http://www.goertzel.org/dynapsyc/2002/AIMorality.htm.

Gordon, John-Stewart. 2018. "What Do We Owe to Intelligent Robots?" *AI & Society*, 1–15.

Gunkel, David J. 2012. *The Machine Question: Critical Perspectives on AI, Robots, and Ethics*. Cambridge, MA: MIT Press.

Gunkel, David J. 2018. "The Other Question: Can and Should Robots Have Rights." *Ethics and Information Technology* 20(2): 87–99.

Gunkel, David J. 2018. *Robot Rights*. Cambridge, MA: MIT Press.

Hall, J. Storrs. 2001. "Ethics for Machines." *KurzweilAI.net*. July 5, 2001. http://www.kurzweilai.net/ethics-for-machines.

Haraway, Donna J. 2008. *When Species Meet*. Minneapolis, MN: University of Minnesota Press.

Hart, H.L.A. 1961. *The Concept of Law*. Oxford: Oxford University Press.

Heidegger, Martin. 1977. *The Question Concerning Technology and Other Essays*. Translated by William Lovitt. New York: Harper & Row.

Heider, Fritz and Marianne Simmel. 1944. "An Experimental Study of Apparent Behavior." *The American Journl of Psychology* 57(2): 243–259.

Henderson, Mark. 2007. "Human Rights for Robots? We're Getting Carried Away." *The Times*. April 24, 2007. https://www.thetimes.co.uk/article/human-rights-for-robots-were-getting-carried-away-xfbdkpgwnov.

Hohfeld, Wesley. 1920. *Fundamental Legal Conceptions as Applied in Judicial Reasoning*. New Haven: Yale University Press.

Hume, David. (1739–1740) 1980. *A Treatise of Human Nature*. New York: Oxford University Press.

Isa, Mariam. 2018. "Should robots be granted human rights?" *Finweek*. June 22, 2018. https://www.fin24.com/Finweek/Featured/should-robots-be-granted-human-rights-20180622.

Johnson, Barbara. 1981. "Translator's Introduction." In Jacques Derrida, *Disseminations*, vii–xxxiii. Chicago: University of Chicago Press.

Johnson, Deborah G. 2006. "Computer Systems: Moral Entities but not Moral Agents." *Ethics and Information Technology* 8: 195–204.

Kahn, Peter H., Hiroshi Ishiguro, et al. 2012. "'Robovie, You'll Have to go into the Closet Now': Children's Social and Moral Relationships with a Humanoid Robot." *Developmental Psychology* 48(2): 303–314.

Kant, Immanuel. 1983. *Grounding for the Metaphysics of Morals*. Translated by James W. Ellington. Indianapolis, IN: Hackett Publishing.

Kramer, Matthew H. 1998. "Rights Without Trimmings." In *A Debate Over Rights: Philosophical Enquiries*, edited by Matthew H. Kramer, N.E. Simmonds and Hillel Steiner, 7–112. Oxford: Clarendon Press.

Kramer, Matthew H., N.E. Simmonds and Hillel Steiner. 1998. *A Debate Over Rights: Philosophical Enquiries*. Oxford: Clarendon Press.

Levinas, Emmanuel. 1969. *Totality and Infinity: An Essay on Exteriority*. Translated by Alphonso Lingis. Pittsburgh, PA: Duquesne University.

Levy, David. 2005. *Robots Unlimited: Life in a Virtual Age*. Boca Raton, FL: CRC Press.

Marks, Paul. 2007. "Robot Rights." *New Scientist Blog*. April 24, 2007. https://www.newscientist.com/blog/technology/2007/04/robot-rights.html.

Marx, Johannes and Christine Tiefensee. 2015. "Of Animals, Robots and Men." *Historical Social Research* 40(4): 70–91.

Miller, Lance Fleming. 2015. "Granting Automata Human Rights: Challenge to a Basis of Full-Rights Privilege." *Human Rights Review* 16(4): 369–391.

Prescott, Tony J. 2017. "Robots are not Just Tools." *Connection Science* 29(2): 142–149.

Reeves, Byron and Clifford Nass. 1996. *The Media Equation: How People Treat Computers, Television, and New Media Like Real People and Places.* Cambridge: Cambridge University Press.

Rosenthal-von der Pütten, Astrid M., et al. 2013. "An Experimental Study on Emotional Reactions Towards a Robot." *International Journal of Social Robotics* 5: 17–34.

Schurz, Gerhard. 1997. *The Is-Ought Problem: An Investigation in Philosophical Logic.* Dordrecht: Springer.

Searle, John R. 1980. "Minds, Brains and Programs." *Behavioral and Brain Sciences* 3(3): 417–424.

Sigfusson, Lauren. 2017. "Do Robots Deserve Human Rights?" *The Crux*. December 5, 2017. http://blogs.discovermagazine.com/crux/2017/12/05/human-rights-robots/#.XLtvg9h7nyM.

Singer, Peter and Agata Sagan. 2009. "When robots have feelings." *The Guardian*. December 14, 2009. https://www.theguardian.com/commentisfree/2009/dec/14/rage-against-machines-robots.

Sparrow, Robert. 2004. “The Turing Triage Test.” *Ethics and Information Technology* 6(4): 203–213.
Steiner, Hillel. 1998. “Working Rights.” In *A Debate Over Rights: Philosophical Enquiries*, edited by Matthew H. Kramer, N.E. Simmonds and Hillel Steiner, 113–232. Oxford: Clarendon Press.
Stone, Christopher D. 1974. *Should Trees Have Standing? Toward Legal Rights for Natural Objects*. Los Altos, CA: William Kaufmann.
Suzuki, Yutaka, Lisa Galli, Ayaka Ikeda, Shoji Itakura and Michiteru Kitazaki. 2015. “Measuring Empathy for Human and Robot Hand Pain Using Electroencephalography.” *Scientific Reports* 5: 15924.
de Tocqueville, Alexis. 1899. *Democracy in America*. Translated by Henry Reeve. New York: The Colonial Press.
Velmans, Max. 2000. *Understanding Consciousness*. London, UK: Routledge.
Wenar, Leif. 2005. “The Nature of Rights.” *Philosophy & Public Affairs* 33(3): 223–252.
Wenar, Leif. 2015. “Rights.” *Stanford Encyclopedia of Philosophy*. https://plato.stanford.edu/entries/rights/.
Žižek, Slavoj. 2006. *The Parallax View*. Cambridge, MA: MIT Press.

CHAPTER 3

Machine Moral Standing: In Defence of the Standard Properties-Based View

Kęstutis Mosakas

1 Introduction

Should artificially intelligent robots be recognised as bearers of moral rights? While the academic contributions on the question of machine moral agency (i.e. programming ethics into machines) have been considerably more plenteous compared to those dealing with the question of rights, the recent years have been marked by a rise of scholarly interest towards the latter issue as well (Gunkel 2015, 2018; 2012; Gordon 2018; Coeckelbergh 2014; 2010; Torrance 2014; Levy 2009; Miller 2015). Given the seemingly inevitable emergence of intelligent machines in the relatively near future and their expected involvement in various spheres of human life, it seems to be absolutely necessary to resolve the related ethical issues before these machines actually come into being. For this reason, I find it to be greatly valuable that the question of machine rights is not only being tackled by proponents of analytic moral philosophy but also by ingenious thinkers who lean towards the continental tradition, such as David J. Gunkel (2018) and Mark Coeckelbergh (2010), who have both done a great deal to expand the scope of the arguments in the ongoing debate on machine rights. Both Gunkel and Coeckelbergh have professed to be in favour of what has been referred to as "the relational turn" in recent literature (Gerdes 2015; Coeckelbergh 2010; Gunkel 2018) – the idea that we should consider moral status not in terms of certain ontological features possessed by entities (e.g. consciousness, rationality, etc.) but rather in terms of relations between entities.[1] By endorsing the relational turn, we would thus be required, first and foremost, to consider the relation between ourselves and a given entity, rather than asking what morally relevant features a given entity does in fact possess.

1 This is especially true of Coeckelbergh, who has specifically argued for what he calls a "social-relational" justification of moral consideration (Coeckelbergh 2010). Gunkel too – while rather cautious in his book *The Machine Question* (2012) – appears to be sympathetic to this line of reasoning in his later works, although he prefers to frame his approach in Levinasian terms of entities "taking up a face" in our personal encounters with them (see Gunkel 2015, 2018; 2012).

 | DOI:10.1163/9789004437876_005

My primary goal in this paper is to examine some of the reasons why one may think that the standard properties-based view is deficient in addressing the question of machine moral standing, as well as to consider the relational approach in its own right to see if it makes for a suitable alternative. To achieve this goal, I shall first outline some of the background issues concerning both artificial intelligence (henceforth AI) and moral theory. In doing so, I argue that the traditional arguments that deem the question of machine rights absurd have virtually been rendered obsolete by our background knowledge, thereby showing that the question is both complex and philosophically significant. Then, I turn to the dispute between the standard view and the relational approach, which have thus far emerged as the most promising candidates for addressing the question of machine moral standing. I henceforth examine some of the main arguments presented by Gunkel (2018) and Coeckelbergh (2010) against the standard view and comment on the relational approach itself in the process. Ultimately, I conclude that while the proponent of the standard view may be forced to concede certain points by Gunkel and Coeckelbergh, it does not appear to be the case that the standard view is as deficient as its critics suggest; neither does it seem to be clear that the relational approach provides us with a plausible account of moral consideration.

2 What Kind of Entities Deserve Moral Consideration?

2.1 *Machine Moral Standing: The Changing Attitudes*

The question of whether machines should be considered to have moral standing may strike one as rather bizarre if not outright absurd – are they not just dumb mechanisms that we have developed to achieve certain ends for ourselves? What moral standing can they possibly have? Similar attitudes can be spotted by taking a glimpse at the historical development of moral rights – a notable example here could be René Descartes, who argued against moral rights for animals on the grounds that they could neither reason, nor feel pain. He thought of them as mindless entities with certain behavioural dispositions – much like today's machines that are not conscious but are able to take certain actions that were programmed into them (Descartes 2006, 47–49). What is interesting about this example is that it makes an appeal to particular properties that machines possess in order to deny moral status for animals – it presupposes that machine-like beings are not the kinds of entities to which moral rights could be meaningfully ascribed. In contrast to human beings whom Descartes thought to have immaterial souls and thereby minds, purely mechanistic entities, he would argue, cannot have any properties that are morally relevant.

Since the times of Descartes, the underlying assumptions of his argument have been challenged time and time again. Possibly the greatest challenge has been posed by Charles Darwin, who made it rational to believe that, in his words, "the difference [...] between man and the higher animals, great as it is, certainly is one of degree and not of kind" (Darwin 1981, 105). What Darwin's theory of evolution has shown is that the distinction between human beings and animals, which was considered to be clear and immovable not that long ago, has turned out to be rather blurry – it seems to be the case that we have all descended from a common ancestor, and presumably, through the same evolutionary mechanisms. Darwin's conclusion of there being quantitative rather than qualitative differences between human beings and animals has since received considerable empirical support – as Gunkel points out: "Everything that had divided us from them is now up for grabs: language, tool use, and even reason. Recent discoveries in various branches of the biological sciences have had the effect of slowly dismantling the wall that Descartes and others had erected between the human and the animal other" (Gunkel 2012, 109). As a result, any argument that attempts to exclude animals from moral consideration on the grounds that they lack something substantial that is possessed by human beings is bound to appear philosophically suspect.[2]

Similarly, we should be very careful about "it's just a machine" kind of arguments. Can we be certain that human beings are not just "biochemical machines," as it were? Today, the views that go under the names of "physicalism," "naturalism," or "materialism" are indeed quite respectable positions in the domain of philosophical anthropology. However, if there is no immaterial aspect to the man, as the mentioned views entail, what definite distinction can be drawn between a human being and a machine?[3] Indeed, one of the earliest French materialists, Julien Offray de La Mettrie, titled what is now his best-known work *L'Homme Machine* (1912), which translates into "Man a Machine." Since de La Mettrie denied the existence of a soul separate from the body, he simply extended the Cartesian animal-machine analogy onto human beings. More recently, an interesting analogy between the man and the machine was drawn by Keith E. Stanovich in his *Robot's Rebellion* (2004), who appeals to evolutionary biology to make his case. The description on the back of the book is particularly illuminating:

2 In addition to this, one could also consider in this context the on-going research on human-animal hybrids, also known as "chimaeras"(Hübner 2018; McNamee 2015).

3 Perhaps one could try to appeal to the distinction between "organic" and "mechanistic." Here, however, I share my scepticism with Gunkel as to whether it is at all plausible that this distinction is significant or even a tenable one (see Gunkel 2012).

> The idea that we might be robots is no longer the stuff of science fiction; decades of research in evolutionary biology and cognitive science have led many esteemed scientists to the conclusion that, according to the precepts of universal Darwinism, humans are merely the hosts for two replicators (genes and memes) that have no interest in us except as conduits for replication. Richard Dawkins, for example, jolted us into realising that we are just survival mechanisms for our own genes, sophisticated robots in service of huge colonies of replicators […]. (Stanovich 2004)

Moreover, one should also take into consideration the transhumanist scenario that envisions the future with human beings so enhanced and modified that they do not resemble "human beings" as we understand them anymore (see Sutton 2009; Bostrom 2014, 36–44). The apparent possibility of this scenario has laid support to the idea that human nature is not a static essentialist notion but rather something that is dynamic and ever-changing. Given these considerations – perhaps only with the exceptions of certain religious views – it seems that the boundaries between humans, animals, and machines have been shaken to the core, which should lead us to questioning the traditional moral arguments based on them. These considerations, I think, make it evident that the question of machine moral standing is considerably more complex than it may appear at first glance.

To complicate matters further, we need to consider the moral status of the so-called social robots, such as robotic care-givers – the anthropomorphic, yet non-conscious autonomous machines that will be taking care of the elderly and the infirm in the near future.[4] Despite being neither rational, nor conscious, some have argued that these machines should still be taken into moral consideration (Coeckelbergh 2010; Gunkel 2015) or at least be granted some form of legal protection (Darling 2016), given their social presence. John Sullins, for example, has (perhaps rather ambitiously) remarked that "even the modest robots of today can be seen to be moral agents of a sort under certain, but not all, levels of abstraction and are deserving of moral consideration" (Sullins 2011, 160). Regardless of whether one agrees with the positions defended by the mentioned scholars, it should be clear that the idea of robots possessing moral rights is one that is worth taking seriously. The long-held view that there is a sharp line dividing "us" from "them" has indeed become quite difficult to defend, whereas the expected engagement of artificially intelligent

4 Relatively simple versions of such robotic carers (and, indeed, pets) are already in operation. While at the moment the scope of their functions is rather narrow, the industry in general is looking very promising (see Petrecca 2018; Hurst 2018).

entities in our socio-political life is bound to raise new challenges that need to be met as soon as possible.[5]

2.2 *The Expanding Circle of Moral Concern*

The considerations on whether a new kind of entity should be included in the so-called moral community immediately brings one back to what is known as the expanding circle analogy. Present as early as 1809 in William E.H. Lecky's *History of European Morals* (1809), the image of moral progress as that of an expanding circle has been picked up and elaborated on by a number of different authors (see Singer 2011; Godlovitch 1998; Jamieson 2002). As Musschenga and Meynen note, this widely accepted image attempts to capture the idea that moral concern first began with our own family or tribe, and since then has been "expanding over time to include larger groups, nations, families of nations, all humans and perhaps even nonhuman animals" (Musschenga and Meynen 2017, 6). The idea here is that moral progress has so far been marked with inclusion – we have come to change our attitudes towards certain previously disadvantaged groups and include them in the moral community. Indeed, as John-Stewart Gordon points out, this process has been strongly shaped by five significant movements that have greatly contributed to the expansion of the boundaries of moral concern:

1. The feminist movement, stressing moral equality between the sexes
2. the gay and lesbian movement, stressing moral equality between people of different sexual orientation;
3. the animal rights movement, stressing the importance of sentience as requiring moral behaviour towards non-human living beings;
4. the disability rights movement, stressing moral equality among persons regardless of physical or mental ability;
5. the environmental rights movement, stressing the importance of non-human natural life. (Gordon 2018, 5)

In general, I view the history of moral rights with optimism. While it is certainly the case that moral progress has been a difficult and non-linear process, we have come a long way since the ancient times when rights in the fullest sense were only ascribed to relatively small groups of privileged men.

5 While in this paper I approach the philosophical aspects of researching machine rights, it also needs to be mentioned that the topic is of great practical significance. Given the generally optimistic predictions of the future of strong AI (see Grace et al. 2017; Kruel 2012; Müller and Bostrom 2014), it seems to be absolutely necessary to resolve the related problems before such AI is built. And while it is true that there is a number of sceptical arguments regarding strong AI, I do not think that any of them is of the right sort to undermine the importance of the question of machine moral standing.

This leads us back to the question of machine moral standing – how do we determine what kind of entities should be included in the circle of moral concern? Moreover, if we were to agree that intelligent robots do have moral standing, should we see them as moral agents – beings that are capable of understanding and acting upon moral reasons (and thereby having moral obligations); or merely as moral patients – beings that are not capable of taking genuinely moral (or immoral) actions but can nevertheless be harmed or wronged in some way? The imagery of the expanding circle may tempt one to affirm simply that the logical thing to do would be to stretch the boundaries of moral concern to include intelligent machines in one way or another. However, this argument surely is a fallacious one – one cannot derive normative judgements from mere historical facts (or indeed any other empirical observations), for moral doctrines need to be demonstrated as norms, not facts (see Fagan 2003). That is to say, in order to address the question of machine moral standing, we need to engage in moral philosophy.[6]

Moreover, as the Brill referee points out, it seems to be question-begging to appeal to other previously disadvantaged groups in the context of machine rights, because we already know they were "short-changed." Thus, if one makes an analogy between robots and the groups, one presupposes precisely what is in question, namely, that it would be wrong to exclude machines from moral consideration. I may point out that I am not making an argument from analogy in this case. Rather, the previously disadvantaged groups (as well as the expanding circle imagery) are mentioned only for illustration purposes, i.e., to provide some background on moral consideration in practice and highlight how our moral attitudes have been influenced by these different rights movements. I may also add that the purpose of this chapter is not to advocate

6 It needs to be mentioned that there is nevertheless an important empirical dimension to the problem of machine moral standing, having to do with the question of the future existence of "moral" machines. For example, Daniel Dennett takes the optimistic view that there is no principal reason why machines should not become moral agents in the future (Dennett 1998). Selmer Bringsjord, on the other hand, takes the pessimistic view that actions taken by machines will never be truly moral or immoral, on the grounds that machines cannot have free will (Bringsjord 2008). An interesting argument is also made by Emile Nadeau, who maintains that only robots and not human beings can be genuine moral agents because human rationality is all too often hindered by emotions (Nadeau 2006). The question of the future of moral machines is both ethical and empirical. It requires one to consider, first and foremost, what constitutes moral status – e.g. what properties one needs to possess in order to enter the moral dimension; and second, whether this threshold of moral status, whatever it may be, is a feasible one for the future engineers who will be manufacturing intelligent machines. Only the ethical question of machine moral standing is within the scope of the present paper.

machine rights as such, although some of the arguments presented may be utilised in pursuit of that goal as well.

2.3 *The Standard View of Moral Consideration*

Throughout the history of moral philosophy, notions such as "moral standing" and "moral status" have proved to be of vital importance. As commonly understood, an entity has moral standing or status – and I shall be using these terms interchangeably – just in case that entity counts morally in its own right, thereby making it permissible/impermissible to do certain things to it for its own sake.[7] The questions of what exactly it means for one to "count morally" and what kind of entities actually do so have been a matter of heated debates for centuries. According to Bostrom and Yudkowsky, many accounts of moral standing can be fit into two rough categories:

- Sentience: the capacity for phenomenal experience of qualia, such as the capacity to feel pain and suffer.
- Sapience: a set of capacities associated with higher intelligence, such as self-awareness and being a reason-responsive agent. (Bostrom and Yudkowsky 2014, 16)

These categories, while by no means exhaustive,[8] should give us a good idea of what the so-called "standard view" of moral standing consists in. Fundamentally, the idea is that there are certain ontological features that make an entity worthy of moral consideration if the entity in question does in fact possess these features. In his paper titled "Robot rights? Towards a social-relational justification of moral consideration," Coeckelbergh (2010) convincingly argues that all three of the main normative theories of ethics – utilitarianism, Kantianism, and virtue ethics – presuppose that moral status is grounded in certain ontological features or properties that one possesses, which puts them all under the label of the standard view.

With respect to utilitarianism, be it the classical version or the more refined ones that emerged in the twentieth century, the notion of sentience has always been one of the central concepts of the theory. Since according to utilitarians the right actions are the ones that promote happiness or pleasure and the wrong ones are those that cause unhappiness or pain, the theory entails that

7 Based on Nick Bostrom's and Eliezer Yudkowsky's (2014, 5) paraphrasing of Kamm (2007, Chapter 7).

8 Consider, for example, Albert Schweitzer who proposed that all living things have inherent moral value (Schweitzer 1929); environmental ethicists who argued that things of beauty and harmony (e.g. ecosystems, land, etc.) should be taken into moral consideration (Leopold 1949; Callicott 1980); or aestheticians who defend the view that works of art have intrinsic moral worth of their own (Gaut 2007; Sauchelli 2012).

the only possible recipients of moral or immoral actions are beings that are capable of having the mentioned experiences (e.g. being happy or being in pain). As Peter Singer points out, non-sentient beings cannot feel, enjoy, or suffer from anything, and therefore it makes no difference to these beings how they are treated (Singer 2011, 123). It is thus the case that for a utilitarian, sentience is the key property that grounds moral status.

In relation to the category that has been labelled as "sapience" by Bostrom and Yudkowsky, a highly important position to consider is that of Immanuel Kant. Kant's ethics was logocentric – that is to say, he believed that rationality was of utmost importance for moral status. All rational beings, Kant argued, "stand under the law that each of them should treat himself and all others never merely as a means but always at the same time as an end in himself" (Kant 1983, 39). Moreover, as James Rachels points out in his summary of Kant's views,

> Humans have an intrinsic worth i.e., dignity, because they are *rational agents*, that is, free agents capable of making their own decisions, setting their own goals, and guiding their conduct by reason. Because the moral law is the law of reason, rational beings are the embodiment of the moral law itself. [...] It makes no sense, therefore, to regard rational beings merely as one kind of valuable thing among others. They are the beings for whom mere 'things' have value, and they are the beings whose conscientious actions have moral worth. So Kant concludes that their value must be absolute, and not comparable to the value of anything else. (Rachels 2003, 131–132)

Given Kant's notorious exclusion of non-human animals from direct moral consideration (i.e., as proper moral subjects and objects) on the grounds that they do not pass the criterion of rationality, as well as the fact that the only known rational agents are human beings, some have come to believe that Kant's view is anthropocentric – i.e., it exclusively privileges human beings. However, it is important to point out that Kant's criterion does not take into account the origin or composition of a given entity, as Gunkel (2012, 34) rightly notes. What ultimately matters to Kant is the capacity to act autonomously, which is something that only beings that use reason to guide their conduct (i.e., are rational) are capable of. And since it is autonomy that grounds one's dignity and thereby moral worth, according to Kant, it follows that it is rationality that is crucial for moral status, not one's origin, composition, species, etc. (see Kant 1983).[9]

9 Of course, it is not within the scope of this chapter to attempt to cover all of the existing views that fit under the umbrella of the standard view. For our current purposes, it may be

Finally, there is virtue ethics. In simple terms, virtue ethics is a theory that focuses not so much on the consequences of our actions or moral duties but rather on one's moral character. Since for the virtue ethicist the right action is one that a person with excellent moral character would perform, the theory, as Coeckelbergh (2010, 213) rightly remarks, is one that is subject-oriented – on virtue ethics, torturing animals is wrong not because it causes them to suffer but because in doing so we fail to exercise certain virtues (e.g. compassion). Even though virtue ethics shifts the focus from the recipient of the moral action towards the author, it still presupposes the standard property-based view of moral consideration. One of the foundational questions for the virtue ethicist to address is what kind of entities can possess virtues. For instance, we typically have no problem in admitting that this or that human individual is virtuous or, indeed, morally defective; but when we see a housecat brutally toying with a mouse before killing it, we do not consider the cat to be immoral or lacking in virtue. Unsurprisingly, thus, virtue ethics has traditionally been anthropocentric – as Coeckelbergh highlights, in its neo-Aristotelian version, the operational boundaries of virtue ethics are entirely set around man, who is conceived as a rational and political animal that can only flourish in a moral and social community (ibid., 213). Entities other than human beings are simply not within the scope of the theory. Of course, an attempt could be made to adjust various aspects of the theory – perhaps, to reconsider it in terms of different levels of abstraction[10] as to what entities can be virtuous and towards what entities virtues should be exercised. The crucial point here, however, is that these questions can only be addressed by appealing to certain morally relevant ontological properties; and thus I find myself entirely in agreement with Coeckelbergh that all of the three main normative theories of ethics hinge on the plausibility of what has been labelled as the standard account of moral consideration.

---

sufficient to add that the mentioned criteria of moral status can be combined in a number of ways. Unlike Kant, one need not affirm that an entity either has full moral status or no moral status whatsoever. Many have come to believe that moral status comes, in fact, in degrees. For example, Jeff McMahan argues that all sentient beings have *some* degree of moral status, but in order to have *full* moral status, they must also be self-aware (see McMahan 2002). For an outline of a variety of other positions on moral status, see Jaworska and Tannenbaum 2018.

10 The concept of "different levels of abstractions" has been borrowed from Luciano Floridi and J.W. Sanders. As they describe it, "LoA [level of abstraction] is determined by the way in which one chooses to describe, analyse and discuss a system and its context" (Floridi and Sanders 2004, 349).

2.4 *The Relational Approach to Moral Consideration*

Despite the prominence of the standard view in both theoretical and practical ethics, it has been persistently contested by Gunkel and Coeckelbergh in the context of machine moral standing. The mentioned authors advocate for what has been labelled as the relational turn – a shift from the properties-based view of moral consideration towards the relational approach. As I understand it, on the relational account moral significance is grounded in (or constructed through) certain special relations, which is to say that an entity X counts morally just because it stands in a particular relation with other entities (biological, social, cultural, caring, etc.).[11,12] This view, while not held as widely as the standard one, has nevertheless served as a ground for a number of respectable positions in moral and social philosophy, including: the care ethics, which is the view that moral significance rests in caring relationships (Gilligan 1982; Noddings 1986; Tronto 1993); Levinasian ethics (Levinas 1969; possibly Haraway 2008); Marxism and communitarian philosophies; certain types of environmental ethics, such as deep ecology (Naess 1973); as well as types of postmodern feminist ethics (e.g. Braidotti 2006).[13,14] While the mentioned views

11 See Torrance 2014 for an alternative definition. Torrance seems to think that "relationists" generally accept the constructivist view of both moral status and consciousness, in contrast to the "realists" who see both as a matter of objective fact (Torrance 2014, 11–12). However, I do not see why the relationist necessarily needs to commit to constructivism in either case. To give an example, both Gunkel and Coeckelbergh argue that it is in principle not *knowable* to us whether a particular entity is conscious, which is completely different from the claim by Torrance that there simply *is* no "correct" answer to the mentioned question for the relationist.

12 Additionally, as the Brill referee points out, the relational approach should also be distinguished from what is known as "situational ethics." The relational approach – as mentioned earlier – is first and foremost concerned with some sort of relations (whatever they may be), whereas situational ethics emphasises on context-sensitive decision-making, rejecting the idea that there are absolute moral norms (see, for example, Fletcher 1997). While both approaches are clearly distinct, they may nevertheless be compatible. Gunkel, for example, while leaning towards the relational approach by Levinas, seems to affirm the core thesis of situational ethics as well. He claims: "Should machines like AIs, robots, and other autonomous systems be granted admission to the community of moral subjects, becoming what would be recognised as legitimate moral agents, patients, or both? This question cannot be answered definitively and finally with a simple "yes" or "no." The question will need to be asked and responded to repeatedly in specific circumstances" (Gunkel 2012, 215).

13 Moreover, the relational approach has seen success in certain domains of practical ethics, such as disability studies (see Koch 2004).

14 Hybrid approaches to moral consideration are also possible – consider, for example, Mary Anne Warren, who argues in favour of a combination of relational and non-relational criteria (Warren 1997; see also Coeckelbergh 2010, 219 for a critique).

certainly bear interesting implications for the question of machine moral standing, my focus in this paper shall nevertheless be limited to Coeckelbergh and Gunkel, who have argued from very similar perspectives in favour of the relational approach precisely in the context of machine moral status.

3 Anthropocentrism and Exclusion

3.1 *High Thresholds and Marginal Cases on the Standard View*

Some of the main arguments employed by Coeckelbergh and Gunkel against the standard view have to do with the notions of anthropocentrism and exclusion in the context of ethics. Coeckelbergh, for example, argues that not only do the standard theories have high thresholds, but they also face what he calls the problem of marginal cases. The two arguments, while strongly related, will be treated as distinct. Coeckelbergh's first concern of high thresholds has to do with the fact that today's robots are neither rational nor sentient and therefore "this renders arguments based upon such features irrelevant to the problem of how to think about giving moral consideration to currently existing intelligent robots" (Coeckelbergh 2010, 212). His second concern (the problem of marginal cases) is presented in the following way:

> If particular properties are agreed upon as being sufficient [sic] for moral status, and if not all humans share these properties (all the time), does that not imply that these human beings are not worthy of our moral concern (all the time)? Small children or aged adults do not meet that criteria. (Coeckelbergh 2010, 212)

The argument appears to be that once we recognise certain properties as necessary for moral status, then we are obligated to exclude human beings who do not possess these properties from our moral concern. Both of the purported problems seem to arise owing to the standard view's inability to accommodate certain entities in the realm of moral consideration.

Are these objections to the standard view any good? In his criticism of high thresholds, Coeckelbergh suggests that because the standard view excludes today's machines from moral consideration, the arguments based on it are irrelevant to the problem. It seems to me that there are two important responses to that: first and most important, this objection does nothing to undermine the plausibility of the standard view because all it concerns is the outcome of the standard view's application to this particular problem. As such, it tells us nothing about whether the standard view is an adequate way of approaching the question in the first place. Secondly, is it really the case that the standard view ends the discussion by excluding today's robots from moral consideration? I think not – in the argument from high thresholds, Coeckelbergh seems to underestimate the variety of moves available to the proponent of the standard view. For instance, one can still argue against the "mistreatment" of the current robots – particularly the human-like ones (generally known as androids) – by pointing out that it damages the human character. If that is really the case, then it surely constitutes a strong reason for abstaining from violent actions towards the rather primitive anthropomorphic machines of today (e.g. robot care-givers). Indeed, this particular threat to our humanity may be high enough and severe enough to provide machines with legal rights even if the bar for moral rights is currently too high for them, as Kate Darling has argued (see Darling 2016).

Here it may be important to mention that Coeckelbergh actually anticipates the possibility of such indirect justifications in a footnote and promises to discuss them (Coeckelbergh 2010, 212), but ultimately he makes no explicit reference to this particular solution. Instead, he ends up mostly focusing on virtue ethics. Gunkel, on the other hand, does respond to Darling's position. While noting that this way of reasoning is "potentially expedient for developing and justifying new forms of legal protections," it is ultimately implausible, he argues, because it "renders the inclusion of previously excluded others less than altruistic […]. The rights of others, in other words, is not about them; it is all about us" (Gunkel 2018, 95). What strikes me as an obvious problem with this objection is that it appears to be question-begging. It may be true from Gunkel's point of view that this kind of solution is inadequate because he accepts the assumption that today's robots should not be excluded from moral consideration; but Darling does not share that assumption with him. Given her rejection of the mentioned assumption, it is perfectly consistent within Darling's inner moral framework to argue that given that robots do not yet meet the moral criteria, we should rather be concerned with beings that do. Similarly, it was not in any way inconsistent for Kant to deem animal abuse wrong solely on the grounds that it might make human beings more abusive to each other, given his underlying logocentric framework of morals. Thus, what one needs to do in order to challenge Darling or Kant adequately is to refute the underlying presuppositions of their argument – namely, that robots and animals, respectively, have no moral worth of their own. One cannot do so by presenting an argument that is already based on the presumed falsehood of their views.

Coeckelbergh's second argument – namely, his appeal to marginal cases – is undoubtedly an interesting attempt, but nevertheless it is not clear that it succeeds. The thought of an infant or a severely demented person not having high moral worth of their own is indeed quite bothersome, but there may be good reasons to bite the bullet. As Singer has argued, it is "speciesism" to give more weight to the interest of a member of one's own species rather than to the corresponding interest of a member of different species (Singer 1990). However, if we were to accept Coeckelbergh's argument, which appears to be exclusively concerned with human beings who do not pass the threshold, we might have to give up on the idea that morality is impartial species-wise and put extra weight on the interests of mentally underdeveloped or severely impaired human beings[15] just because they are human beings. Thus, there seems to be a considerable trade-off to this idea of providing special moral treatment to these "marginal" human beings, one that comes in the form of speciesism.

3.2 *Universal Ethical Consideration and the Structures of Power*

Curiously enough, while Coeckelbergh appears to be endorsing a degree of anthropocentrism in his objections to the standard view, Gunkel seems to be not just vehemently opposed to it but also highly critical of any kind of "moral centrism." Indeed, following Thomas Birch (1993, 1995), he has argued in favour of what they call "universal ethical consideration" – the idea that nothing can be excluded from moral concern *a priori* and that anything can "take up a face (i.e. be considered in moral terms) (Gunkel 2012, 131; 183; Gunkel 2018, 97). This brings us to the fundamental question of the nature of moral consideration. At very least, it does not appear to be immediately obvious that moral consideration is as totalising and universal as Gunkel believes it is. After all, many entities are thought not to have any intrinsic moral worth. For example, a rock does not have such worth – "we may crush it, pulverise it, or subject it to any treatment we like without any concern for the rock itself" (Bostrom and Yudkowsky 2014, 6). What are then the arguments for universal ethical consideration? One such argument has been presented by Birch, and it goes as follows:

> The institution of *any* practice of *any* criterion of moral considerability is an act of power over, and ultimately an act of violence toward, those others who turn out to fail the test of the criterion and are therefore not permitted to enjoy the membership benefits of the club of *consideranda*. They become 'fit objects' of exploitation, oppression, enslavement, and finally extermination. As a result, the very question of moral

15 Coeckelbergh's use of language could also be questioned here – not even utilitarians believe that "small children" or "aged adults" (see Coeckelbergh 2010, 212) are below the threshold of moral relevance. Usually, the bar is placed considerably lower and one talks about the status of infants/foetuses and that of severely mentally impaired adults.

> considerability is ethically problematic itself, because it lends support to the monstrous Western project of planetary domination. (Birch 1993, 317)

Notice the crucial phrase that "the institution of *any* criterion for moral consideration is an act of violence" – if one thinks that sentience is a necessary criterion, they are being violent towards the non-sentient, whereas if one thinks it is rationality, that is an act of violence against the non-rational. On this view, even the apparently inclusive theories, such as the one by Singer, are in fact exclusive and violent. Moreover, it is argued by Gunkel that moral theorists like Singer are being anthropocentric in terms of the general strategy they adopt – if not in the spirit of their work. Here he once again appeals to Birch, who makes the following argument:

> The nub of the problem with granting or extending rights to others, a problem which becomes pronounced when nature is the intended beneficiary, is that it presupposes the existence and the maintenance of a position of power from which to do the granting. Granting rights to nature requires bringing nature into our human system of legal and moral rights, and this is still a (homocentric) system of hierarchy and domination. The liberal mission is to open participation in the system to more and more others of more and more sorts. They are to be enabled and permitted to join the ranks and enjoy the benefits of power; they are to be absorbed. But obviously a system of domination cannot grant full equality to *all* the dominated without self-destructing. (Birch 1995, 39)

According to Gunkel, what this shows is that the extension of moral rights in the way that the supporters of animal rights do "does not in any way challenge the basic power of anthropocentric structure" and that it "leaves the existing hierarchies and structures of domination intact" (Gunkel 2012, 126).

What should one make of these objections? Consider first the Birch's claim that any attempt to establish a criterion for moral status is already immoral in its own right. The problem with Gunkel's adopting this as an argument against the standard view is that it assumes a peculiar conception of morality rather than argues for it. The claims that those that are excluded from moral consideration are denied the benefits of the club of moral *consideranda* and that it makes them fit objects for exploitation presuppose that these entities – whatever they may be – should be accounted for in moral terms and should not be exploited, but no argument is presented in support of the idea that there are no entities that we could exclude from moral consideration and use solely for our ends. Hence, the idea of universal ethical consideration in this case is affirmed but not argued for. Indeed, the earlier-mentioned example of rock can itself be turned into an argument against universal ethical consideration, for if we affirm that sentience is both a sufficient and necessary condition for moral status, then it follows that, indeed, there are some entities that we can "exploit." And, in order to avoid begging the question, we could argue for sentience as a criterion for moral consideration based on the kind of reasoning similar to that of Singer – since a rock is not sentient, it has no welfare of its own, and therefore it makes no difference to it how it is treated.

The second argument strikes me as a bizarre one. Most important, it unapologetically begs the question against the realist conception of morality. On the realist view, which is indeed the mainstream one among philosophers,[16] morals are not invented but discovered, in which case it simply makes no sense to say that it is human beings who are (literally) doing the granting of moral rights. On moral realism, human beings may choose to *recognise* an entity as a bearer of such rights, but whether an entity does in fact *possess* them is not a matter of anyone's decision.[17] Similarly, the claim that there is an anthropocentric power structure that absorbs into itself those that it chooses to accept seems to rest on a confusion between issues pertaining to moral theory (the way morality is perceived to be) and those related to moral practice (the way it is lived out by human beings). It may be true that the human power structures would hesitate to recognise certain entities as right bearers if they perceived these entities as a threat to their power, but that is simply no argument against the moral theorists who appeal to principles that are impartial species-wise, substance-wise, origin-wise, etc.[18] If one believes, as I do, that moral status comes in degrees and that morality is impartial in the senses mentioned earlier, then there is no reason whatsoever for thinking that the moral worth of human beings cannot be surpassed by other entities, even if no such entity is known to exist at this point in time. Hence, Gunkel's appeal to this argument by Birch does not seem to achieve its purposes.

In sum, the arguments by Coeckelbergh and Gunkel do not appear to hit the standard view particularly hard. The high thresholds of the standard theories seem to be justified and there are other moves and considerations for the

16 According to the PhilPaper Surveys (https://philpapers.org/surveys/results.pl), moral realists outnumber the anti-realists roughly 2 to 1, with 56.4% of the respondents accepting or leaning towards the realism, 27.7% being more sympathetic to anti-realism, and 15.9% accepting other views.

17 See Sayre-McCord 2014 for a detailed outline of moral realism.

18 Consider, for example, the principle of equal consideration of interests by Singer (1993) or the principle of substance non-discrimination as outlined by Bostrom and Yudkowsky (2014).

proponent of the standard view to make even if it is the case that current robots have no moral status. With respect to Gunkel's arguments,[19] no good reason for accepting the so-called universal ethical consideration has been given, whereas the considerations on the structures of power bear no interesting implications for the proponent of moral realism and moral impartiality. Indeed, the very idea of universal ethical consideration should be questioned, given its rejection of the very possibility of there being criteria for moral consideration. Even if one could fit all existing things into the domain of moral concern, there still appears to be a moral hierarchy that needs to be accounted for (e.g. why is it permissible to destroy a rock, but not a human being?) to which universal ethical consideration seems quite oblivious.

4 The Problem of Other Minds as an Argument against the Standard View

Another reason why both Gunkel and Coeckelbergh think that the standard approach is inadequate has to do with what is known in philosophy as the problem of other minds. The idea is that the most typical criteria for moral consideration on the standard view are based on mental properties (consider the examples of sentience and sapience) that can only be accessed privately and subjectively by the individual possessing them. As Gunkel elaborates,

> Because suffering is understood to be a subjective and private experience, there is no way to know, with any certainty or credible empirical method, how another entity experiences unpleasant emotions such as fear, pain, or frustration. For this reason, it appears that the suffering of another <...> remains fundamentally inaccessible and unknowable. (Gunkel 2012, 118)

The implication of this problem is that one cannot establish with certainty whether any entity does in fact possess the internal aspects of mentality that

19 Of course, it is not possible to cover all the arguments put forward by Gunkel in his *The Machine Question* in this chapter. For this reason, an attempt was made to cover only those presented at the subsection 2.3.3., where he attempts to identify the faults of the standard view in principle, which he labels as ethical problems, as opposed to the terminological, epistemological, and methodological ones (see Gunkel 2012, 125–133). In the mentioned subsection, three arguments are presented in total, but two of them are based on universal ethical consideration, and thus I chose not to distinguish between them and simply respond to the main underlying assumption.

are generally thought to be necessary for moral status on the standard view. As Coeckelbergh succinctly summarises, "skepticism tells us that we can never be certain about the internal states of other entities" (Coeckelbergh 2014, 63). As a result, this problem supposedly makes the standard view epistemically opaque. Thus, even if it could be granted that suffering, for instance, is morally relevant, we could never take it into account in our moral decision-making, for it is a private mental event that is completely exterior to anyone aside from the subject experiencing it.

4.1 *Solutions and Scope of the Problem*

There is a number of possible responses to this argument. First, we could attempt to (dis)solve the problem, like Steve Torrance does.[20] According to Torrance, if we grant that what he calls CSS (conscious satisfaction or suffering) properties are not utterly mysterious, then they must be *natural* properties, in which case it would be completely inexplicable why "the other" would be different from the experiencing subject if they are sufficiently similar to each other in terms of their neurological make-up (Torrance 2014, 19). Of course, this does not rule out certain sceptical scenarios: For instance, it remains conceivable that there is only one genuinely conscious experiencing subject while the rest of the world is something like a matrix or a simulation created by the Cartesian Demon. Additionally, if one is a proponent of the hard problem of consciousness, he or she may well argue that it is possible that some of us are philosophical zombies – entities that are physiologically indistinguishable from human beings but are incapable of having any conscious experiences (see Chalmers 1995). The problem with both of these positions is that they are bizarre and asymmetrical. While neither the realist view of other minds, nor the mentioned sceptical scenarios can in principle be proven or disproven, we may still be justified in leaning towards realism on the virtue of its explanatory simplicity.

Moreover, by utilising the other minds problem in the context of moral status Gunkel and Coeckelbergh are taking an epistemological perspective to attack what is fundamentally an ontological position.[21] On the standard view of

20 An interesting proposal could also be found in the context of philosophy of religion by Alvin Plantinga, who argued that certain beliefs, including the belief in the reality of external world and other minds, are properly basic, which is to say we are rational and justified in holding these beliefs even if we have no evidence for them, unless some defeater to them is presented (see Plantinga 1993).

21 Here my line of argumentation diverges from that of Torrance. Torrance goes on to differentiate between "thin" and "thick" conceptions of consciousness, arguing in favour of the latter, which "sees consciousness as conceptually inseparable from the lived,

moral consideration, the central claim is that some features of entities just *are* morally relevant. The argument that we cannot *know* whether any entity in fact possesses these features simply does nothing to challenge this fundamental claim, for it belongs to a different category (epistemology as opposed to ontology). Certainly, as Gunkel and Coeckelbergh argue, the problem of recognising the morally relevant properties in practice is real and significant; however, I am not convinced that it justifies – at least in its own right – the drastic paradigmatic shift that the mentioned authors propose. As long as one is willing to concede that some properties are morally relevant, even on the assumption that it cannot be known with certainty which entities possess them, the leap into the relational approach, as long as it is taken to be at odds with the standard view, is logically unjustifiable. The epistemological challenge raised by Gunkel and Coeckelbergh is thus rather limited in what it can achieve in the context of the question of moral status. What the relationist really needs to be after, I think, is the central claim of the standard view, namely, that some properties are morally relevant, and I have already addressed some of the challenges raised against it in the earlier parts of the paper.

4.2 *"Mindless" Approaches to Moral Consideration*

Finally, I suspect that the problem of other minds (or rather employing it in the context of moral standing) may, indeed, have certain radical implications even for the relationist. Since this paper considers two competing views of moral standing, it is important to see whether what appears to be a problem with one of the views does not also cause problems for its competitor. According to Coeckelbergh, the relationist can avoid the problem of other minds by appealing to phenomenology:

> The alternative approach I propose attempts to avoid the scepticism by replacing the requirement that we have certain knowledge about *real* ontological features of the entity by the requirement that we experience the features of the entity as they *appear* to us in the context of the concrete human-robot relation and the wider social structures in which that relation is embedded. (Coeckelbergh 2010, 214)

---

physiological processes that make up a conscious being's embodied existence" (Torrance 2014, 20). Rather than taking this route, I shall grant the thin Cartesian conception that Gunkel and Coeckelbergh endorse and consider its implications for the question of machine moral standing instead.

Here my first thought is that this "phenomenological turn," as it were, does not seem to be wholly incompatible with the standard view. All that the proponent of the properties approach needs here is the assumption that the appearance of an entity exhibiting property P is a piece of evidence[22] favouring the claim that it does in fact possess P.[23] And while this view does not leave us with certainty, it may still offer us sufficient guidance in our moral decision-making. Let us suppose, however, that the relationist were to deny the mentioned assumption, thereby effectively accepting that there is no reason whatsoever for thinking that the likelihood of human beings possessing consciousness is any higher than rocks possessing it. Since Gunkel and Coeckelbergh ground moral consideration in relations, the question of consciousness and mental properties – at least at first glance – seems to be rendered obsolete. Thus, even if it is the case that we are not justified in maintaining that humans are more likely to be conscious than rocks, the entire question is simply irrelevant – all that matters is how different entities appear to us and what kind of relations we establish with them, based on these appearances.

While I do think that this offers grounds for a coherent account of moral consideration, my main worry has to do with its plausibility. Consider the responses available to the socio-relationist to the following questions:[24]

1. Are social relations possible between non-conscious entities?
2. What makes these relations matter morally?

The first question is closed-ended – either it is the case that some forms of interaction between non-conscious entities can be recognised as genuinely social or it is not. The phenomenologically inclined relationist has to answer this question in the affirmative because what is presented as one of the main selling points of their view is that it does not force us, in Gunkel's words, "[to] get into the head of other entities" (Gunkel 2012, 139–140). Responding in the

22 The notion of evidence that I employ here is a probabilistic one, which is to say that an observation X counts as evidence for a hypothesis H if and only if it makes H more probable than it would be in the absence of X.

23 Machines that were purposely built to mimic conscious behaviour, of course, could still be excluded on the basis of our background knowledge. Even though there is no agreement on what exactly consciousness is, I think a better case could be made for granting rights (or stronger rights, anyway) to a machine that was built according to the neurological architecture of the human brain than to a classical computational one, even if both of them were to exhibit behaviour that would appear equally conscious.

24 I am using the term "socio-relationism" here to refer to the kinds of relational approach that emphasise the moral significance of social relations and appearances, as opposed to other types of relational accounts (e.g. those based on species, family, community, and/or cultural relations, ecological models, etc.) to which the argument that I proceed to develop may not be directly applicable.

negative would simply bring one back to the other minds problem, thereby offering no advantage to relationism over the standard view in this regard. Presumably, then, the interactions between unconscious entities (e.g. two chatterbots) can be considered as genuinely social on the relational account.

The second question concerns the issue of moral justification and allows for a multitude of responses. The way I see it, responding positively to the first question makes it very difficult for the socio-relationist to offer any substantial justification for why moral consideration should be grounded in these nonconscious relations. The standard view, in contrast, is considerably more robust – we feel that it captures something fundamental about what really matters to us and, presumably, other entities as well (pleasure and pain, mental life, reasoning, making life plans, etc.). It needs to be noted that this path is also available to those advocating certain other relational accounts – for instance, a care ethicist could argue that moral relations must necessarily involve at least one conscious entity because unconscious beings are not capable of genuinely caring about anything – at most, they can simulate care. The proponent of other minds problem in ethics, however, does not have that option. For them, the justification of their account of moral consideration cannot include any direct or indirect appeals to consciousness, despite the strong and, I would argue, virtually universal intuition that a difference in one's mental states can be so profound as to decide whether one's life is worth living.[25] In contrast, what is so special about these non-conscious "social" relations? Why think that they can determine one's moral standing? The relationist may object here that they do not necessarily need to maintain that there is no conscious element in social relations that we observe. However, this objection misses the point because if we grant the assumptions made by the relationist that (a) moral consideration emerges through social relations; and (b) social relations are possible between nonconscious entities, then it follows logically that moral consideration can in principle emerge through relations between two nonconscious entities, which

25 This intuition can be firmly supported by a variety of thought experiments. Consider, for example, the idea of "the worst possible misery" as outlined by Sam Harris in his *Moral Landscape*. What Harris puts forward as one of the fundamental ethical intuitions is that a world in which every sentient being suffers as much as they can is worse than one in which they flourish (Harris 2010, 39). Unlike Harris, we need not go as far as to say that ethics is *only* about the well-being of conscious creatures; nevertheless, the intuition that the former world is worse than the latter seems to carry too much weight to be left completely unaccounted for in moral terms (which appears to be the case on relationism). Similarly, we should feel that a world that is inhabited by non-sentient entities that only *appear* to be suffering as much as they can is still by a large margin better than one in which that actually is the case.

is a view that seems to be *prima facie* implausible. It therefore seems to be the case that while socio-relationism appears to overcome some practical problems with the standard view, it seems to be doing so at the expense of its own plausibility.

4.3 *Relations, Moral Consideration, and Objectivity*

Here the relationist may respond that I have not taken certain aspects of their view into account. They may argue that what they are proposing is not merely a new theoretical criterion for what counts morally (i.e. a new kind of "special relationship" that is meant to be morally significant) but also (or, perhaps, instead) an entirely different vantage point. For example, Coeckelbergh and Gunkel stress the precedence of social relations with "others" to any "external" considerations about their moral standing. As Coeckelbergh argues,

> Inspired by Haraway's work, we may wonder if robots will remain 'machines' or if they can become companions. Will people start saying, as they tend to say of people who have 'met their dog,' that someone has 'met her robot'? Would such a person, having that kind of relation with that robot, still feel shame at all in front of the robot? And is there, at that point of personal engagement, still a need to talk about the 'moral standing' of the robot? Is not moral quality already implied in the very relation that has emerged here? (Coeckelbergh 2014, 69–70)

In agreement with Coeckelbergh (whose position he considers to be a "game changer"), Gunkel notes that "morality, conceived of in this fashion, is not determined by a prior ontological determination concerning the essential capabilities, intrinsic properties, or internal operations of these other entities" and that "instead it is determined in and by the way these entities come to face us in social interactions" (Gunkel 2015, 161). Since social relations precede any deliberations about properties of other entities, it appears to give us a reason for thinking that it is the former that is crucial for moral consideration, not the latter.

Nevertheless, it is not clear that this fundamental reconfiguration is helpful. Most important, it seems doubtful that social relations are a necessary condition for moral consideration. For instance, I think we have moral reasons for being concerned about the welfare of future generations or people in distant countries that have been struck by disasters, even though we have no social ties with them. Certainly, the moral validity of this concern could be better explained by some version of the standard view – that is to say, even though we have no ties with these people, we should be morally concerned for them because they are beings with their own desires and welfare. Additionally, the proponent of the standard view can agree with the precedence of social relations and still dispute their being fundamental to moral consideration. To take an example from Coeckelbergh's argument above, a companion robot may be an object of its owner's care and concern prior to any considerations about its ontological properties, but based on that alone, it does not seem to be immediately obvious that the robot has moral standing of its own. It could be the case that its value is not intrinsic, but instrumental – for instance, breaking the robot might be bad only because it hurts the owner. If that is the case, then the robot would still need to be taken into account in our moral decision-making, despite not being a proper object of moral concern itself. Thus, the view that relations precede things related does not seem to bear a convincing argument in favour of the relational approach, for it appears that the proponent of the standard view has sufficient explanatory tools at their disposal to account for the apparent moral quality in the relation.

Same goes, I think, for Coeckelbergh's and Gunkel's analysis of "the conditions of possibility for relations," which they carry out in their paper on the moral status of animals. In the analysis, they highlight the importance of a number of factors in our moral "facing and defacing" (to use the Levinasian terminology that they adopt) of animals, such as our use of language (Coeckelbergh and Gunkel 2014, 11–12), material-technological context, and geography (ibid., 13). I think we could all agree on the importance of discussing and analysing the reasons why we may be more inclined to treat certain entities as deserving moral concern and others not. For the proponent of the standard view, such analyses would probably fall under the purview of moral psychology and anthropology and could be greatly useful in determining and finding remedies for the moral deficiencies of human beings. Despite their importance, however, it is simply not clear that these conditions of possibility have to do with moral status *per se* and not merely with human moral perceptions.

The issue here, I suspect, is Coeckelbergh's and Gunkel's constructivist view of moral status. On their account, moral status is dynamic – it is ascribed through social "facing and defacing" of entities, which means it is bound to vary in different social contexts. There are considerable difficulties with this view. To give an example of what appears to be one of the key issues with socio-relationism (and other similar approaches), in the words of Agnieszka Jaworska and Julie Tannenbaum, is that it "do[es] not truly offer an account of moral status, but only of particular agents' reasons vis-à-vis the individual at issue" (Jaworska and Tannenbaum 2018). Consider once again the example by Coeckelbergh discussed earlier – the owner of the pet and the robot may have reasons for treating them well, but she does not seem to have any comparable reasons for doing the same with entities with which she has no social ties. The problem here seems to be that the grounding for moral status on the relational approach is not firm enough, given that moral status is generally thought to be universally binding – i.e., it has to be taken into account by *all* entities capable of moral action.

Finally, if what has been labelled by Jaworska and Tannenbaum as particular agents' reasons vis-à-vis the individual at issue is all there is to the question of moral status, the problem of relativism seems to come barging in. As Simon Kirchin explains, "the key relativistic thought is that the something that acts as a standard will be different for different people, and that all such standards are equally authoritative [...]" (Kirchin 2012, 15). Particularly problematic is the extreme version, which denies there being any moral judgments or standards that could be objectively true or false (in contrast to moderate versions that do admit of a certain degree of objectivity) (Moser and Carson 2001, 3). Given the apparent rejection of any such standard by Coeckelbergh and Gunkel, they seem to be hard-pressed to explain how the radically relational ethics (to use Coeckelbergh's own term (Coeckelbergh 2010, 218)) that they are advocating avoids the extreme version. If that cannot be done, one faces a view that is not only unintuitive but also plausibly incoherent (see Pojman 2015).[26] Ultimately, I choose to remain agnostic about whether an adequate solution to this problem can be offered by the socio-relationist; nevertheless, I think it should be evident that socio-relationism has a long way to go in order to become a plausible model.

26 To my knowledge, only Gunkel has attempted to respond to this charge of relativism. He rightly raises the question that "if moral status is relational [...], are we not at risk of affirming an extreme form of moral relativism?" (Gunkel 2015, 161). However, rather than attempting to discharge the objection, he chooses to follow Slavoj Žižek's strategy of "fully endorsing what one is accused of" (ibid., 162). The problem is that in order to make this strategy work, one needs to show that the indictment that one is prepared to endorse is in fact unproblematic, whereas Gunkel does not seem to do that. He points out that his version of ethics, although relativistic, is carefully articulated (ibid., 162); but he does not respond to any of the specific problems posed by the extreme form of metaethical relativism that made it the bogeyman of moral philosophy (Mackie 1977; Kirchin 2012; Hales 2011). Moreover, Gunkel maintains that a universally accepted response would reply to the charge of relativism (Gunkel 2015, 161–162), but that seems to rest on a confusion between a universally accepted standard and an objectively valid one. If Gunkel wishes to endorse the relativism of beliefs about moral judgments, then he is free to do that, but that offers no response to the particular implications of the extreme form of metaethical relativism, which are severe enough to merit some consideration.

5 Conclusions

In conclusion, this chapter offers a response to the main arguments put forward by the proponents of relational ethics in the context of machine moral standing. Ultimately, I think both the relationist and the proponent of the standard view should agree that the traditional "it's just a machine" kind of arguments do not do much justice to the complexities of the problem. Putting this agreement aside, despite its apparent flexibility and definite merits in stimulating a further discussion, the relational approach seems to fall short as an approach by not offering substantial grounds for moral status, denying the moral significance of mental states, and inviting the problem of relativism. Similarly, the arguments put forward by the relationists against the standard view have not been found to have enough force to warrant the rejection of the standard view. For these reasons, while nonetheless remaining inclined to support the initiative of "thinking otherwise" by Coeckelbergh and Gunkel, I would put an even stronger emphasis on what they have been arguing against – finding "yet another, more refined criterion and argumentation concerning moral standing" (Coeckelbergh and Gunkel 2014, 16).

References

Birch, Thomas H. 1993. "Moral Considerability and Universal Consideration." *Environmental Ethics* 15(4): 313–332.

Birch, Thomas H. 1995. "The Incarnation of Wilderness: Wilderness Areas as Prisons." In *Postmodern Environmental Ethics*, edited by Max Oelschlaeger, 137–162. Albany, NY: SUNY Press.

Bostrom, Nick. 2014. *Superintelligence: Paths, Dangers, Strategies*. Oxford: Oxford University Press.

Bostrom, Nick, and Eliezer Yudkowsky. 2014. "The Ethics of Artificial Intelligence." In *Cambridge Handbook of Artificial Intelligence*, edited by Keith Frankish and William Ramsey, 316–334. New York: Cambridge University Press. https://intelligence.org/files/EthicsofAI.pdf.

Braidotti, Rosi. 2006. *Transpositions: On Nomadic Ethics*. UK: Polity Press.

Bringsjord, Selmer. 2008. "Ethical Robots: The Future Can Heed Us." *AI & Society* 22(4): 539–550.

Callicott, J. Baird. 1980. "Animal Liberation, A Triangular Affair." *Environmental Ethics* 2(4): 311–338.

Chalmers, David J. 1995. "Facing Up to the Problem of Consciousness." *Journal of Consciousness Studies* 2(3): 200–219.

Coeckelbergh, Mark. 2010. "Robot Rights? Towards a Social-Relational Justification of Moral Consideration." *Ethics and Information Technology* 12(3): 209–221.
Coeckelbergh, Mark. 2014. "The Moral Standing of Machines: Towards a Relational and Non-Cartesian Moral Hermeneutics." *Philosophy & Technology* 27(1): 61–77.
Coeckelbergh, Mark, and David J. Gunkel. 2014. "Facing Animals: A Relational, Other-Oriented Approach to Moral Standing." *Journal of Agricultural and Environmental Ethics* 27(5): 715–733.
Darling, Kate. 2016. "Extending Legal Protection to Social Robots: The Effects of Anthropomorphism, Empathy, and Violent Behavior Towards Robotic Objects." In *Robot Law*, edited by Ryan Calo, A. Michael Froomkin and Ian Kerr, 213–231. Northampton: Edward Elgar.
Darwin, Charles. 1981. *The Descent of Man, and Selection in Relation to Sex*. Princeton: Princeton University Press.
De La Mettrie, Julien O. 1912. *Man a Machine*. Chicago: Open Court.
Dennett, Daniel. 1998. "When HAL kills, Who's to Blame? Computer ethics." In *Hal's Legacy: 2001's Computer as Dream and Reality*, edited by David Stork, 351–365. Massachusetts: The MIT Press.
Descartes, René. 2006. *A Discourse on the Method*. Translated by I. MacLean. Oxford: Oxford University Press.
Fagan, Andrew. 2003. "Human Rights." In *Internet Encyclopedia of Philosophy*. https://www.iep.utm.edu/hum-rts/.
Fletcher, Joseph. 1997. *Situation Ethics: The New Morality*. Louisville, KY: Westminster John Knox Press.
Floridi, Luciano, and JW Sanders. 2004. "On the Morality of Artificial Agents." *Mind Mach* 14(3): 349–379.
Gaut, Berys. 2007. *Art, Emotion, and Ethics*. Oxford: Oxford University.
Gerdes, Anne. 2015. "The Issue of Moral Consideration in Robot Ethics." *ACM SIGCAS Computers & Society* 45(3): 274–280.
Gilligan, Carol. 1982. *In a Different Voice*. Cambridge: Harvard University Press.
Godlovitch, Stan. 1998. "Morally We Roll Along: (Optimistic) Reflections on Moral Progress." *Journal of Applied Philosophy* 15(3): 271–286.
Gordon, John-Stewart. 2018. "What Do We Owe to Intelligent Robots?" *AI & Society*, 1–15.
Grace, Katja, et al. 2017. "When Will AI Exceed Human Performance? Evidence from AI Experts." https://arxiv.org/abs/1705.08807.
Gunkel, David J. 2012. *The Machine Question: Critical Perspectives on AI, Robots, and Ethics.* Cambridge, MA: MIT Press.
Gunkel, D.J. 2015. The Rights of Machines: Caring for Robotic Care-Givers. In *Machine Medical Ethics*, edited by S.P. van Rysewyk and M. Pontier. Heidelberg/New York/Dordrecht/London: Springer, 151–166.

Gunkel, David J. 2018. "The Other Question: Can and Should Robots Have Rights." *Ethics and Information Technology* 20(2): 87–99.

Hales, Steven D., ed. 2011. *A Companion to Relativism*. Chichester: Wiley-Blackwell.

Haraway, Donna J. 2008. *When Species Meet*. Minneapolis, MN: University of Minnesota Press.

Harris, Sam. 2010. *The Moral Landscape: How Science Can Determine Human Values*. New York: Free Press.

Hübner, Dietmar. 2018. "Human-Animal Chimeras and Hybrids: An Ethical Paradox behind Moral Confusion?" *The Journal of Medicine and Philosophy: A Forum for Bioethics and Philosophy of Medicine* 43(2): 187–210.

Hurst, Daniel. 2018. "Japan Lays Ground for Boom in Robot Carers." *The Guardian*. February 6, 2018. https://www.theguardian.com/world/2018/feb/06/japan-robots-will-care-for-80-of-elderly-by-2020.

Jamieson, Dale. 2002. *Morality's Progress: Essays on Humans, Other Animals, and the Rest of Nature*. Oxford: Oxford University Press.

Jaworska, Agnieszka, and Julie Tannenbaum. 2018. "The Grounds of Moral Status." *The Stanford Encyclopedia of Philosophy*: https://plato.stanford.edu/archives/spr2018/entries/grounds-moral-status.

Kamm, Frances M. 2007. *Intricate Ethics: Rights, Responsibilities, and Permissible Harm*. New York: Oxford University Press.

Kant, Immanuel. 1983. *Grounding for the Metaphysics of Morals*. Translated by James W. Ellington. Indianapolis, IN: Hackett.

Kirchin, Simon. 2012. *Metaethics*. Great Britain: Palgrave Macmillan.

Koch, Tom. 2004. "The Difference that Difference Makes: Bioethics and the Challenge of 'Disability.'" *Journal of Medicine and Philosophy* 29(6): 697–716.

Kruel, Alexander. 2012. "A list of Q&A Style Interview Series on AI and Risks Associated with It." *Less Wrong*. https://wiki.lesswrong.com/wiki/Interview_series_on_risks_from_AI.

Lecky, William E.H. 1809. *History of European Morals*. New York: D. Appleton & Co.

Leopold, Aldo. 1949. *A Sand County Almanac*. Oxford: Oxford University Press.

Levinas, Emanuel. 1969. *Totality and Infinity: An Essay on Exteriority*. Translated by Alphonso Lingis. Pittsburgh, PA: Duquesne University.

Levy, David. 2009. "The Ethical Treatment of Artificially Conscious Robots." *International Journal of Social Robotics* 1(3): 209–216.

Mackie, John L. 1977. *Ethics: Inventing Right and Wrong*. London: Harmondsworth.

McMahan, Jeff. 2002. *The Ethics of Killing: Problems at the Margins of Life*. Oxford University Press.

McNamee, Shane M. 2015. "Human-Animal Hybrids and Chimeras: What's in a Name?" *JAHR – European Journal of Bioethics* 6(11): 45–66.

Miller, Lantz F. 2015. "Granting Automata Human Rights: Challenge to a Basis of Full-Rights Privilege." *Human Rights Review* 16(4): 369–391.

Moser, Paul K., and Thomas L. Carson, eds. 2001. *Moral Relativism: A Reader*. New York: Oxford University Press.

Müller, Vincent C. and Nils Bostrom. 2014. "Future Progress in Artificial Intelligence: A Survey of Expert Opinion." In *Fundamental Issues of Artificial Intelligence*, edited by V.C. Müller. 553–570. Berlin: Springer.

Musschenga, Albert W., and Gerben Meynen. 2017. "Moral Progress: An Introduction." *Ethical Theory and Moral Practice* 20(1): 3–15.

Nadeau, Joseph E. 2006. Only Androids can be Ethical. In *Thinking About Android Epistemology*, edited by Ken Ford and Clark Glymour, 241–248. Cambridge: MIT Press.

Naess, Arne. 1973. "The Shallow and the Deep, Long-Range Ecology Movement. A Summary." *Inquiry* 16: 95–100.

Noddings, Nel. 1986. *Caring: A Relational Approach to Ethics and Moral Education*. Berkeley: University of California Press.

Petrecca, Laura. 2018. "Technology is Transforming Caregiving." *AARP*. October 15, 2018. https://www.aarp.org/caregiving/home-care/info-2018/new-wave-of-caregiving-technology.html.

Plantinga, Alvin. 1993. *Warrant and Proper Function*. New York: Oxford University Press.

Pojman, Louis. 2015. "Who's to Judge?" In *Ethics: Theory and Contemporary Issues*, edited by Barbara MacKinnon and Andrew Fiala, 56–63. Stamford, CT: Cengage Learning.

Rachels, James. 2003. *The Elements of Moral Philosophy*. New York: McGraw-Hill Higher Education.

Sauchelli, Andrea. 2012. "Ethicism and Immoral Cognitivism: Gaut vs. Kieran on Art and Morality." *Journal of Aesthetic Education* 46(3): 107–118.

Sayre-McCord, Geoff. 2014. "Metaethics." *The Stanford Encyclopedia of Philosophy* (Summer 2014 Edition). https://plato.stanford.edu/archives/sum2014/entries/metaethics/.

Schweitzer, Albert. 1929. *Civilization and Ethics: The Philosophy of Civilization*. London: A & C Black.

Singer, Peter. 1990. *Animal Liberation*. New York: New York Review/Random House.

Singer, Peter. 1993. *Practical Ethics*. Cambridge: Cambridge University Press.

Singer, Peter. 2011. *The Expanding Circle: Ethics, Evolution and Moral Progress*. New Jersey: Princeton University Press.

Stanovich, Keith E. 2004. *The Robot's Rebellion: Finding Meaning in the Age of Darwin*. University of Chicago Press.

Sullins, John P. 2011. "When Is a Robot a Moral Agent?" In *Machine Ethics*, edited by Michael Anderson and Susan Leigh Anderson, 151–161. Cambridge: Cambridge University Press.

Sutton, Sean D., ed. 2009. *Biotechnology: Our Future as Human Beings and Citizens*. Albany, NY: State University of New York Press.

Torrance, Steve. 2014. "Artificial Consciousness and Artificial Ethics: Between Realism and Social Relationism." *Philosophy and Technology* 27(1): 9–29.

Tronto, Joan. 1993. *Moral Boundaries: A Political Argument for An Ethics of Care*. New York: Routledge.

Warren, Mary. 1997. *Moral Status: Obligations to Persons and Other Living Things*. Oxford: Oxford University Press.

CHAPTER 4

Ascribing Rights to Robots as Potential Moral Patients

Janina Loh

1 Introduction

Since the middle of the 20th century at the latest, robots have found their way into more and more areas of human life. Starting with industry and its first "official" robot *Unimate*[1] via military and warfare up to the service sector in care, medicine, and household – robots are doing various jobs and often those that are usually considered "dull, dangerous, and dirty" either today or presumably in the foreseeable future (Marr 2017). Now, nothing that humans do is purely descriptive or evaluatively neutral, because human action is always (whether consciously or unconsciously) determined by norms and values. It is precisely intentionality that distinguishes an action in the conventional sense from instinct or mere behavior.[2] Through the intention, values "enter" the action. People choose between different alternatives for actions based on reasons. And those reasons were previously implicitly or explicitly weighed against other reasons. Therefore, technics in the sense of technologies, i.e. technological artefacts and thus products of human action, is not neutral, and thus neither are robots as specific technologies.[3] The philosophical discipline of robot ethics deals with the normative criteria that play a role in the construction of robots, in dealing with robots, and in the relationship between humans and robots.

1 A composite of "Universal" and "Automation," built by Joseph Engelberger in 1961; Ichbiah 2005, 207. Toby Walsh (2017) offers a good overview of the history of artificial intelligence and robotics.

2 Cf. Sombetzki (now Loh) 2014, 47. Classical works of philosophical action theory include Elisabeth Anscombe's monograph *Intention* (1957) and Donald Davidson's *Essays on Actions and Events* (1980).

3 The following explanations are based on a broad understanding of technology which includes artefacts (technologies, material technology), well-rehearsed, standardized practices, methods and processes (technologies, process technology), other forms of technology such as materials (real technologies), signs (intellectual technologies), executions (social technologies) and attributes (e.g. affine in technics).

 | DOI:10.1163/9789004437876_006

In the following, I first present the discipline of robot ethics and then give an overview of its three fields, namely (1) robots as potential moral agents, (2) robots as potential moral patients, and (3) inclusive approaches that question the classical subject-object dichotomy. In the fourth section, I will deal with the attribution of (moral) rights with regard to the second robot-ethical field, which (as will be shown) in turn is closely related to the first field. Since David Gunkel's theory of a "thinking otherwise" in this anthology is in my opinion to be understood as representative of the inclusive approaches,[4] this robot-ethical field will be dealt with in an excursus in the fifth section.

With regard to robots as potential moral patients some thinkers advocate the possibility of attributing (some) robots rudimentary or even fundamental rights. Apart from academic philosophical discourses, this question is of great relevance for political agents and institutions. The European Parliament, for instance, is currently working on a concept that would allow (some) robots to be given the status of "electronic persons." And in October 2017 the robot Sophia was the world's first robot to be granted full citizenship in Saudi Arabia (Wootson Jr. 2017). In this chapter, however, the focus lies on fundamental moral rights, rather than on fundamental legal rights.

2 What Is Robot Ethics?

Robot ethics is a sub-category of machine ethics and represents a relatively new form of applied ethics within the so-called Western cultural area (Allen et al. 2006; Anderson and Anderson 2011; 2007; Bendel 2018a; Edgar 2003; Misselhorn 2018a; Moor 2006; Rath et al. 2019).[5] All robots are machines, but not all machines are robots (Loh 2018). This classification of robot ethics as an applied ethics sub-category of machine ethics is based on an understanding of ethics as a category of action that is initially specific to humans. Thus, the

4 In his paper "Thinking otherwise" (2007) Gunkel uses the term "inclusive" (as different from my own understanding) for those centrist approaches (e.g. anthropocentrism, pathocentrism etc.) that are sometimes willing to attribute intrinsic value to nonhuman beings, but nevertheless still cling to the inherent challenges of essentialism and paternalism that beset every centrist approach due to their underlying subject-object dichotomy. He discusses Michael Anderson's, Susan Leigh Anderson's, and Chris Armen's "machine ethics," Robert Sparrow's "android ethics," and Luciano Floridi's "information ethics" as examples for such inclusive thinking.

5 Another sub-category of machine ethics is computer ethics. Machine ethics, again, is a sub-category of the sub-discipline ethics of technology which – in turn – is a field in philosophy of technology.

philosophical discipline of ethics, since the classical definition of Aristotle, has been concerned with human customs, traditions and habits, with the good life, has carried out the scientific reflection of human practice and judged the criteria of good and bad action (Aristotle 2006). A lot of people assume that only human beings are acting beings, whose doings are not subject to blind instinct and drive, but are formed by intentions, norms and reasons.

I differentiate two types of applied ethics: One concerns those ethical systems that deal with a nonhuman counterpart. In addition to machine and robot ethics, these ethical systems also include animal, plant, environmental, computer ethics, and, in general, ethics of technology. Similar to robot ethics, animal ethics, for instance, focuses on the normative criteria that play a role in the breeding, domestication and keeping of animals, in general in dealing with animals and in the relationship between humans and animals (Beauchamp and Frey 2011; Cohen 1986; deGracia 1996; Donaldson and Kymlicka 2013; Regan 1983; Schmitz 2014; Singer 2015; Wolf 2004, 2008; regarding plant ethics cf. Coccia 2018). The robot ethicist David Gunkel has in *The Machine Question* (2012) already stated that robot ethics is related to animal ethics in particular, in so far as "[t]he machine question [...] is the other side of the question of the animal" (Gunkel 2012, 5). René Descartes, as Gunkel understands him, initially attributed the same ontological status to animals and machines (Gunkel 2012, 3). It was not until the 20th century that this ontological equality between animals and machines was abolished in favor of animals.

The other group of applied ethics assembles ethical systems for special domains of human life, where values are represented, norms are enforced, and rules are formulated that are usually attributed a different status in people's everyday lives. Medical ethics, ethics of humanitarian interventions, ethics of war and of economics, and ethics of international relations can be cited as examples of this form of applied ethics. This second type of applied ethics is only marginally addressed, if at all, in this paper.

In German-speaking countries, robot ethics is not yet a generally recognized discipline within academic philosophy, even though interest in interdisciplinary collaborations, including philosophers, is growing. In comparison with the English-speaking world, where the ethical examination of artificial systems has produced a canon of classical literature since the middle of the last century (Asaro 2006; Bekey 2005; Brey et al. 2008; Capurro and Nagenborg 2009; Lin et al. 2017; Lin et al. 2012; Lin et al. 2011; Stahl 2004; Sullins 2006; Turing 1950; Versenyi 1974; Wallach and Allen 2009), the German-language discourse is, although increasing, still relatively small (Hilgendorf 2014; Loh 2017; Loh 2019; Sombetzki 2016; Mainzer 2010; Misselhorn 2018b; Remmers 2018).

Here, robot ethics from the ranks of academic philosophers occasionally has to put up with the accusation that it is not ethics in the genuine sense at all or that it has no specific object, since ethics (at least in its traditional, hence Aristotelian, sense; see above) only deals with human action and not with the inanimate. But even if, after prior examination, it turns out that robots cannot be subjects of moral action and are therefore not capable of morally acting themselves, they should be assigned a place in the moral universe. After all, all sorts of beings and entities – such as landscapes, ecosystems, animals, plants, houses, cars or smartphones – are objects of moral actions, and we are willing to attribute values, even rights, to a whole range of nonhuman and (partially) inanimate entities. What kinds of values are involved in the case of specific robots, whether purely instrumental values or intrinsic values, remains to be discussed and depends on the underlying approach. But where, if not in ethics, would be the appropriate space for such a discussion? Comparable to the other applied ethics, which deal with nonhuman beings and entities, robot ethics on the one hand thinks about the extent to which the counterpart in question is a carrier of values and perhaps even rights, whether one should behave towards it in a certain way, whether it is entitled to a specific treatment, and to what extent in general it is first and foremost an object of moral action and thus falls into the category of so-called moral patients, "as entities that can be acted upon for good or evil." On the other hand, it is asked to what extent robots themselves have to be interpreted as moral agents and thus as subjects of moral action, in other words to what extent they are "entities that can perform actions" (Floridi and Sanders 2004, 349). Furthermore, robot ethics discusses alternatives to this classical distinction between subjects and objects of moral action – these are what I call inclusive approaches.

Robot ethics poses traditional ethical questions with regard to robots such as, for instance, which competencies we generally regard as fundamental for moral agency, which moral (and other) values we should implement in artificial systems, what kind of moral self-image can be inferred when we treat robots "badly," and in which areas – industrial, military, medical, service robotics and others – we want to rely exclusively or at least to a significant extent on human and not on artificial expertise.

Robot ethics was above referred to as sub-form of machine ethics, since all robots are machines, but not all machines are robots. A machine is an artificial structure that consists of individual parts moved by a drive system (motor, wind, water, etc.) and converts energy (Canguilhem 2012; Strandh 1980). Robots are special machines. Historically, the term "robot" goes back to the Czech word "robota" for work, compulsory service, and forced labor, and was coined in 1920 by the artist Josef Čapek. His brother Karel Čapek used it in the play

R.U.R. Rossum's Universal Robots (1921) for humanoid apparatuses that render services to humans. In this original understanding is anchored the idea of robots as artificial slaves (from the Slavic root "rab" for "slave"; Jordan 2017, 50) and assistance systems that relieve humans of tiresome, boring (e.g. repetitive) and dangerous work. The first established human domain where robots were deployed to undertake these dull, dangerous and dirty jobs, namely in industry, reflects this vision, which Čapek creates in the above-mentioned play. It is also the core of the so-called industry 4.0, the technological transformation of the human working environment through digitization and automation. Čapek is, however, no blind technology enthusiast and in *R.U.R.* deals with numerous challenges which accompany the creation of robots. The plot of his play ultimately amounts to a revolt of robots who are no longer willing to submit to their human creators, to turn against humans in order to take over world domination. But Čapek also raises many other philosophical questions, such as the nature of "the" human being, the responsibility of scientists for their artificial creatures, and what it means to form an emotional bond with another being. His play ends with the prospect of a love relationship that begins to develop between two robots. Thus, in the historical understanding of the robot founded by Čapek, a broad basis is laid for the discussions that should arise in the following decades.

The following reflections are based on an extended version of the definition proposed by Catrin Misselhorn, according to which a robot is an electro-mechanical machine that (a) has some form of independent body, (b) consists of at least one processor, (c) with sensors that collect information about the world, (d) and effectors or actuators that translate signals into mechanical processes. The behavior of a robot (e) is or at least appears to be autonomous and it can (f) influence its environment (Misselhorn 2013, 43; cf. Bekey 2005, 2; Jordan 2017, 41–59; Mataric 2007, 2; Remmers 2018, 28–31). This understanding is not unproblematic, since some of the conditions cited (such as embodiment, autonomy and influence) are ambiguous and therefore at least in need of explanation. It also excludes those artificial systems, such as computers, chatbots, medical assistance systems and drones, which have so far often been referred to as robots. These machines open up the large grey area which deserves to be illuminated by philosophers of technology and robot ethicists. Isaac Asimov has already suspected this robot-related sphere in which we also meet these other relatives and acquaintances of robots (Asimov 1982, 53). In fact, according to the definition given here, something is not a robot in the strict sense if one of the mentioned conditions (a) to (f) is not fulfilled: Computers do not meet condition (f) and whether they possess an independent body – condition (a) – remains to be discussed. The specific kind and manner of embodiment, on the other hand, plays no role, robots exist in every conceivable shape, the humanoids among them are called androids (Drux 1988). In a metaphorical sense, computers represent the "brain" of a robot, just as algorithms can metaphorically be regarded as their "nervous system", or even better as their "mental behavior patterns" and "learned processes," but not as the actual robot itself. Artificial systems such as the surgical assistance system DaVinci and drones lack condition (e) (Jordan 2017, 31, 59, 262–266). The conditions (a) to (f) are necessary in detail and together sufficient to provide a complete definition of "robot" in the strict sense anticipated in this paper.

3 The Three Fields of Robot Ethics

Traditionally, research in robot ethics distinguishes two fields. In one area, it is discussed to what extent robots are to be understood as moral patients, i.e. as having value and perhaps even rights, but in general above all as objects of moral action. The other field deals with the question of the extent to which robots themselves have to be interpreted as moral agents and thus as subjects of moral action. The two fields of work are not necessarily exclusive; they complement each other. As will be shown, both face – thanks to their first ethical (Aristotelian) premises – some challenges that the third (and younger) field intends to meet. These will be discussed in more detail in the following.

Within the field in which robots are regarded as moral agents, the question is asked to what extent robots are capable of acting morally, what competencies they have to be equipped with, and to what extent, if they are to be able to act morally. Depending on the underlying understanding of agency, morality and the competences that are necessary for its realization, this field of robot ethics focuses, for instance, on the attribution of freedom and autonomy as a condition for moral agency, on cognitive competences (e.g. thinking, mind, reason, judgement, intelligence, consciousness, perception and communication), but also on empathy and emotions. It should be noted, however, that the robots that exist so far are candidates for the attribution of moral patiency and not of moral agency.

Within the field in which robots are discussed as moral patients, as objects of moral action, the issue is how to deal with artificial systems, what value they have, even if one holds that they are not capable of morally acting themselves. Here robots are consistently understood as tools or complements of human beings. Topics within this robot-ethical field include the formulation of codes of ethics in companies, the desirability and possibility of relationships with robots, the "enslavement" of robots or the assessment of the use of robots for therapeutic purposes. Some thinkers also advocate or discuss the possibility of attributing (some) robots rudimentary or even fundamental rights. Similar to Immanuel Kant in § 17 of the second part of his *Metaphysics of Morals*, which argues against cruelty to animals because this leads to a brutalization of humans, Kate Darling (Darling 2012), for instance, argues for robot rights, since, according to her, humans are then more likely to succeed in remaining "human." The European Parliament, too, is currently working on a concept that would allow (some) robots to be given the status of "electronic persons." Within this area of robot ethics, the moral competence (i.e. the decision on the behavior of a robot, which happens through the personalization of a respective artificial system by its users) remains consistently with the human owners. The competence-competence (i.e. the decision on any framework, values and principles which are specified in the programming of the respective robot and cannot be changed by the users) remains with the manufacturers and distributors or with the law. Within this field of robot ethics, it is therefore only humans who decide on the morality of their creatures and who are responsible in the event of an accident.

The group of moral agents is generally much smaller than that of moral patients, because usually we only distinguish people with moral abilities in the genuine sense. However, a whole series of beings and things is attributed a moral value or even (rudimentary) rights, so they are moral patients – at least to the extent that these entities are morally worthy of consideration, even if they may not have any intrinsic value, but only a high instrumental value. As a moral agent, a being is at the same time a carrier of values; conversely, not all carriers of values are moral agents equally. The attribution of moral values to living beings and objects depends, for example, on the perspective taken in each case. An anthropocentric position argues that only humans have an intrinsic value. Anthropocentrism means on the one hand that people have a moral advantage and thus special status over all other beings (moral anthropocentrism), and on the other that only people have the capacity for knowledge and can develop a capacity for judgement (epistemic anthropocentrism) (Krebs 1997, 343). An alternative to anthropocentrism is, for instance, pathocentrism, which morally ascribes an intrinsic value to beings capable of suffering and epistemically advocates the position that values come into the world through beings capable of suffering. As a result, they must also be objectively accepted outside the human cognitive horizon. Biocentrism, which considers all living morally and epistemically, and physiocentrism, which considers all nature (holistic physiocentrism) or everything in nature (individualistic physiocentrism) morally and epistemically, can also be classified in the circle of centrist approaches, as they are called below. The inclusion of robots in the horizon of things endowed with an intrinsic value could perhaps open up a further perspective, a mathenocentrism, for instance (from the Greek "matheno," "learning"), which measures all this with an intrinsic value that is controlled or programmed or capable of learning in a specific way.

The above-mentioned understanding of ethics in general, which goes back to Aristotle, and the resulting status of applied ethics, as well as these and all other conceivable centrist approaches, are now accompanied by some serious philosophical challenges, which the inclusive robot-ethical approaches intend to meet. For the starting point within the classical two fields of robot ethics (robots as potential moral agents and moral patients) is always first of all "the" human being who defines the ethical commonsense, depicts the "normal case" or the genuine and "ideal type" of ethical agency. Thus every ethic (in the Aristotelian understanding) that defines "man" as the pivot rests, at least implicitly, on an anthropological foundation.

The anthropological disciplines in biology, cultural studies, ethnology and not least philosophical anthropology each take their own paths in order to define the human being and to differentiate humans with the greatest possible clarity from all other beings. In doing so, they usually proceed in an essentialistic way, i.e. they seek to describe "the" human being by means of an individual attribute or a series of characteristics. The *Historical Dictionary of Philosophy* presents in the entry "Mensch" a total of four common anthropological approaches, all of which are ultimately essentialist (Grawe et al. 1971–2007, 1072): The *Alkmaion topos* defines "the" human being by a single attribute. Alkmaion, a pre-Socratic from the late 6th or early 5th century BC, distinguishes "the" human from other living beings for the first time via a single competence. He defines "man" "by the fact that he alone understands, while the remaining [*creatures*] perceive, but do not understand" (my translation; Diels, 1966, 24, B 1a). On the other hand, the *compositional topic* ("Kompositionsformel"; my translation) captures "the" human being not only by one property, but by two or more attributes. The *microcosm-macrocosm approach* understands "the" human being as what the cosmos is on a small scale. The *horizon topic* ("Horizontformel"; my translation) finally determines "the" human being "through the spheres into which he projects upwards and downwards" (my translation; Grawe et al. 1971–2007, 1073).

This essentialism, which is taken for granted in philosophical anthropology and other (anthropological) disciplines, poses certain challenges, such as moral status attributions, which suggest a specific treatment of animals, machines and other alterities. Even radically exclusive positions such as speciesism, racism and sexism often argue essentialistically by denying the excluded beings certain characteristics. (Anthropological) essentialism continues to be accompanied by an epistemic uncertainty as to whether the being in question actually possesses the attributed qualities. Our attribution of certain competences, such as freedom of will, is often based on a metaphysical foundation. Actually, we do not only not know what it is like to be a machine or an animal, e.g. a bat – to quote the title of a famous paper by Thomas Nagel (1974). But we also don't really know what it is like to be another human. Because it cannot be determined with unambiguity whether humans are actually equipped with freedom of will and similar abilities. We cannot clearly prove them empirically. The only difference lies in the fact that we are willing to make an additional assumption with humans, namely that at least the beings we call humans have the competences in question (Churchland 1999; Coeckelbergh and Gunkel 2014).

Those applied ethics approaches that are (implicitly or explicitly) based on Aristotelian, anthropological, and centrist principles address a respective and also essentialist defined counterpart to "man," who has a moral status due to certain characteristics. Within a pathocentric approach, for example, all beings to whom sentience can be attributed are given a place in the moral universe. Furthermore, these positions have to react not only to (moral) centrism and (anthropological) essentialism, but also to the consequences of an at least implicitly assumed subject-object dichotomy. This subject-object dichotomy underlies Aristotelian ethics as well as centrist approaches in general and emerges, on closer examination, as a consequence of epistemic centrism. For instance, epistemic anthropocentrism recognizes a specific *subject*, namely "the" human being, who, endowed with cognitive capacity, judgement and reason (or comparable competences), attributes values to all other beings as the *objects* of "his" cognitive practice and is the only being that brings values into the world in the first place. The same applies to other morally centrist approaches, such as moral pathocentrism, in combination with epistemic anthropocentrism. But also other epistemic centrisms remain committed to the paternalism implicit in the subject-object dichotomy by defining a respective cognitive subject that assigns or denies values, abilities and competences to an imaginary counterpart as the object.

The inclusive approaches of robot ethics present in quite different ways alternatives to the Aristotelian understanding of ethics, the anthropological essentialism that accompanies it, the moral and epistemic anthropocentrism that is often embedded in it, and face the philosophical challenges of centrist approaches in general. Here, too, the debate revolves around the attribution of moral agency and moral patiency. With regard to the ascription of moral agency, however, the traditional understanding of the moral agent is questioned in order to extend it to nonhuman beings. On the other hand, competencies that have been ascribed essentialistically to individual subjects of actions within the framework of the usual approaches should now be understood relationally and processually as realizing themselves in the interaction of different agents and non-agents. With a view to robots as objects of moral actions, the debate in this third area of robot ethics also revolves around, for instance, the anthropomorphization of nonhuman beings and the possibility of entering into relationships with them.

In their concern, the inclusive approaches overlap in many respects with post-feminist, post-structuralist, post-colonial and critical-posthumanist positions. The unifying intention of all these perspectives is an inclusive program (morally and epistemically) that neither emphasizes the position of "the" human being (whatever may be hidden behind this label in detail) over other beings or negatively devalues "the" human to the "level" of all other beings. Exclusive theories (with regard to the topic of this chapter approaches on robots as potential moral agents or moral patients) either value "the" human being as special and better than all other beings (this position could be called an exclusive authoritarianism) or exclude "the" human being from the genuine realm of morality (call this exclusive relativism[6] a position less frequently defended than the first). In contrast to these exclusive theories, the inclusive approaches seek to place all other beings on the same "level" as human beings and thus include them as morally equal companions with accompanying fundamental moral rights in the moral universe in which human beings are located.[7] It must, however, be pointed out that it is not the aim of the inclusive approaches of robotic ethics to put an end to essentialism or to the distinction between subjects and objects in general. A critical awareness of the morally highly questionable consequences of some concrete essentialisms as well as of some specific differentiations between subjects and objects is needed.

6 A position that can perhaps be assumed with Joseph Emile Nadeau (2006), who only attributes genuine moral ability to androids. John Danaher (2017) discusses in his text "The rise of the robots and the crisis of moral patiency" the thesis that the development of artificial systems "could compromise both the *ability* and *willingness* of humans to act in the world as responsible moral agents."

7 To illustrate the distinction between in- and exclusive approaches I will use two current examples: The German AfD represents a morally exclusive authoritarianism, since it discriminates against some people morally, i.e. excludes them from the circle of the morally best qualified persons. US President Donald Trump and his advisor Kellyanne Conway can be interpreted with the phrases "fake news" and "alternative facts" as the implicit precursors of an epistemically exclusive relativism, which in the final analysis threatens to abolish any distinction between "right" and "wrong," "fact" and "fiction" and, until then, first tries to discredit a number of journalistic media.

4 Ascribing Rights to Robots as Potential Moral Patients

In the second area of robot ethics artificial systems are considered as potential carriers of values and rights. Here, it is discussed whether one should behave towards them in a certain way, whether they are entitled to a specific treatment, and to what extent they are generally initially the object of moral actions. This field of robot ethics probably comprises the majority of currently existing robots. Nevertheless, this area is not at all its own field of work in the same way as the one that deals with robots as potential moral agents. For in most cases moral patiency is regarded as dependent on moral agency, as a pure negative to it, which does not require any independent consideration.[8] Every moral agent, to briefly summarize this train of thought, which Luciano Floridi and J.W. Sanders call the "standard position" (2004, 350), is at the same time also an object of moral action, in that it is precisely the status of moral agency that requires this being to be treated in a specific (morally appropriate) way.[9] Moral agents and objects of moral actions coincide under this perspective. The circle of moral agents is small, since the usually defined conditions of moral agency are demanding and in need of interpretation. In addition, some advocates of artificial moral agency (such as John P. Sullins, James H. Moor, Wendell Wallach and Colin Allen, and Catrin Misselhorn) have already shown that, although one might be willing to grant (some, specific) robots moral agency, there is still a difference to be made between humans on the one hand and all other nonhuman beings on the other. From their perspective, robots are perhaps (rudimentary) moral agents, but aren't in principle capable of acting in the same way as humans and therefore require a different (morally appropriate) behavior than humans in the sense of their moral patiency.

It is all the less surprising, then, that other thinkers implicitly or explicitly reject any reduced or quasi-agency of nonhuman and, in particular, artificial beings in general. However, this does not necessarily mean (as in the case of Joanna Bryson) that these beings are at least given the status of a moral object of action, i.e. may be moral patients. Rather, completely in the spirit of the standard position, with the loss of the possibility of being accepted into the community of moral subjects, they are also denied the hope of being able to demand a specific (morally appropriate) mode of treatment as objects of

8 Gunkel considers this, too (2012, 94 f.), also cf. Mane Hajdin (1994, 180) and Tom Regan (1983, 152).

9 Oliver Bendel, on the other hand, sees things differently, identifying (some) robots as "new, alien, strange subjects of morality," but not as moral patients or "objects of morality" (my translations; Bendel 2018b, 34).

moral actions. With these entities people may indeed do as they please, no behavior towards them can be considered morally reprehensible in any way. From the point of view of the representatives of the standard position, moral patiency does not exist independently, but is fundamentally bound to moral agency. If the latter does not exist, the former does not exist either. Against the background of the standard position, an introduction to robot ethics would only require a discussion of the question of whether robots can be moral agents. If one denies this, as for example Joanna Bryson, the necessity of dealing with the question of a possible moral patiency of the machines in question or artificial systems in general would be omitted. Thus, the entire discipline of robot ethics itself would become obsolete sooner or later. As Gunkel has in *The Machine Question* (2012) already pointed out, Bryson is in fact not only consistently opposed to possible undertakings to allow robots to play a moral role. But from this "'fact'" (that robots *are* not moral agents; Gunkel 2012, 95) she also derives a rejection of their possible moral patiency. As with any other tool, as with any other object, Bryson says we should be allowed to treat robots as we choose – there exists no morally inappropriate or reprehensible behavior toward them, just as there is no morally inappropriate or reprehensible behavior toward "a car, piano, or couch" (Bryson 2010, 66; cf. Bryson 2018).

But not everyone takes this position. In contrast, many would intuitively agree that at least some (special, important) artificial systems can claim a specific handling, even if they are not capable of independently (morally) acting themselves, even if they exist only for human users to fulfill their needs. According to this intuition we must not simply deal with these (special, important) robots as we see fit. There may be some beings that may be objects of moral actions, and that is to say, require a specific (morally appropriate) approach without themselves being moral agents, not even in a rudimentary or quasi-sense. These artificial beings, or some of the philosophical approaches that correlate with this view, are presented and discussed in the following, with particular reference to a possible attribution of rights.

That artificial moral patients in fact make up the majority of the currently existing robots is obvious against the background of what has been said so far. Because either the authority that determines the (moral) status of a respective artificial non-agent and that is thus allowed to decide whether the robot in question is entitled to a certain (morally appropriate) treatment has not been defined yet. Or because there is no universal and objective authority of this kind. Therefore, this chapter starts from the premise that theoretically all robots can be considered as potential moral objects. Restrictions to this maximally inclusive view, which is potentially prepared to grant every artificial being a place in the community (after all) of moral patients, are only valid and justifiable within the framework of the respective approaches discussed. However, all those theories that do not necessarily exclude the possibility of artificial moral agency per se fall within this robot-ethical field of work of moral patiency, since they simply do not have robots as potential moral agents in mind, but deal with them exclusively as tools or supplements of human beings. This, too, is a reason for assuming that the field of robots as potential moral objects of actions encompasses (as far as possible) all currently existing artificial systems.

Although some controversial approaches are currently being discussed in the field of jurisprudence, such as the European Parliament, which is currently working on a concept that would allow (some) robots to be given the "status of electronic persons,"[10] or the case of the robot Sophia, the world's first robot to be granted full citizenship in Saudi Arabia in October 2017 (Wootson Jr. 2017), the present study has to be limited to the *moral* status of robots and to their fundamental *moral* rights for reasons of space (cf. for the legal debate Beck 2010; Calverley 2011, 2006; Gunkel 2018; McNally and Inayatullah 1998; Robertson 2014; Tavani 2018; Teubner 1998). Kate Darling offers a comprehensible transition between the debates on legal rights and moral rights for (some) artificial systems. Although she presents her position with a view to "legal rights"

---

10 European Parliament, „Entwurf eines Berichts mit Empfehlungen an die Kommission zu zivilrechtlichen Regelungen im Bereich Robotik. (2015/2103(INL))," 2014–2019, http://www.europarl.europa.eu/sides/getDoc.do?pubRef=-//EP//NONSGML%2BCOMPARL%2BPE-582.443%2B01%2BDOC%2BPDF%2BV0//DE. Some thinkers, such as Tavani (2018, 4), assume that the concept of the electronic person is very similar to that of the legal person. According to Eric Hilgendorf, however, this debate is "less about rights for robots than about how to avoid or close liability gaps in the case of damage to humans by machines. Let's assume that a perfectly functioning self-learning system has learned 'the wrong thing' and therefore causes material damage. If one assumes that the manufacturer cannot be blamed for any fault here, there is a danger that the injured party will not receive any compensation. The concrete 'injuring party' is a machine and can therefore neither be sued nor held responsible in any other way. According to the prevailing view since the Enlightenment, legal or moral subjects of responsibility cannot be things, but only other people, i.e. 'natural persons.' After all, the law recognized the figure of the 'legal person' more than a hundred years ago, so that, for example, companies can be sued. If this possibility did not exist, many aggrieved parties would remain sitting on their damages, because in view of the considerable complexity of a large company, it is often not recognizable who exactly caused the damage. The same thought could lead to the recognition of electronic persons. Manufacturers would have to allocate a certain amount of assets to their robots (e.g. via insurance), and injured parties would then have the opportunity to sue the machine directly and, if they are justified, to be compensated out of the machine's assets." My translation; Beck et al. 2019, 247 f. According to Hilgendorf, cases of this kind are at stake in the debate on the electronic person.

(Darling 2012),[11] she justifies it by appealing to Immanuel Kant. Similar to Kant's objection to cruelty to animals in § 17 of the second part of his *Metaphysics of Morals*, as this leads to a moral brutalization of humans (Kant 1797, 443), Darling argues for robot rights, as according to her humans would then be more likely to remain "human." Her argument remains limited to the field of social robots, since they are in intimate and emotional relationships with humans and therefore need a special treatment.[12]

Also interpreting Kant, Anne Gerdes also attests rights to social robots, since "having indirect duties *with regard to* non-human entities and animals rest upon our [direct] duties *to* ourselves" (2017, 277). According to Kant, human beings as rational beings in the true sense have only direct duties towards themselves. All further duties towards nonhuman beings are derived as indirect duties from the direct duties: "Even gratitude for the long services rendered by an old horse or dog (as if they were housemates) belongs indirectly to the duty of man, namely in respect of these animals, but directly it is always only man's duty towards himself." However, since Kant in §§ 16 and 17 of the *Metaphysics of Morals* does not speak exclusively of the "living, although unreasonable part of creatures," i.e. of animals, but also includes the "lifeless in nature" in his considerations, according to Gerdes the indirect duties can and must consistently extend also to robots (my translations; Kant 1797, 442 f.).

Both Darling and Gerdes, in their extension of Kant's train of thought to social robots, adopt what some scholars interpret as his implicit

11 Migle Laukyte also allows a similar train of thought to Darling in that she uses an argument for criminal or social rights for robots as "group agents" (2017, 1) to allow implications about identifying them as moral agents with moral rights.

12 In the area of social robotics, artificial systems are being developed that are intended to help people in their everyday lives. And not only that, according to Kate Darling, social robots have "been designed to act as […] companion[s] to humans" (2012). Whether the vacuum cleaner robot *Roomba* (iRobot), lawn mower robots like the *Automower* (Husqvarna), the sales assistant *Paul*, who guides customers through the corridors of Saturn's electrical retailer, or the household assistance systems and entertainment robots like *Pepper* (Aldebaran Robotics SAS in cooperation with SoftBank Mobile Corp.) – they all represent the growing group of social robots that are used in the near human environment and therefore need social skills, depending on what their task is and to what extent they interact directly with humans. Darling argues that people can and should identify themselves with social robots used in elderly care and private homes, because these machines are intimately related to their owners and otherwise users would not be able to engage sufficiently with the artificial systems. With, for example, mine detection robots, however, people should probably not enter into a close emotional relationship in order not to mourn too much the probable loss of the machine in question and thus perhaps even risk no longer being able to adequately fulfil their own role as soldiers.

anthropocentrism.[13] For machines are not identified as moral objects of actions for their own sake, but rather on the basis of their utility for human beings. They have no intrinsic value. Robots are therefore moral patients who can claim morally appropriate behavior, but for instrumental reasons for the protection of the morally appropriate human character.

Herman T. Tavani, on the other hand, formulates a physiocentric approach with reference to Hans Jonas that grants social robots moral rights and thus the status of moral objects of action. According to him, an orientation towards Jonas has two advantages over other philosophical positions: On the one hand, our duties towards robots would be different than within the horizon of Kant's thinking not only indirect, but "direct," since "our obligations to social robots are direct (e.g., because social robots are a significant part of our technological-world to which we, as humans, already have direct moral obligations)." On the other hand, "Jonas's ethical framework explicitly includes non-human entities as 'objects' eligible for moral consideration" (Tavani 2018, 12). Tavani's considerations are primarily based on Jonas' main work *Das Prinzip Verantwortung*, in which he demands a radical change in human morality and perception of responsibility due to "modern technology" (my translation; Jonas 1984, 11). Against the background of a sharp critique of the "previous ethics," which does not deal with a possible "effect on non-human objects," since it attaches moral significance exclusively to the "direct contact of man with man" and is thus throughout "*anthropocentric*" (my translations; Jonas 1984, 22) Jonas formulates a new Categorical Imperative: "Act in such a way that the effects of your action are compatible with the permanence of genuine human life on earth" (my translation; Jonas 1984, 36). Jonas's imperative differs from Kant's original Categorical Imperative, which in its basic formula reads "Act only according to that maxim whereby you can at the same time will that it should become a universal law" (my translation; Kant 1785, 421), among other things by including the context of human action, the earth, and thus the nonhuman and also inanimate nature directly in the moral evaluation.

Without having the space at this point to undertake a detailed comparison of Kant's and Jonas's approaches, it can at least be stated at first glance that Jonas makes the explicit claim to break out of the anthropocentric world view of the traditional, Aristotelian ethic traditions, above all Kant's ethics. His

13 Other Kant scholars, however, would disagree with this understanding of Kant's thinking and would argue that Kant is prone to a logocentrism and not anthropocentrism (Altmann 2011; I thank John-Stewart Gordon for this advice). However, within the framework of this interpretation, Kant's approach would also have to meet the challenges of centrist approaches in general.

new Categorical Imperative establishes a physiocentrism that also attests an intrinsic value to nature (animate and inanimate; cf. Krebs 1997, 352, 358). Consequently, as Tavani correctly observes, robots, as part of nonhuman and inanimate nature, would also have a place in the moral universe, being valuable moral patients for their own sake. Usually (see above) one distinguishes between a holistic physiocentrism which morally and epistemically considers *all nature* and an individualistic physiocentrism which morally and epistemically considers *everything in nature*. It cannot be clearly determined which physiocentric position Jonas (and with him Tavani) holds, but in principle he does not define responsibility as an unambiguous yardstick that would allow the moral value of individual nonhuman entities to be determined. Thus, the question remains whether each robot has an intrinsic value in itself, whether only a group of special robots (such as the social robots on which Darling and Gerdes focus) or not even a single artificial system, but only "the" robot in general (as part of the technical side of human beings and nature) or the robots as species.

Furthermore, Tavani's position, which allows the recognition of "some kinds of social robots [to] qualify for moral consideration as moral patients" (Tavani 2018, 1), remains within the horizon of centrist approaches and thus exposed to the philosophical challenges they generally face. Basically, as outlined above, there are three difficulties of centrist approaches, namely the morally questionable consequences of centrism, the (anthropological) essentialism inherent in centrism, and the implicit paternalism that accompanies the subject-object dichotomy embedded in an (epistemic) centrist approach. Even an individualistic physiocentrism, in order to remain capable to act, must ascribe an instrumentalistic value to some entities, i.e. a value weaker than an intrinsic one, or define a hierarchy of values in some other way. And even within a physiocentric ethics system, the morally significant entities are identified essentialistically in so far as everything that belongs to nature (and therefore may understand itself endowed with a number of attributes depending on the authority of definition) has an intrinsic value. Finally, within the physiocentric horizon, the difference between recognizing subjects and (recognized) objects also remains, even if physiocentrism is applied not only on the moral level but also on the epistemic level and thus attests to "the" nature or "all parts" of nature cognitive faculty and power of judgement.

5 Excursus: Inclusive Approaches in Robot Ethics

For all of these reasons given above Mark Coeckelbergh, David Gunkel, Luisa Damiano and Paul Dumouchel, Donna Haraway, Lucy Suchman and other inclusive thinkers each grant in their own way moral rights to robots. The approaches discussed within this third area of robot ethics question, on the one hand, the traditional understanding of the moral agent in order to extend it to nonhuman beings, too. On the other hand, competencies that have been ascribed essentialistically to individual agents within the framework of the exclusive approaches should now be understood relationally as realizing themselves in the interaction of different human and nonhuman agents.

As explained above, the unifying intention of all these perspectives is an inclusive program (morally and epistemically) that neither emphasizes the position of "the" human being over other beings or negatively devalues "the" human to the "level" of all other beings. In contrast to these exclusive theories, which either value "the" human being as special and better than all other beings (exclusive authoritarianism) or exclude "the" human being from the genuine realm of morality (exclusive relativism), the inclusive approaches seek to place all other beings on the same "level" as human beings and thus include them as morally equal companions (having the same fundamental moral rights) in the moral universe in which human beings are located. Rather, inclusive positions seek to locate all beings on the same "level" as humans and thus include them as morally equal companions in the moral universe. In other words, inclusive approaches neither deny human beings morally relevant qualities or fundamental rights nor restrict their rights as human beings. Including theories claim to draw attention to the (at times) questionable images of human beings, which are the basis of morally and legally important institutions such as human rights and humanism.

In the following, I will give one example for an inclusive approach: the anthropologist, feminist, and scientist of Science and Technology Studies, Lucy Suchman, has been working for many years on human-computer interaction and especially on military technologies. In an updated and expanded new edition of her book *Plans and Situated Actions* (1987) – *Human-Machine Reconfigurations* (2007) – she develops an approach in the context of the theories of Donna Haraway, Bruno Latour, Judith Butler, and others, but especially with Karen Barad's critical-posthumanist "relational ontology" (my translation; Barad 2012, 18). Suchman attemts to reconfigure (as she puts it) the ways in which we relate to other humans and nonhumans (especially technologies) and understand ourselves as self-sufficient beings.

With her agential realism, Barad takes a path towards overcoming anthropocentrism by "recognizing the dynamic power of matter" (my translation; Barad 2012, 40 f.). According to Barad, matter is neither simply passive, but discursive, nor is discourse alone active, but also material. This view ultimately makes it possible not only to recognize new nonhuman agents as such, but also to radically question the dichotomy of humans and nonhuman beings (2012, 30–32). In her conversation with Jennifer Sophia Theodor, Barad vividly explains the procedure of "an agential-realist analysis" (my translation; Barad 2015, 184), using the examples of the Californian raisin and the fetus. When we eat a Californian raisin, according to Barad, we also bite into the "material-discursive apparatuses" (my translation; 2015, 185) that were involved in their production "such as capitalism, colonialism, and racism" (my translation; 2015, 185), and many more. As a reaction to essentialism, Barad formulates a relational approach that focuses less on the relata, the subjects and objects, and more on the relation, that is, what lies between them, in a nonessentialist way (Coeckelbergh & Gunkel 2014). Thus, she designs a relational ontology in which no singular, autarkic agents exist, but these are merely recognized within their mutual entanglements:

> This relational ontology is the basis for my posthumanist performative account of material bodies (both human and nonhuman). This account refuses the representationalist fixation on words and things and the problematic of the nature of their relationship, advocating instead *a relationality between specific material (re)configurings of the world through which boundaries, properties, and meanings are differently enacted* [...] *and specific material phenomena* [...]. (Barad 2007, 139; cf. 2007, 33, 93; 2012, 18)

With her relational ontology Barad provides the critical-posthumanist foundation for an anthropology and ethics to be based on it. In the last chapter of *Human-Machine Reconfigurations* Suchman makes a proposal to her readers for its outline. The chapter is preceded by a quote of Barad's paper "Posthuman Performativity: Toward an understanding of how matter comes to matter": "Agency is not an attribute but the ongoing reconfigurings of the world" (2003, 818). Suchman, with a view to computer technologies and robots, tackles the project of a new understanding of the "traditional humanist" (Suchman 2007, 270) agent and the "human-machine boundaries" (2007, 260). She comes to the conclusion that not all differences per se become obsolete. However, we can learn to make other distinctions than those suggested by the Western and humanistic image of the human being. Moreover, and this is what Suchman is concerned with here, questioning the classical understanding of agency supports us in rethinking the "nature of difference" (2007, 260) itself. Not only *what* differences we encounter is of anthropological and ethical concern, but we must also visualize and discuss *how* we ourselves want to understand difference, what difference *is* (2007, 260).

Along numerous examples from the technical sciences, robotics and human everyday life, which is fundamentally shaped and structured by technologies, Suchman proposes a relational alternative to the human-machine relationship, which up to now has been interpreted in an essentialist way and which rests on clearly defined boundaries and attributable (and thus always deniable) characteristics. According to her, the "interface" (2007, 263) must not necessarily be seen "as a priori or self-evident boundary between bodies and machines but as a relation enacted in particular settings and one, moreover, that shifts over time" (2007, 263). There is no predetermined way of reconfiguring the agent's understanding, the dichotomies of human and machine, and the categories in which we can define our relationships to human and nonhuman beings in terms of fundamental moral rights. The way in which we understand difference and make specific distinctions raises "questions of alignment or dislocation, relation or alienation" (2007, 266) that can be answered in different ways. None of these answers is inherent in any form of human-machine interaction or precedes the relation, the relationship itself. "[T]hey [the distinctions we make; J.L.] are effects lived and experienced within multifaceted subject-object assemblages" (2007, 266).

Suchman bases her ethics of reconfiguration in Barad's tradition on the assumption of "*intra-action*" and thus associates with this in particular the renegotiation of the subject-object dichotomy, as exemplified in her paper "Subject objects" (2011) using the example of humanoid robots (2011, 121). Intra-action, unlike "interaction" (2011, 267), does not presuppose (at least) two autonomous "entities" (2011, 267), which then enter into a kind of relationship and "exchange" (2011, 267) with each other. But intra-action is based on the idea of interference:

> *Interference* is the physical term for the superposition of waves. It thus describes a form of relationship that cannot be translated into common notions of difference and identity, in which isolated, stable entities relate to one another. In interference, the wave operates by becoming something else, another wave, several, smaller, larger waves, or possibly no waves at all. Waves cannot meet without interfering, without becoming something else. In this sense, interference stands for a non-binary, non-linear and yet relational form of generating otherness, difference, which, however, does not merge into these differences. (my translation; Bath et al. 2013, 7)

From the concepts of intra-action and thus implicitly interference, Suchman derives the mutually constitution of "humans and artifacts" (2007, 268). This "constitution of humans and artifacts does not occur in any single time and place, nor does it create fixed human-artifact relations or entities" (2007, 268). Suchman arrives at a processual idea of humans and nonhumans, who constantly recreate each other in their constant interferences (Suchman 2007, 278 f.). "Agencies [...] reside neither in us nor in our artifacts but in our intra-actions" (2007, 285).

With her ethics of the mutual constitution of humans and nonhumans, Suchman does not intend to abolish differences in general. However, we should consider how much and what exactly we want to continue to accept from our heavy humanistic heritage as the basis of our view of human beings and the world. According to her, "we need a story that can tie humans and nonhumans together without erasing the culturally and historically constituted differences among them" (2007, 270). We must "avoid the twin traps of categorical essentialism and the erasure of differences that matter" (2011, 137). However, "[m]utualities" are "not necessarily symmetries" (2007, 269), but in order to be able to include nonhuman forms of otherness in our moral universe, we must be able to endure and support forms of "asymmetry" (2007, 269) or better still "dissymmetry" (2007, 269) that are constantly being renegotiated as realizations of the interferences between humans and nonhumans who are constantly redesigning themselves in a processual way. Suchman therefore concludes her paper "Subject objects" with the request to "extricat[e] ourselves from a tradition in which our interest in nonhumans is for either their reflective or contrastive properties vis-à-vis (a certain figure of) our own, in favour of an attention to ontologies that radically – but always contingently – reconfigure the boundaries of where we stop, and the rest begins" (2011, 138).

6 Concluding Remarks

Further above, reference was made to the heterogeneous strategies according to which moral patiency is to be granted or denied. As long as it is not determined which robots to look at, it must be assumed that potentially all robots are entitled to a certain morally appropriate behavior. This seems to be the case with the approaches of Gerdes and Tavani. Both run the risk of unduly expanding the circle of potentially morally considerable beings and thus becoming incapable of action. With Gerdes, this challenge arises from the fact that in the papers under discussion she does not narrow down "robots" as potential moral patients. Tavani, on the other hand, in his interpretation of Jonas, represents a physiocentrism that has not been further concretized, and which needs to be more precisely executed in order to be able to prove its ethical viability. Darling narrows the range of artificial systems worthy of moral consideration down to that of social robots. Following her considerations, it is comparatively easy to identify which robots are to be identified as moral patients and which are not. Furthermore, it could be shown that the recognition of a robot as a moral object of action often depends on the fact that it is also seen as a moral agent (the standard position).

Inclusive approaches generally accuse exclusive theories, i.e. those that identify robots either as moral agents or as moral patients, of not adequately dealing with the exceptionally problematic (implicitly discriminatory, sexist, heteronormative, racist, etc.) foundations of their thinking. Thus, these first premises of their arguments are at least implicitly confirmed and perpetuated. Sometimes even the attempt is made to justify these basic assumptions as objective and universal. The exclusive positions, on the other hand, usually accuse the inclusive thinkers – such as Suchman, whose approach was exemplarily outlined in the fifth section of this chapter – of a general blurring of their actual concerns, a darkening of the circumstances by an unnecessary and confusing softening of conceptual categories and species boundaries. After all, not even the decision-making authority is set over inclusion and exclusion into or out of the moral universe.

How the attribution of fundamental moral rights is undertaken in the field of robot ethics, in which robots are interpreted as potential moral patients, was discussed in this chapter. In particular, the limits of these theories, which in my opinion are questioned by inclusive thinkers, were shown.

References

Allen, Colin, Wendell Wallach, and Iva Smit. 2006. "Why Machine Ethics?" *Intelligent Systems IEEE* 4: 12–17.

Altmann, Matthew C. 2011. *Kant and Applied Ethics: The Uses and Limits of Kant's Practical Philosophy*. New Jersey: John Wiley & Sons Inc.

Anderson, Michael, and Susan Leigh Anderson, eds. 2011. *Machine Ethics*. Cambridge: Cambridge University Press.

Anderson, Michael, and Susan Leigh Anderson. 2007. "Machine Ethics: Creating an Ethical Intelligent Agent." *AI Magazine* 4: 15–26.

Anscombe, Elizabeth. 1957. *Intention*. Oxford: Basil Blackwell.

Aristotle. 2006. *Nikomachische Ethik*. Translated by Ursula Wolf. Reinbek: Rowohlt.

Asaro, Peter M. 2006. "What Should We Want From a Robot Ethic?" *International Review of Information Ethics* 6(12): 9–16.

Asimov, Isaac. 1982. *The Complete Robot. The Definitive Collection of Robot Stories*. London: Harper Collins Publications.

Barad, Karen. 2003. "Posthuman Performativity: Toward an Understanding of How Matter Comes to Matter." *Signs. Journal of Women in Culture and Society* 28(3): 801–831.

Barad, Karen. 2007. *Meeting the Universe Halfway. Quantum Physics and the Entanglement of Matter and Meaning*. Durham, London: Duke University Press.

Barad, Karen. 2012. *Agentieller Realismus*. Berlin: Suhrkamp.

Barad, Karen. 2015. *Verschränkungen*. Berlin: Merve.

Bath, Corinna, Hanna Meißner, Stephan Trinkaus, and Susanne Völker. 2013. "Einleitung." In *Geschlechter Interferenzen. Wissensformen – Subjektivierungsweisen – Materialisierungen*, edited by Corinna Bath, Hanna Meißner, Stephan Trinkaus, and Susanne Völker, 7–25. Berlin: LIT Verlag.

Beauchamp, Tom L., and R.G. Frey, eds. 2011. *The Oxford Handbook of Animal Ethics*. New York: Oxford University Press.

Beck, Susanne, et al. 2019. "Brauchen wir eine Roboterethik? Eine Kontroverse mit Susanne Beck, Bruno Gransche, Eric Hilgendorf, Janina Loh, Catrin Misselhorn und Thomas Zoglauer." *Jahrbuch Technikphilosophie* 5: 231–267.

Beck, Susanne. 2010. "Roboter, Cyborgs und das Recht – von der Fiktion zur Realität." In *Aktuelle Herausforderungen der Life Sciences. Band 1 der Reihe „Recht der Lebenswissenschaften"*, edited by Tade M. Spranger, 95–120. Berlin: LIT.

Bekey, George. 2005. *Autonomous Robots. From Biological Inspiration to Implementation and Control*. Cambridge: MIT Press.

Bendel, Oliver, ed. 2018a. *Handbuch Maschinenethik*. Wiesbaden: Springer.

Bendel, Oliver. 2018b. "Überlegungen zur Disziplin der Maschinenethik." *Aus Politik und Zeitgeschichte* 68(6–8): 34–38.

Brey, Philip, Adam Briggle, and Katinka Waelbers, eds. 2008. *Current Issues on Computing and Philosophy*. Amsterdam: IOS Press.

Bryson, Joanna. 2018. "Patiency is not a Virtue: The Design of Intelligent Systems and Systems of Ethics." *Ethics and Information Technology* 20(1): 15–26.

Bryson, Joanna J. 2010. "Robots Should Be Slaves." In *Close Engagements with Artificial Companions: Key Social, Psychological, Ethical and Design Issues*, edited by Wilks Yorick, 63–74. Amsterdam: John Benjamins.

Calverley, David J. 2011. "Legal Rights for Machines." In *Machine Ethics*, edited by Michael Anderson and Susan Leigh Anderson, 213–227. Cambridge: Cambridge University Press.

Calverley, David J. 2006 "Android Science and Animal Rights. Does an Analogy Exist?" *Connection Science* 4: 403–417.

Canguilhem, Georges. 2012. "Maschine und Organismus." In *Die Erkenntnis des Lebens*. Translated by Till Bardoux, Maria Mühle und Francesca Raimondi, 183–232. Berlin: August Verlag.

Capurro, Rafael, and Michael Nagenborg, eds. 2009. *Ethics and Robotics*. Heidelberg, Amsterdam: IOS Press.

Churchland, Paul M. 1999. *Matter and Consciousness*. Cambridge: MIT Press.

Coccia, Emanuele. 2018. *Die Wurzeln der Welt*. München: Hanser.

Coeckelbergh, Mark, and David Gunkel. 2014. "Facing Animals: A Relational, Other-Oriented Approach to Moral Standing." *Journal of Agriculture and Environmental Ethics* 27(5): 715–733.

Cohen, Carl. 1986. "The Case for the Use of Animals in Biomedical Research." *New England Journal of Medicine* 315(14): 865–870.

Danaher, John. 2017. "The Rise of the Robots and the Crisis of Moral Patiency." *AI & Society*. https://www.researchgate.net/publication/321150508_The_rise_of_the_robots_and_the_crisis_of_moral_patiency.

Darling, Kate. 2012. "Extending Legal Protection to Social Robots." *IEEE Spectrum*, September 10, 2012. https://spectrum.ieee.org/automaton/robotics/artificial-intelligence/extending-legal-protection-to-social-robots.

Davidson, Donald. 1980. *Essays on Actions and Events*. New York: Oxford University Press.

deGracia, David. 1996. *Taking Animals Seriously. Mental Life and Moral Status*. New York: Cambridge University Press.

Diels, Hermann. 1966. *Die Fragmente der Vorsokratiker*, edited by Walther Kran. Hildesheim: Weidmann.

Donaldson, Sue, and Will Kymlicka. 2013. *Zoopolis. Eine politische Theorie der Tierrechte*. Frankfurt am Main: Suhrkamp.

Drux, Rudolf, ed. 1988. *Menschen aus Menschenhand. Zur Geschichte der Androiden. Texte von Homer bis Asimov*. Stuttgart: Metzler.

Edgar, Stacey L. 2003. *Morality and Machines. Perspectives on Computer Ethics*. Boston, Toronto, London, Singapore: Jones & Bartlett Learning.

European Parliament. 2014–2019. *Entwurf eines Berichts mit Empfehlungen an die Kommission zu zivilrechtlichen Regelungen im Bereich Robotik*. (2015/2103(INL)). http://www.europarl.europa.eu/sides/getDoc.do?pubRef=-//EP//NONSGML%2BCOMPARL%2BPE-582.443%2B01%2BDOC%2BPDF%2BV0//DE.

Floridi, Luciano, and J.W. Sanders. 2004. "On the Morality of Artificial Agents." *Minds and Machines* 14(3): 349–379.

Gerdes, Anne. 2017. "The Issue of Moral Consideration in Robot Ethics." *ACM SIGCAS Computers & Society* 45: 247–279.

Grawe, Christian, et al. 1971–2007. "Mensch." In *Historisches Wörterbuch der Philosophie. Band: L–Mn*, edited by Joachim Ritter, 1059–1105. Basel: Schwabe.

Gunkel, David. 2018. *Robot Rights*. Cambridge, London: MIT Press.

Gunkel, David. 2012. *The Machine Question. Critical Perspectives on AI, Robots, and Ethics*. Cambridge: MIT Press.

Gunkel, David. 2007. "Thinking Otherwise: Ethics, Technology and Other Subjects." *Ethics and Information Technology* 9: 165–177.

Hajdin, Mane. 1994. *The Boundaries of Moral Discourse*. Chicago: Loyola University Press.

Hilgendorf, Eric, ed. 2014. *Robotik im Kontext von Recht und Moral*. Baden-Baden: Nomos.

Ichbiah, Daniel. 2005. *Roboter. Geschichte – Technik – Entwicklung*. Munich: Knesebeck.

Jonas, Hans. 1984. *Das Prinzip Verantwortung. Versuch einer Ethik für die technologische Zivilisation*. Frankfurt am Main: Suhrkamp.

Jordan, John. 2017. *Roboter*: Translated by Manfred Weltecke. Berlin: Berlin University Press.

Kant, Immanuel. 1797. *Die Metaphysik der Sitten*, AA VI. https://korpora.zim.uni-duisburg-essen.de/kant/aa06/443.html.

Kant, Immanuel. 1785. *Grundlegung zur Metaphysik der Sitten*, AA IV. https://korpora.zim.uni-duisburg-essen.de/kant/aa04/421.html.

Krebs, Angelika. 1997. "Naturethik im Überblick." In *Naturethik. Grundtexte der gegenwärtigen tier- und ökoethischen Diskussion*, edited by Angelika Krebs, 337–379. Frankfurt am Main: Suhrkamp.

Laukyte, Migle. 2017. "Artificial Agents Among Us: Should We Recognize Them as Agents Proper?" *Ethics and Information Technology* 19: 1–17.

Lin, Patrick, Keith Abney, and George Bekey, eds. 2012. *Robot Ethics. The Ethical and Social Implications of Robotics*. Cambridge, London: MIT Press.

Lin, Patrick, Keith Abney, and George Bekey. 2011. "Robot Ethics: Mapping the Issues of a Mechanized World." *Artificial Intelligence* 175(5–6): 942–949.

Lin, Patrick, Ryan Jenkins, and Keith Abney, eds. 2017. *Robot Ethics 2.0. From Autonomous Cars to Artificial Intelligence*. New York: Oxford University Press.

Loh (née Sombetzki), Janina. 2019. *Roboterethik. Eine Einführung*. Berlin: Suhrkamp.

Loh (née Sombetzki), Janina. 2018. "Maschinenethik und Roboterethik." In *Handbuch Maschinenethik*, edited by Oliver Bendel. Wiesbaden: Springer: 75–93.

Loh (née Sombetzki), Janina. 2017. "Roboterethik. Janina Loh über eine noch junge Bereichsethik." *Information Philosophie* 1: 20–33.

Mainzer, Klaus. 2010. *Leben als Maschine. Von der Systembiologie zur Robotik und Künstlichen Intelligenz*. Paderborn: Mentis.

Marr, Bernard. 2017. "The 4 Ds Of Robotization: Dull, Dirty, Dangerous and Dear." *Forbes*, October 16, 2017. https://www.forbes.com/sites/bernardmarr/2017/10/16/the-4-ds-of-robotization-dull-dirty-dangerous-and-dear/#346ac3c93e0d.

Mataric, Maja J. 2007. *The Robotics Primer*. Cambridge: MIT Press.

McNally, Phil, and Sohail Inayatullah. 1998. "The Rights of Robots: Technology, Law and Culture in the 21st Century." *Futures* 20: 119–136.

Misselhorn, Catrin. 2018a. *Grundfragen der Maschinenethik*. Stuttgart: Reclam.

Misselhorn, Catrin 2018b. "Maschinenethik und 'Artificial Morality': Können und sollen Maschinen moralisch handeln?" *Aus Politik und Zeitgeschichte* 68: 29–33.

Misselhorn, Catrin. 2013. "Robots as Moral Agents?" In *Ethics in Science and Society. German and Japanese Views*, edited by Frank Rövekamp and Friederike Bosse, 42–56. München: Iudicium.

Moor, James H. 2006. "The Nature, Importance, and Difficulty of Machine Ethics." *Intelligent Systems* IEEE 4: 18–21.

Nadeau, Joseph Emile. 2006. "Only Androids Can Be Ethical." In: *Thinking about Android Epistemology*, edited by Kenneth Ford, Clark Glymour, and Patrick J. Hayes, 241–248. Menlo Park, Cambridge, London: MIT Press.

Nagel, Thomas. 1974. "What is it Like to be a Bat?" *The Philosophical Review* 4: 435–450.

Rath, Matthias, Friedrich Krotz, and Matthias Karmasin, eds. 2019. *Maschinenethik. Normative Grenzen autonomer Systeme*. Wiesbaden: Springer.

Regan, Tom. 1983. *The Case for Animal Rights*. Berkeley, CA: University of California Press.

Remmers, Peter. 2018. *Mensch-Roboter-Interaktion*. Berlin: Logos.

Robertson, Jennifer. 2014. "Human Rights vs. Robot Rights: Forecasts from Japan." *Critical Asian Studies* 4: 571–598.

Schmitz, Friederike, eds. 2014. *Tierethik. Grundlagentexte*. Berlin: Suhrkamp.

Singer, Peter. 2015. *Animal Liberation. Die Befreiung der Tiere*. Translated by Claudia Schorcht. Erlangen: Harald Fischer.

Sombetzki (now Loh), Janina. 2016. "Roboterethik." In: *Zur Zukunft der Bereichsethiken. Herausforderungen durch die Ökonomisierung der Welt*, edited by Matthias Maring, 355–379. Karlsruhe: KIT Scientific Publishing.

Sombetzki (now Loh), Janina. 2014. *Verantwortung als Begriff, Fähigkeit, Aufgabe. Eine Drei-Ebenen-Analyse*. Wiesbaden: Springer VS.

Stahl, Bernd Carsten. 2004. "Information, Ethics, and Computers: The Problem of Autonomous Moral Agents." *Minds and Machines* 14(1): 67–83.

Strandh, Sigvard. 1980. *Die Maschine. Geschichte – Elemente – Funktion*. Freiburg im Breisgau: Herder.

Suchman, Lucy. 2011. "Subject Objects." *Feminist Theory* 12(2): 119–145.

Suchman, Lucy. 2007. *Human-Machine Reconfigurations. Plans and Situated Actions*. Cambridge: Cambridge University Press.

Sullins, John P. 2006. "When Is a Robot a Moral Agent?" *International Review of Information Ethics* 6(12): 23–30.

Tavani, Herman T. 2018. "Can Social Robots Qualify for Moral Consideration? Reframing the Question about Robot Rights." *Information* 9. http://www.mdpi.com/2078-2489/9/4/73.

Teubner, Gunther. 1998. "Rights of Non-Humans? Electronic Agents and Animals as New Actors in Politics and Law." *Journal of Law and Society* 33(4): 497–521.

Turing, Alan. 1950. "Computing Machinery and Intelligence." *Mind* 59(236): 433–460.

Versenyi, Laszlo. 1974. "Can Robots be Moral?" *Ethics* 84(3): 248–259.

Wallach, Wendell, and Colin Allen. 2009. *Moral Machines. Teaching Robots Right from Wrong*, Oxford, New York: Oxford University Press.

Walsh, Toby. 2017. *Android Dreams. The Past, Present and Future of Artificial Intelligence.* London: C. Hurst & Co Publishers Ltd.

Wolf, Ursula. 2004. *Das Tier in der Moral.* Frankfurt am Main: Vittorio Klostermann Verlag.

Wolf, Ursula, ed. 2008. *Texte zur Tierethik*, Stuttgart: Reclam.

Wootson, Cleve R., Jr. 2017. "Saudi Arabia, which Denies Women Equal Rights, Makes a Robot a Citizen." *The Washington Post*, October 29, 2017. https://www.washingtonpost.com/news/innovations/wp/2017/10/29/saudi-arabia-which-denies-women-equal-rights-makes-a-robot-a-citizen/?noredirect=on&utm_term=.191afe283aa8.

CHAPTER 5

Fundamental Rights and Smart Health Technologies

Adam Poulsen and Anwaar Ulhaq

1 Introduction

This chapter addresses the fundamental rights in healthcare involving smart health technologies. To be able to explore their implementation in healthcare, those rights must first be declared. The fundamental rights in care can be examined in the light of "*The Universal Declaration of Human Rights* (*United Nations General Assembly 1948*)" and "*The Charter of Fundamental Rights of the European Union* (*European Convention 2012*)." We have compiled the following list comprising fundamental mental rights in healthcare based on these two essential documents as follows:

- Right to life, liberty, and personal security
- Right to equality
- Right to dignity
- Right to recognition as an individual
- Right to justice
- Right to privacy
- Right to protection of personal data
- Right to rest and leisure
- Right to adequate living standards, social security, social assistance, and health care
- Freedom of movement
- Freedom from torture and inhuman or degrading treatment or punishment
- Freedom from discrimination
- Freedom of belief, thought, conscience, and religion
- Freedom of opinion and information

Later in this chapter, these rights will be reflected on in relation to smart health technologies. The following sections will describe the importance of these rights in healthcare, the description of how these rights are supported and regulated, and how caregivers practically go about respecting these rights.

 | DOI:10.1163/9789004437876_007

1.1 *Framing Fundamental Rights in Healthcare Law and Codes*

Fundamental human rights in care are regulated and supported by healthcare laws in each country individually. The principles and standards set out by laws are then passed down into codes of conduct and ethics which enforce the law itself on a smaller scale. Codes outline the rules which caregivers and other healthcare professionals must follow when providing care. For instance, the United Kingdom Human Rights Act of 1998 adopted 16 of the fundamental human rights from The European Court of Human Rights (1950) which in turn sourced many of its rights from The Universal Declaration of Human Rights (United Nations General Assembly 1948). The Human Rights Act of 1998 is the central healthcare law in the UK for regulating the healthcare domain and those persons within it. To provide more specific guidelines to caregivers, the standards and fundamental rights within the Act are then translated into codes of conduct and ethics. One of these codes is the Nursing and Midwifery Council's (2002) Code of professional conduct which gives some general recommendations on how to respect the fundamental rights of care recipients. Here we present an example of some exemplary fundamental human rights and how they are supported by the code with nursing guidelines.

> Right to recognition as an individual
>
> "As a registered nurse, midwife or health visitor, you must respect the patient or client as an individual. You must recognise and respect the role of patients and clients as partners in their care and the contribution they can make to it."
>
> Right to dignity, freedom from discrimination, freedom of belief, thought, conscience, religion, and freedom of opinion and information
>
> "You are personally accountable for ensuring that you promote and protect the interests and dignity of patients and clients, irrespective of gender, age, race, ability, sexuality, economic status, lifestyle, culture and religious or political beliefs."
>
> Right to privacy, right to protection of personal data, and right to rest and leisure
>
> "You must, at all times, maintain appropriate professional boundaries in the relationships you have with patients and clients. You must ensure that all aspects of the relationship focus exclusively upon the needs of the patient or client."
>
> Right to equality and right to adequate living standard, social security and social assistance, health care

> You must promote the interests of patients and clients. This includes helping individuals and groups gain access to health and social care, information and support relevant to their needs.
>
> Nursing and Midwifery Council 2002

Consider another example: examine the Nursing and Midwifery Board of Australia's code of conduct for nurses (hereon in referred to Australian nursing code). To follow are some exemplary fundamental rights in care and caregiver guidelines as found in the Australian nursing code.

> Right to recognition as an individual, right to privacy, and right to protection of personal data
>
> "Nurses engage with people as individuals in a culturally safe and respectful way, foster open and honest professional relationships, and adhere to their obligations about privacy and confidentiality."
>
> Freedom from discrimination and freedom of belief, thought, conscience, and religion
>
> "Culturally safe and respectful practice requires nurses and midwives to challenge bias and beliefs based on assumption, such as assumptions based on gender, disability, race, ethnicity, religion, sexuality, age or political beliefs."
>
> Right to life, liberty, personal security and right to rest and leisure
>
> "accept that the person has the right to refuse treatment, or to request withdrawal of treatment, while ensuring the person receives relief from distress."
>
> Nursing and Midwifery Board of Australia 2018

1.2 *The Problem with Codes for Caregivers*

There is an inherent problem with healthcare codes, such as those in the UK and Australia. That is, while codes do support fundamental rights in care, unfortunately, there is still little guidance for caregivers as to how one should practically go about ensuring the rights of care recipients in each unique situation. Take the right to dignity as a use case, to follow is everything the Australian nursing code tells caregivers about the assurance of care recipient dignity.

> "[A caregiver must] create a positive, culturally safe work environment through role modelling, and supporting the rights, dignity and safety of others, including people and colleagues."

> "Each person is treated as an individual with the aim of respecting people's ownership of their health information, rights and preferences while protecting their dignity and empowering choice."
>
> "The nurse is sensitive to a person's situation and purposefully engages with them using knowledge and skills with respect, compassion and kindness. In the relationship, the person's rights and dignity are recognised and respected."
>
> "In the [therapeutic] relationship the person's rights and dignity are recognised and respected."
>
> Nursing and Midwifery Board of Australia 2018

It should be evident that while these recommendations in the Australian nursing code are good, unfortunately, they are mostly useless in telling one how to practically go about providing good care which ensures that fundamental rights are respected. Achieving good care which aims to protect fundamental rights requires more than laws and codes, it needs context and training. That is, a caregiver's experience and training, as well as responsiveness to each unique care situation, are essential elements to providing good care which respects rights.

1.3 *Good Care is Ethical, Responsive Care*

The context plays an important role in defining care. Lacking practical, realistic care instructions from nursing codes, one might instead look to nursing use cases which provide situational context and describe effective nursing processes and procedures. Use cases give specific care scenarios and possible caregiver actions based on the elements of the scenarios. The reality is that the only real objective guidance to be found is in specific situations such as those presented in use cases. Examine the following use case provided by Australia's NSW Department of Health (2009) describing how a caregiver ought to care for a critically ill, near-to-death care recipient in order to respect their right to dignity.

> "Remember that dying patients and the critically ill have very special needs. The care of a dying patient and the critically ill is an extremely intimate and private event occurring at a time when a person is at their most vulnerable, and deserves to be treated with the utmost respect.
>
> Dying patients have priority for single rooms. If this is not able to be provided due to the clinical needs of other patients, please consider other alternatives.
>
> Keep environmental noise around the dying patient and the critically ill to a minimum.

> Talk to support staff (domestic staff, food service staff) so they know whether or not to enter the room.
>
> Cultural, religious or faith tradition perspectives on death and dying are available through: Aboriginal Liaison Officers, Multicultural Health Services, Accredited Hospital Chaplains, etc."
>
> NSW Department of Health 2009

Taking a quick look at this use case, one can get a picture of the scenario, those involved, and the proper direction as to how to deal with the scenario as a caregiver. Objectively, use cases are a good source of guidance for caregivers. However, just like rights in isolation, use cases in isolation are still unhelpful in a generalisable sense. Use cases are important in that one can learn different ways to deal with different variations in the scenario. But, the real lesson to be learned is responsiveness, one must learn to deal with each situation as it comes and get a positive result. That is, training oneself to be responsive to each dynamic care situation is essential in being a good caregiver. For example, in the use case above, one can see that the caregiver is told to be responsive – "please consider other alternatives." Moreover, the use case suggests that the caregiver seek advice from expert peers on how to deal with each care recipient's cultural, religious, and faith-based expectations and adapt their care delivery to respect them. *Caregiver responsiveness*, to both the situation and care recipients' unique needs, is the key to good care.

1.4 *Good Care and Smart Health Technologies*

As a summary of the above, in general, fundamental rights in care tell us the end goal of care delivery. However, they do not determine caregiver acts in each different, unique situation. Instead, a caregiver's responsiveness, experience, and training are key to knowing how to go about caring for someone to ensure their rights are upheld. Caregivers must make decisions in practice to best uphold both a care recipient's rights and good care in general, and smart health technologies should do the same.

Just as human caregivers interpret their environment and make dynamic decisions based on situational information, smart health technologies do the same. Additionally, just as human caregivers adapt their care delivery to both the care context and the unique needs of care recipients, smart health technologies should do the same. As a broad and simple example, a care robot should understand that a care recipient wants privacy when changing clothes and thus it leaves the room to respect that right. As another example, an ambient intelligence system such as a fall detection sensor network should be able to differentiate between a person doing yoga on the floor and someone falling over. Thus, if it can make this determination and withhold sounding the alarm which indicates that someone has fallen, then it is respecting one's right to an adequate living standard.

Smart health technologies make room for the human touch in technology-delivered care; they add caregiver responsiveness and provide care recipient freedoms to otherwise standardised care. If they are smart, then healthcare technologies are taking in the situational context and using contextual data to provide a responsive and personalised level of care which furthers technology-delivered care towards the goal of upholding the fundamental rights in care. Smart health technologies do this by making good, dynamic care decisions which are responsive, scenario-derived, and person-centred. This will be expanded on in Sections 4 and 5. But, before that, smart health technologies themselves will be examined.

Section 2 looks at current smart health technologies either commercially available or in development. In that section, we will provide a brief overview of the technical aspects of smart health technologies.

Section 3 looks to the future of smart health technology by exploring the design considerations of some of the most common smart health technologies, ambient intelligence and care robots. There, we will allude to some of the fundamental rights in care that smart health technologies must uphold.

Section 4 will introduce the topic of care ethics. There, we will go deeper into demonstrating the importance of smart health technologies capable of responsive, care ethics-based decisions for ensuring fundamental rights. Within, we will give detailed scenarios showing the positive benefits of those types of smart technologies in the care setting, comparing them to the responsiveness of human caregivers. Furthermore, we will reflect on the positive and negative impacts that health technologies have on fundamental rights in care.

Section 5 we will present and evaluate design approaches which can be applied to the creation of smart health technologies. We will reflect on fundamental rights in care and care ethics while examining those design approaches. Then, finally, we will make recommendations on how to best prepare for the future of smart health technology design.

2 Smart Health Technologies

The term "smart technology" is popular for its self-monitoring, analysis, and reporting supporting feature that enables the dumb objects around us to interact with us and even guide our behaviour. Smart technology is an umbrella term that encapsulates various forms of technologies which capture data from its surroundings, store and analyse the data, and has the means to automate actions based on the data input. The term "smart health technology" is the combination of "smart technology" and "healthcare," i.e. the use of smart technologies for healthcare. Smart health technologies are, therefore, gradually building their capacity to record valuable health information using increasingly advanced sensors, storing and computing meaningful health information from available data, and delivering automated health services. Some notable examples of smart health technologies include smart pills, medicine dispensation, smart surgeries, wearables, ambient sensors, and care robots.

Internet-of-things (IoT) plays a significant role in the implementation of healthcare systems. The IoT adds ubiquitous and faster communication, internet connectivity, and efficient computing along with ambient intelligence. With internet-of-medical-things (IoMT), smart technologies can now provide more connected healthcare solutions. These connected healthcare solutions further improve the quality and efficiency of healthcare by facilitating self-care and complementing it with remote care. While smart healthcare solutions can operate in an autonomous fashion, connected healthcare provides the added benefit of supplying users with feedback from clinicians (Sundaravadivel et al. 2018). Cloud computing extends the ubiquitous notion of patient-centric, data-driven healthcare services by providing highly available, elastic, and scalable infrastructure.

The diversity of smart devices, faster internet. and cloud highly scalable infrastructure, however, increases volume, velocity, or variety of data. The abundance of smart sensors, activities on social networks, web services, and usage of mobile devices are estimated to generate more than 2.5 quintillion bytes per day (Mohanty, Choppali, and Kougianos 2016). The Big Data and its remote processing can introduce delays which can be costly in a healthcare system. Fog computing is introduced to coup with this delay by pushing the computing capacity toward the edges of the network (Ni et al. 2018). Server-less computing and blockchain are strengthening the security and privacy of health information (Kuo, Kim, and Ohno-Machado 2017). Today, this valuable integration of smart items – fog, cloud and blockchain technologies – are leading smart health technologies research.

These smart technologies also assist users to share data to better understand health and wellbeing. The Quantified Self, for example, is the popular movement that gives people access to a large data about themselves (e.g. physiological, mood changes, behavioural patterns, and environmental impacts) from mobile self-tracking apps (Luxton, Hansen, and Stanfill 2014) and other wearable devices. It also provides them with the opportunity to compare their data to the data of millions of other people (Wolf 2011; Lupton 2013) with the objective of self-monitoring and self-reflection to facilitate change or health improvement.

Smart care systems can be classified based on the types of devices used, their technologies, management framework, availability of services, applications areas, and end user profile. Figure 5.1 (adapted from Sundaravadivel et al. 2018) provides a classification of smart healthcare systems. The applications of smart health technologies vary from fitness monitoring to vital-sign monitoring in hospitals (Xu et al. 2014; Istepanian et al. 2011) and in-body or intrabody sensing to smart surroundings (Mayer et al. 2016; Sprint et al. 2016). Intrabody sensing refers to the monitoring of multiple health-related features. For instance, smart watches for fitness tracking can be used for heart-rate monitoring alongside counting the steps taken and active hours. Interbody sensing can be used to monitor the internal organs with nano-smart devices like a pill camera. A pill camera can be swallowed like a pill to capture high-resolution images of the internal organs (Kiourti and Nikita 2017). Smart environment sensing may facilitate in establishing communication between the hospital and the patient like calling ambulance assistance, evacuation strategies for disaster management, or accurate patient tracking through RFID tags.

Today, various commercial and scientific organizations are working constantly on the advancement of smart health technologies. Intel is coming up with market-ready solutions like Ambient AWARE™ based on innovative technologies for data analytics and assistive technology. IBM's Watson, an artificially intelligent computer system, can more effectively use the patient's health record and relevant medical information to provide crucial health-care solutions. Apple, Medtronic, Johnson & Johnson, Qualcomm, Amazon, and various other organizations are investing in their smart health solution research and development on a large scale. Google has established a life-sciences division fully dedicated to the development of novel technologies in healthcare.

In this chapter, we consider two important types of smart health technologies: (1) Ambient Intelligence (AmI) and (ii) Care Robotics. These two smart health technologies are being considered because both are common, easily relatable, and have direct interactions with persons which impact on fundamental human rights in a demonstrable way. Ambient Intelligence (AmI) originated in The Netherlands at Royal Philips in 1998 when a group of individuals, including Eli Zelkha and Brian Epstein (who coined the term "Ambient Intelligence" along with Simon Birrell), described a future in which homes have a distributed intelligent network of devices that provide information, communication, and entertainment (Aarts and Wichert 2009). Robots in healthcare have two main functions. Those which aid in the overall systems such as in hospitals and those which assist an individual, such as a sufferer of disease, to

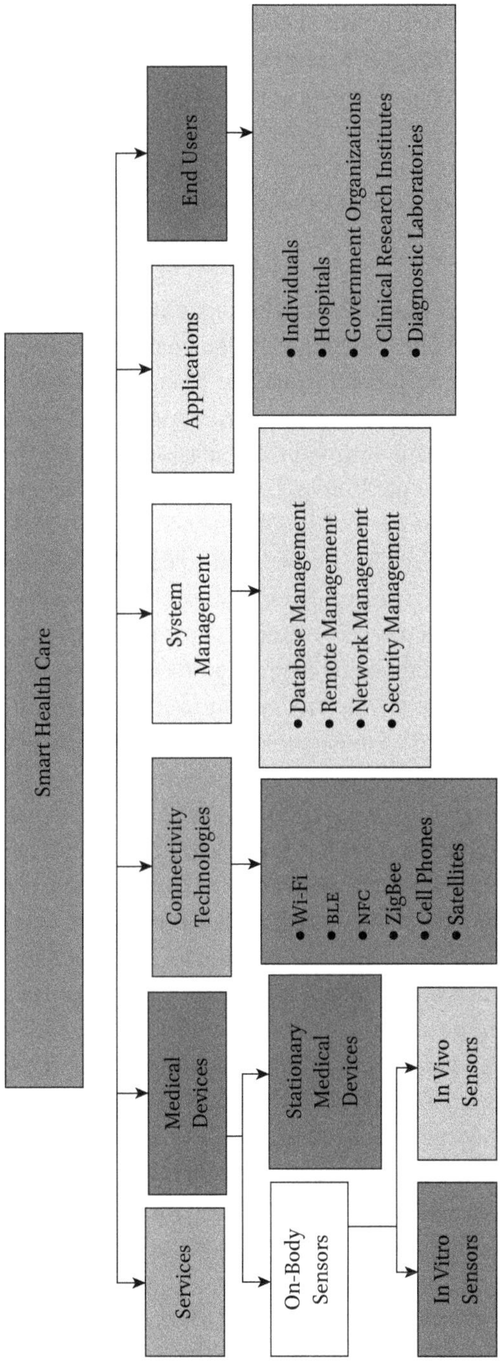

FIGURE 5.1 A classification of smart health care. BLE: bluetooth low energy; NFC: near-field communications
ADAPTED FROM SUNDARAVADIVEL ET AL. (2018)

keep a constant check on whether someone is safe and well (Cresswell, Cunningham-Burley, and Sheikh 2018). The latter type of robotics in healthcare is known as care robotics. In the following sections, we discuss the design considerations of smart health technologies to guide future design goals.

3 Looking Forward: Design Considerations of Smart Health Technologies

The design considerations for smart health technologies can broadly be classified into functional and non-functional design goals. We capture these design goals in the POPV design model. Functional design goals are targeted at the specific Processing and Operational (PO) requirements of the smart health systems. For example, accurate and real-time sensing, efficient data collection, processing, and storage are all PO requirements. Functional requirements also include ubiquity, anticipatory, and intelligence. The functional goals change with the application at hand. Non-functional goals are comparatively generic and may apply in a broader context. Such goals capture a system's Performance and Value (PV) requirements. For example, cost-effectiveness, performance metrics, adaptivity, and value sensitivity are all PV requirements meeting non-functional goals. Performance metrics can further be based on software and hardware evaluation metrics. For example, efficient smart care systems require low power, quality of service, higher efficiency, high availability, high resilience, and better storage and connectivity (Mieczkowski et al. 2019; Ikuta, Ishii, and Nokata 2003; Vandemeulebroucke, de Casterle, and Gastmans 2018; Lupton 2013). The value requirements include ethics, value sensitivity, and personalization. The POPV model is presented in Figure 5.2.

In ambient intelligence design, the sensors are the key elements. Sensors have different classifications (Olivier et al. 2009). Sensors can be active or passive, pervasive and non-pervasive, wired or wireless, and so on. Therefore, the sensor network design is an important consideration. Other important considerations include context awareness (AmI can recognize the user and their situational context), personalization (AmI can be tailored to user needs), adaptiveness (AmI can change in response to user) and anticipatory (AmI can anticipate user needs without conscious mediation) and natural interfaces (AmI can provide access to a user interface to communicate needs and preferences).

Similarly, care robots can provide physical assistance such as rehabilitation support and mobility support. For example, the "My Spoon Robot" can help someone with eating problems and the "Sanyo Bath Robot" can ensure hygienic

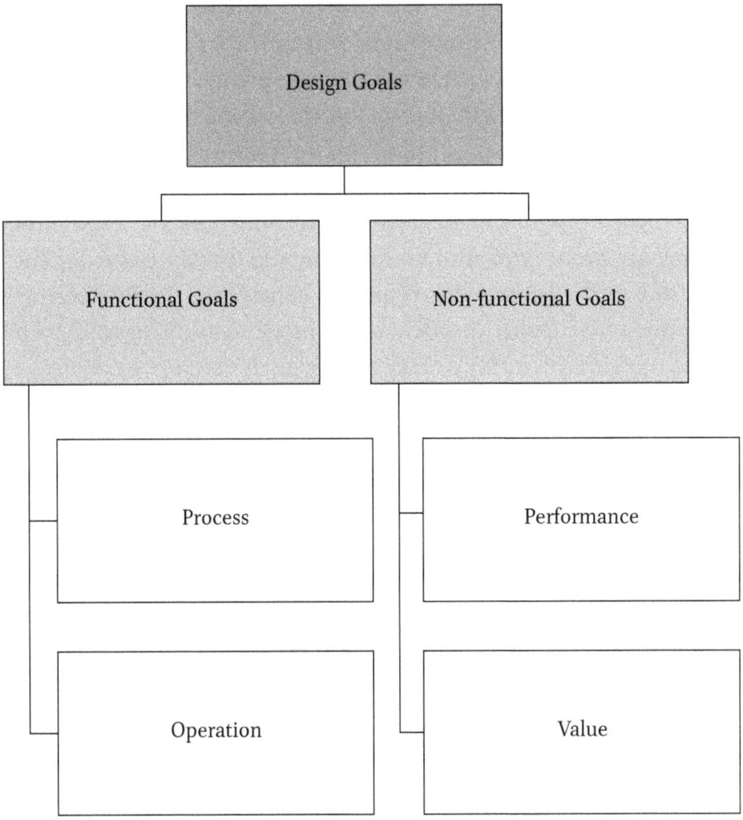

FIGURE 5.2 Classification of design goals for smart health technologies

care to elderly (Bedaf, Gelderblom, and de Witte 2015). Care robots can also provide social assistance through social interaction with users. This assistance can be provided through friendly conversation, gesture understanding, task collaboration and so on. For example, the dog-like robot AIBO and the seal-like robot Paro (Bemelmans et al. 2012) can provide social support in a way an animal would. Some care robots can provide both functionalities like the human-like robot Robovie and the robot Pearl (Kachouie et al. 2014).

The social interaction element of healthcare is arguably the most intricate and impactful on fundamental human rights due to the complexity of human behaviours. The design considerations of a socially competent smart health technologies, such as care robots, are therefore the most challenging. Care robots should be capable of high-level dialogue-based communication, automated recognition of natural cues (e.g. face, action, and gesture), emotional expression (perception and demonstration), recognition of other related agents like objects, an exhibition of distinctive personality and character, and user-friendliness. While providing social assistance, a care robot must respect fundamental human rights such as the rights to privacy and protection of personal data, as well as the freedom from discrimination and freedoms of belief, thought, conscience, religion, opinion, and information.

However, one could also argue that the physical interactions and the relevant design considerations of smart health technologies are more important. That is, they are more impactful on fundamental human rights for the same reason i.e. due to the complexity of human behaviours. A physically-assistive care robot must be capable of safely handling persons, recognizing a person, adapting to the physical metrics of users, foreign or obstructive object recognition and avoidance, fall detection, and so on. When providing physical assistance, a care robot must be designed to respect fundamental human rights in care such as the rights to dignity, personal security and privacy, as well as the freedom of movement and freedoms from inhuman or degrading treatment.

In the context of the discussion in this chapter, we particularly focus on the value requirements (captured as non-functional goals in the POPV design model) and value sensitive design considerations of smart health technologies. This is because the value requirements are those which highlight the importance of caregiver responsiveness in smart health technologies.

4 Beyond Fundamental Rights

In the introduction, smart health technologies were compared to human caregivers. This comparison lies in the shared capacity for smart technologies and human caregivers to be aware and responsive to the care context. Both gather information about the situation and use it to make dynamic, person-centred decisions that ultimately work towards achieving fundamental rights in care. Here, this shared ability will be shown with scenarios which highlight the responsiveness of smart health technologies. The goal here is not to say that smart health technologies are better than human caregivers at achieving, but to try and show how smart health technologies are on par with humans at delivering care in narrow situations. In no way is the claim that smart health technologies are as well-rounded as human caregivers, instead, the claim is that in some cases carefully tailored smart technologies can match the awareness and responsiveness of human caregivers.

How can one compare smart health technologies to a human caregiver? To answer this, the idea of what it is to be a good care provider with *caregiver responsiveness* needs to be expanded upon to ensure that it is clear how smart health technologies and human caregivers are relatable. Here, we are focusing on the non-functional goals and value requirements of smart technologies (see Figure 5.2). Caregiver responsiveness is grounded in the *care ethics* tradition. In care ethics, good care decisions are made by responsive caregivers reacting dynamically to both the situation and the interests of those receiving care. Care ethics states that good care decisions, including those which positively impact fundamental rights of care recipients, are made in situ by the caregiver. To explain it in another way, good care decisions come from a caregiver's subjective interpretation of each situation and a care recipient's unique needs. That interpretation determines how a caregiver ought to behave to ensure those needs are met (Upton 2011; Vallor 2011; Vanlaere and Gastmans 2011).

4.1 *Care Ethics*

Care ethics presents principles that describe moral behaviour and action when taking care of someone (Vanlaere and Gastmans 2011). A good caregiver must make decisions affecting those in their care during the practice of caring (Upton 2011). In care ethics, it is considered ethical if a moral decision arises from the "good" which is internal to practice, rather than external normative moral criteria or principles (Vallor 2011). Care ethics are determined by a carer's interpretation of a patient's individual needs as observed in the carer-patient relationship. That is to say, care ethics are determinative in practice (Beauchamp 2004). Expanding on this, for care to be good one needs to be aware of the entire scenario and those involved. Moreover, a good caregiver must have the appropriate knowledge (including training and experience) to react to each scenario appropriately. It is for these reasons that caregiver responsiveness is essential to care. Moreover, it is for these reasons that healthcare technologies require responsiveness also and why the "smart" part of smart health technologies is essential to good technology-delivered care. Just as human caregivers require the appropriate knowledge to respond to the situation and care recipient needs correctly, smart health technologies also require such knowledge. Section 5 will discuss how smart health technologies can be designed with that knowledge, along with the capabilities to gather further required knowledge about the situation at run-time.

Fundamental rights are perfectly explicit, but they are also perfectly broad. Look at the right to dignity. Broadly, the right to dignity declares that each person has an inherent, fundamental right to dignity. But, specifically, what is dignity? Dignity is a difficult concept to define (Jacobson 2007, Gallagher et al. 2008, Holmerova et al. 2007, Horton 2004, Marmot 2004). Answering this question would require a whole book itself and so we will not answer it here, especially since it has been answered by many before us, such as Kant (1996),

Sacks (2009), Jacelon et al. (2004), and Caygill (1990), with varying definitions. In defining dignity, Gordon (2018) reasons that the question we should be asking is "what is indignity?." It is easier to think about dignity in terms of the possible cases of indignity which should be avoided, rather than trying to positively define dignity.

Here, the reason we present the question of "what is dignity?" is to demonstrate both the difficulty of defining dignity itself as well as the difficulty of putting forward instructions for caregivers on how to ensure the dignity of care recipients. How is a caregiver supposed to interact with a care recipient to assure their dignity? This is the importance of care ethics, answering this question depends on the situation and the persons involved. Just as human caregivers must observe good care ethics and make responsive, dynamic decisions in situ to ensure fundamental rights of care recipients, smart health technologies must do so at run-time.

4.2 *Ambient Intelligence as a Smart Health Technology*

Consider the following scenario. Margaret is 85, she lives alone in a senior living community in her own private apartment. Margaret's apartment is equipped with ambient intelligence technology, specifically a sensor network which uses thermal sensors to monitor the apartment and detect falls. If the system detects a human-sized motionless thermal signature on the floor, then it presumes that it is a motionless, collapsed Margaret and sends an emergency alert to the on-site caregivers so they may intervene.

This sensor network sounds very beneficial, however, there are some potentially negative impacts on fundamental human rights in care. What if Margaret has a large dog which sets off the emergency alarm when it is sleeping on the floor? Or, as an extreme case to demonstrate the importance of value sensitive systems identifying seemingly insignificant details to improve the overall quality of life of users, what if Margaret does yoga and sets off the alarm when she is holding a ground pose for an extended period? Having a caregiver knocking on the door, or potentially charging inside without notice, violates Margaret's rights to privacy, dignity, rest and leisure, and adequate living standard, as well as her freedom from degrading treatment. These violations occur because the sensor network lacks the appropriate awareness and responsiveness of a good caregiver.

So, how do we make sure that there are no negative impacts on fundamental human rights? The answer is smart health technologies. Transforming the sensor network in Margaret's apartment into healthcare technology which is "smart" – i.e. having the capacity to learn from its environment and respond to the users appropriately (caregiver responsiveness) – can be done as follows so it protects Margaret's fundamental rights in those exemplary instances mentioned above.

If the network detects a motionless, floor-bound thermal signature it should do the following two checks before sounding the alarm.

1. Being aware that Margaret has a large dog which it might be detecting it searches for an additional thermal signature which is moving, standing, or raised off the floor (e.g. a person who is sitting on a couch, at a table, or lying in bed). If it finds such a signature, it assumes that this is Margaret and that the floor bound signature is Margaret's dog. Thus, it does not sound the alarm. As an additional check, the system emits high-frequency sound that only a dog could hear. If the floor-bound thermal signature moves at that moment, then the system confirms its assumption that it is Margaret's dog reacting to the high-frequency sound. Knowing that the dog is in that exact area showing a motionless, floor bound thermal signature, it will not need to emit the high-frequency sound again to check if it is a fallen Margaret. If the system did not adhere to this logic, then it would keep disturbing the dog which in turn might negatively impact Margaret's right to rest and leisure if she has a distressed dog. If the dog moves (e.g. it wakes up), this check will be reset and made ready for future use.
2. Being aware that Margaret does floor yoga which might cause her to appear as though she has fallen if she is staying in one pose for too long, the network system will attempt to learn Margaret's yoga routine. In doing so, it will be aware of the shapes of the yoga poses and the time intervals between pose changes. Knowing this information, the system can be more flexible, extending the time it waits to call for emergency human intervention after detecting a motionless, grounded thermal signature.

As an example of the additional benefits of having a smart sensor system which upholds the fundamental rights of care recipients, consider the medical advantages. As an additional feature, the system could be used to detect lower or higher-level than normal body temperature which might indicate a medical issue Margaret is having. This functionality works towards ensuring Margaret's fundamental right to healthcare. When it detects rapid fluctuations or long-term changes in body temperature, the system passes that information onto the patient directly through an email service advising them to bring the issue up with her doctor. The system communicates directly with Margaret to protect her privacy, as opposed to with the doctor which might be a meaningful breach of privacy (obviously this is a point which is up for debate in the medical ethics field, not a decision for smart technology to make). Regardless of the debatable privacy issue, directly communicating with the doctor instead of

Margaret is certainly interfering with both the right to protection of personal data and the right to an adequate living standard. In terms of responsiveness, however, if Margaret indicates that she is happy for the system to pass the health information straight onto the doctor, then this should be done by the system. This is done in response to her desire for the system to uphold her right to rest and leisure without having the system disturb her with constant updates on her fluctuating temperature.

Here, we have looked at fundamental rights in care and caregiver responsiveness as they relate to ambient intelligence technology. Next, we will go into depth about those care requirements pertaining to smart care robots. Before that, however, we will identify traditional design approaches being used to create care robots and demonstrate how they fail to meet caregiver responsiveness. We will then introduce a recently developed care robot design approach and demonstrate how it can be used to achieve smart care robots which are responsive to care recipient needs and values and inherently good at respecting fundamental rights in care.

5 Recommendations for the Future

Now that we know what caregiver responsiveness looks like in smart health technologies, the question is, how do we *design* smart health technologies to provide that responsiveness which in turn works towards ensuring fundamental rights in care? Here, we propose using Values in Motion Design (VMD) (Poulsen, Burmeister, and Kreps 2018), formerly the attentive framework (Poulsen and Burmeister 2019), to design those technologies. Before we explore VMD, let's look at what current design approaches are being used and how they can be used to create smart health technologies. To do this, we will use care robots as a running example.

Design approaches describe how a designer ought to undergo the practice of design ethically. Each approach provides practical design practices for one or more of the following for the design process: data gathering, data analysis, discussion of data, and application of data to the design process.

5.1 *User-Centred Design*

User-centred design (UCD) is a design approach which brings users into the design process (Duarte and Guerra 2012; Vredenburg et al. 2002), by doing so it hopes to create products which are focused around improving the user experience (Strömberg 2005; Abras 2004). UCD could be seen as a subset of human-centred design which "puts human needs, capabilities, and behaviour first, then designs to accommodate those needs, capabilities, and ways of behaving" (Norman 2013). UCD intends to offer practitioners design processes to create a design which is shaped by the influence of users, placing them at the centre of the system design and thus making the system easier to use and more useful (Duarte and Guerra 2012, Putnam et al. 2016, Brunner et al. 2017).

UCD also intends to capture details about users to directly determine the final product, it does this by involving the users in the design process (Duarte and Guerra 2012; Vredenburg et al. 2002) and focusing on the user experience (Strömberg 2005; Abras 2004). The involvement of users is key to creating a clear understanding of user and task requirements (Vredenburg et al. 2002); this process is commonly referred to as participatory design. UCD emerges from the need for designers to have a deep understanding of a user's goals and needs, as well as the users themselves (Chammas, Quaresma, and Mont'Alvão 2015).

Users fit into UCD in three broad ways during care robot design. First, users are needed for the data gathering and conceptual stages of care robot design to determine user requirements which can then be used to inform the first care robot design and physical prototype (Abras 2004). Second, users can participate in usability testing to determine if the robot design suitably meets user requirements (Abras 2004). Third, users are involved in the design process, acting as design partners next to designers who are to make sure that user requirements are integral to the design (Abras 2004). In this way, the care robot is designed to be user-centric.

UCD has been utilised for designing socially assistive long-term care robots (Di Nuovo et al. 2018), elderly care service robot interactions (Mast et al. 2012), and every day, domestic robots in a care setting (Sung 2011).

5.2 *Value Sensitive Design*

Value sensitive design (VSD), another design approach, aims to produce systems which are sensitive to indirect and direct stakeholder's values by involving them in the design process (Friedman, Kahn, and Borning 2006). VSD accounts human values in the design process of information systems (van Wynsberghe 2016a; Cummings 2006; Friedman 1996; Friedman, Kahn, and Borning 2006; Friedman, Nathan, and Yoo 2016; Manders-Huits 2011; van Wynsberghe 2013a). Friedman, Kahn, and Borning define a value as "what a person or group of people consider important in life" (2006). They go on to say that a value is substantiated by the interests and desires of human beings within a cultural milieu, that they cannot be motivated by an empirical account of the external world (Friedman, Kahn, and Borning 2006); the focus on the subjective interpretations of design stakeholder values is core to VSD.

Umbrello and De Bellis (2018) explain that VSD is a unique design approach because it is proactive in such a way that it encourages predicting emerging values and issues and realizing solutions in designs. Another advantage of VSD is that it invites a multidisciplinary approach to better address the diverse complexities of design with the involvement of philosophers, ethicists, social scientists, behavioural scientists, computer scientists, and designers (Friedman 2015). VSD realises values and incorporates them into design via its tripartite methodology consisting of conceptual, empirical, and technical investigations.

VSD has also been applied to care robot design. It has been used for designing seizure detection systems (van Andel et al. 2015), conducting a retrospective analysis of therapeutic care robots (Melson et al. 2005), and in humanoid care robot design (Cheon and Su 2016).

5.3 *Care Centred Value Sensitive Design*

UCD and VSD are traditional approaches often used in the design of technologies, including care robot design. However, there are approaches found in the literature that are tailored to care robot design due to the failings of traditional approaches. VSD is often criticised for lacking an explicit normative ethical theory that a designer can call upon to deal with value trade-offs (Manders-Huits 2011; van Wynsberghe 2013a). Additionally, Manders-Huits (2011) also finds fault with the traditional VSD approach as it relies on broad values. Value trade-offs have been discussed in the engineering, philosophical, and design literature, but little of that discussion has so far addressed care robots (Spiekermann 2015; Friedman, Hendry, and Borning 2017; van den Hoven 2017; van Gorp 2007).

To deal with the faults of VSD, as well as to make VSD more suited to the design of care robots, van Wynsberghe developed care centred value sensitive design (CCVSD) (2013a, b, 2016b, 2016a). CCVSD is a care robot design approach which expands upon the VSD approach, grounding it in normative care ethics. In doing so, designers have normative ethical grounds from which to make value trade-offs, as well as a narrowing of broad values to care values. It aims to explicitly and systematically incorporate care ethics into the design process of robots (van Wynsberghe 2016a). The CCVSD approach to care robot design brings together VSD and the care ethics perspective to answer the following, "how can values be included in the design of (care) robots and which values ought to be included?" (van Wynsberghe 2016a). In answering that question, van Wynsberghe provides designers with two tools, the care centred (CC) framework and the CCVSD methodology. These tools encourage the design of care robots by focusing on the required attentiveness, competencies, components, and responsiveness of robot caregivers.

5.4 *Values in Motion Design*

UCD, VSD, and CCVSD are all adequate approaches for designing care robots. However, the robots created are not what would be described as "smart" care robots. That is, care robots created with these design approaches are not smart health technologies inherently. Traditional design approaches can design for fundamental rights in care, but, they lack explicit design considerations for caregiver responsiveness (Poulsen, Burmeister, and Tien 2018). One design approach that is purposefully catered to designing smart care robots for ensuring caregiver responsiveness and fundamental rights in care is Values in Motion Design (VMD).

VMD is an innovate design approach, derived from CCVSD, which suggests how to design for fundamental rights and caregiver responsiveness in care robots. VMD grounds itself in fundamental rights, professional ethics (e.g. caregiver ethics), and care ethics. For design purposes, fundamental rights and professional ethics are represented as intrinsic care values, whereas care ethics are captured as extrinsic care values. Intrinsic care values are the normative, end goals of care (e.g. rights in care). Extrinsic care values are the unique needs and wants of each care recipient, these tell a caregiver how they ought to deliver care to best reach the end goals for each person in their care; extrinsic care values can only be captured subjectively with caregiver responsiveness. As extrinsic care values are grounded in care ethics, they also capture care ethical values such as those influenced by one's culture. As an example, LGBTIQ+ elder care values should be accounted for in the design of elderly care robots (Poulsen, Fosch-Villaronga, and Søraa 2020).

Value sensitive design features (hardware and software) to respect intrinsic values should be unchanging and guarantee fundamental rights and professional ethics. Moreover, those design features to respect extrinsic values should adapt and respond to the user and respect care ethics.

To create smart, responsive care robots, VMD takes extrinsic values to be constantly changing and recommends that they are respected dynamically in situ. A care recipient's values change over time as they age, have interactions with new technologies, and so on (van Wynsberghe 2013a; Sharkey 2014). Thus, care robots should be responsive and adaptive, and this could be done by designing them to interpret extrinsic values and value priorities. This could be achieved with a graphical user interface, voice recognition, or artificial intelligence (Poulsen and Burmeister 2019). Accounting for each care recipient's value interpretations and different and dynamic value priorities, as well as the differences in value types, is central to VMD. Moreover, accounting for user values is essential to ensuring that a care robot is capable of caregiver responsiveness and respecting fundamental rights in care.

To follow is a demonstration of how VMD works in practice to create a smart care robot capable of caregiver responsiveness and respecting fundamental rights in care.

Take the care practice "bath care recipient" for example. Without VMD, using traditional design approaches, designers may program a robot to close the curtains around the bath to respect to care recipient's dignity. However, what if the care recipient feels threatened by that, or they feel alone with a robot they do not entirely understand or trust, or they do not like enclosed spaces, or they feel like they are endangered because they are hidden from a human caregiver seeing them if they were drowning, or perhaps they would like to talk to their partner in the bathroom doorway.

Such a care robot is not very "smart" nor is it being responsive to the needs and wants of care recipients or their fundamental rights. Instead what if intrinsic and extrinsic values were declared as is done in VMD. One could examine the fundamental rights in care or nursing codes of ethics to reveal the relevant rights and intrinsic values in bathing a care recipient. For example, one can identify the right to an adequate living standard which encapsulates the intrinsic values of hygiene and safety. Then if caregivers and care recipients are consulted, the following extrinsic values could be discovered: privacy, dignity, social contact, emergency help, happiness, touch, eye contact, independence, and comfort.

The intrinsic values should be embedded into the design of the care robot, the assurance of them should not be conflicted by any other possible action. In this case, to ensure the value of hygiene the robot could have a cleaning function to wash the care recipient, have cameras to see any dirt on the care recipient, arms and hands to hold a sponge, and so on. To ensure safety, the cameras might observe the care recipient to ensure their head is above water and to see any distress, one arm could be at the ready to grab the care recipient under their arm to pull them up if they are drowning, that arm could be covered in a soft material so it cannot hurt the care recipient, and so on.

Now that hygiene and safety have been implicitly embedded into the design, one can consider the extrinsic values. Such values should be represented explicitly in the robots physical and information systems. One can design any number of functions/behaviours/features (that can even conflict with each other) and provide them to the care robot for it to determine during run-time which is the most appropriate behaviour to respond to the care recipient's desires i.e. the care robot has caregiver responsiveness and makes dynamic decisions based on its awareness of user preferences. For privacy, the robot could avert its anthropomorphic eyes and could delete any related video recording files (perhaps to ensure the intrinsic value of personal security, those video files might be kept for a few hours after the bath just in case the care recipient is hurt, and the recordings are needed for liability). The robot ought to be able to sense when an injury occurs to account for that instance. For dignity, the robot could shut a curtain around the bath. For happiness, the robot could turn on the radio or tv, or even talk to the care recipient. And so on. Such design features may conflict because the care robot is to observe care recipients and decide which behaviour is more suited to their needs and values during run-time. See more design considerations in Table 5.1 below.

TABLE 5.1 A care recipient bathing, smart care robot design built with VMD

Value	Value sensitive design feature
Intrinsic value	
Hygiene	A cleaning function to clean the care recipient. Cameras to see any dirt on the care recipient. Arms and hands to hold a sponge.
Safety	Cameras could observe the care recipient to ensure their head is above water and to see any distress. One arm could be at the ready to grab the care recipient under the arm to pull them up if they begin to drown. Arms could be covered in soft material, so it cannot hurt the care recipient.
Extrinsic values	
Privacy	Avert anthropomorphic eyes. Additionally, delete video recordings (a few hours after the bath just in case the care recipient is hurt, and the recordings are needed for liability. The robot ought to be able to sense when an injury occurs to account for this).
Dignity	Shut a curtain around the bath.
Happiness	Turn on the radio or tv. Talk to the care recipient.
Eye contact	Feature anthropomorphic eyes that follow the care recipients when they are talking.
Touch	Have a hand which is made of soft material that can gently hold the care recipient.
Independence	If it is deemed safe, ask the care recipient if they want to perform the care practice on their own. Stay with them and watch them to ensure they do not drown and that the care practice is done sufficiently.

6 Summary

This chapter reviewed the fundamental rights in care as they are impacted by the use of smart health technology. We have provided a discussion about these rights and how caregivers go about respecting them with *caregiver responsiveness*. We further elaborated on smart health technologies with focus on two common implementations: ambient intelligence and care robots. Following that, we framed the important design considerations of smart health technologies and identified how caregiver responsiveness is captured in the proposed POPV model. In addition, smart health technology design approaches with the case of smart care robots were presented. It included three contemporary design approaches: user-centred design, value sensitive design, and care centred value sensitive design. To address the failure of these design approaches to create care robots which are "smart," we introduced Values in Motion design (VMD). In the end, we demonstrated how VMD is the most feasible design approach capable of creating smart care robots i.e. those which have caregiver responsiveness and respect fundamental rights in care in situ.

References

Aarts, Emile, and Reiner Wichert. 2009. "Ambient intelligence." In *Technology Guide*, edited by Hans-Jörg Bullinger, 244–249. Berlin, Heidelberg: Springer.

Abras, Chadia, Diane Maloney-Krichmar, and Jenny Preece. 2004. "User-Centered Design." In *Encyclopedia of Human-Computer Interaction*, edited by William Sims Bainbridge, 1023–1027. Thousand Oaks: Sage Publications.

Beauchamp, Tom L. 2004. "Does Ethical Theory Have a Future in Bioethics?" *The Journal of Law, Medicine & Ethics* 32(2): 209–217.

Bedaf, Sanda, Gert Jan Gelderblom, and Luc de Witte. 2015. "Overview and Categorization of Robots Supporting Independent Living of Elderly People: What Activities Do They Support and How Far Have They Developed." *Assistive Technology* 27(2): 88–100.

Bemelmans, Roger, Gert Jan Gelderblom, Pieter Jonker, and Luc de Witte. 2012. "Socially Assistive Robots in Elderly Care: A Systematic Review into Effects and Effectiveness." *Journal of American Medical Directors Association* 13(2): 114–120.

Brunner, Julian, Emmeline Chuang, Caroline Goldzweig, Cindy L. Cain, Catherine Sugar, and Elizabeth M. Yano. 2017. "User-Centered Design to Improve Clinical Decision Support in Primary Care." *International Journal of Medical Informatics* 104: 56–64.

Caygill, J. 1990. "Dignity in Dementia." *New Zealand Nursing Journal* 83(2): 18–20.

Chammas, Adriana, Manuela Quaresma, and Cláudia Mont'Alvão. 2015. "A Closer Look on the User Centred Design." *Procedia Manufacturing* 3: 5397–5404.

Cheon, EunJeong, and Norman Makoto Su. 2016. "Integrating Roboticist Values into a Value Sensitive Design Framework for Humanoid Robots." The Eleventh ACM/IEEE International Conference on Human Robot Interaction, Christchurch, New Zealand.

Cresswell, Kathrin, Sarah Cunningham-Burley, and Aziz Sheikh. 2018. "Health Care Robotics: Qualitative Exploration of Key Challenges and Future Directions." *Journal of Medical Internet Research* 20(7).

Cummings, Mary L. 2006. "Integrating Ethics in Design Through the Value-Sensitive Design Approach." *Science & Engineering Ethics* 12(4): 701–715.

Di Nuovo, Alessandro, et al. 2018. "The Multi-Modal Interface of Robot-Era Multi-Robot Services Tailored for the Elderly." *Intelligent Service Robotics* 11 (1): 109–126.

Duarte, Joana, and Arminda Guerra. 2012. "User-Centered Healthcare Design." *Procedia Computer Science* 14: 189–197.

European Convention. 2012. "Charter of Fundamental Rights of the European Union." *Official Journal of the European Union* 2012/C 326/02: European Union.

European Court of Human Rights. 1950. "European Convention on Human Rights." Council of Europe. https://www.echr.coe.int/Documents/Convention_ENG.pdf.

Friedman, Batya, David G. Hendry, and Alan Borning. 2017. "A Survey of Value Sensitive Design Methods." *Foundations and Trends® in Human–Computer Interaction* 11(2): 63–125.

Friedman, Batya., David G. Hendry, Alina Huldtgren, Catholijn M. Jonker, Jeroen van den Hoven, & Aimee van Wynsberghe. 2015. "Charting the Next Decade for Value Sensitive Design." *Aarhus Series on Human Centered Computing* 1(1): 4.

Friedman, Batya. 1996. "Value-Sensitive Design." *Interactions* 3(6): 17–23.

Friedman, Batya, Peter H Jr Kahn, and Alan Borning. 2006. "Value Sensitive Design and Information Systems." In *Human-Computer Interaction and Management Information Systems: Foundations*, edited by Ping Zhang and Deniis Galletta, 348–372. New York: Routledge.

Friedman, Batya, Lisa P Nathan, and Daisy Yoo. 2016. "Multi-Lifespan Information System Design in Support of Transitional Justice: Evolving Situated Design Principles for the Long(er) Term." *Interacting with Computers* 29(1): 80–96.

Gallagher, Ann, Sarah Li, Paul Wainwright, Ian Rees Jones, and Diana Lee. 2008. "Dignity in the Care of Older People – A Review of the Theoretical and Empirical Literature." *BMC Nursing* 7(1).

Gordon, John-Stewart. 2018. "Indignity and Old Age." *Bioethics* 32(4): 223–232.

Holmerova, Iva, Bozena Juraskova, Z. Kalvach, E. Rohanova, M. Rokosova, and Hana Vankova. 2007. "Dignity and Palliative Care in Dementia." *Journal of Nutrition Health and Aging* 11(6): 489–494.

Horton, Richard. 2004. "Rediscovering Human Dignity." *Lancet* 364(9439): 1081–1085.

Ikuta, Koji, Hideki Ishii, and Makoto Nokata. 2003. "Safety Evaluation Method of Design and Control for Human-Care Robots." *The International Journal of Robotics Research* 22(5) :281–297.

Istepanian, Robert S.H., Songlin Hu, Nada Y. Philip, and Ala Sungoor. 2011. "The Potential of Internet of M-Health Things 'm-IoT' for Non-Invasive Glucose Level Sensing." *IEEE Engineering in Medicine and Biology Society* 2011: 5264–5266.

Jacelon, Cynthia S., Thomas W. Connelly, Rana Brown, Kathy Proulx, and Thuy Vo. 2004. "A Concept Analysis of Dignity for Older Adults." *Journal of Advanced Nursing* 48(1): 76–83.

Jacobson, Nora. 2007. "Dignity and Health: A Review." *Social Science and Medicine* 64(2): 292–302.

Kachouie, Reza, Sima Sedighadeli, Rajiv Khosla, and Mei-Tai Chu. 2014. "Socially Assistive Robots in Elderly Care: A Mixed-Method Systematic Literature Review." *International Journal of Human–Computer Interaction* 30(5): 369–393.

Kant, Immanuel. 1996. *The Metaphysics of Morals*. Translated by Mary J. Gregor. Cambridge, England: Cambridge University Press.

Kiourti, Asimina, and Konstantina S. Nikita. 2017. "A Review of In-Body Biotelemetry Devices: Implantables, Ingestibles, and Injectables." *IEEE Transactions on Biomedical Engineering* 64(7): 1422–1430.

Kuo, Tsung-Ting, Hyeon-Eui Kim, and Lucila Ohno-Machado. 2017. "Blockchain Distributed Ledger Technologies for Biomedical and Health Care Applications." *Journal of the American Medical Informatics Association* 24(6): 1211–1220.

Lupton, Deborah. 2013. "Quantifying the Body: Monitoring and Measuring Health in the Age of mHealth Technologies." *Critical Public Health* 23(4): 393–403.

Luxton, David D., Ryan N. Hansen, and Katherine Evans Stanfill. 2014. "Mobile App Self-Care Versus In-Office Care for Stress Reduction: A Cost Minimization Analysis." *Journal of Telemedicine and Telecare* 20(8): 431–435.

Manders-Huits, Noëmi. 2011. "What Values in Design? The Challenge of Incorporating Moral Values into Design." *Science and Engineering Ethics* 17(2): 271–287.

Marmot, Michael. 2004. "Dignity and Inequality." *The Lancet* 364(9439): 1019–1021.

Mast, Marcus, et al. 2012. "User-Centered Design of a Dynamic-Autonomy Remote Interaction Concept for Manipulation-Capable Robots to Assist Elderly People in the Home." *Journal of Human-Robot Interaction* 1(1): 96–118.

Mayer, Simon, Ruben Verborgh, Matthias Kovatsch, and Friedemann Mattern. 2016. "Smart Configuration of Smart Environments." *IEEE Transactions on Automation Science and Engineering* 13(3): 1247–1255.

Melson, Gail F., Jr. Peter H. Kahn, Alan M. Beck, Batya Friedman, Trace Roberts, and Erik Garrett. 2005. "Robots as Dogs? Children's Interactions with the Robotic Dog

AIBO and a Live Australian Shepherd." CHI '05 Extended Abstracts on Human Factors in Computing Systems, Portland, OR, USA.

Mieczkowski, Hannah, Sunny Xun Liu, Jeffrey Hancock, and Byron Reeves. 2019. "Helping Not Hurting: Applying the Stereotype Content Model and BIAS Map to Social Robotics." March 14, 2019. ACM/IEEE International Conference on Human-Robot Interaction (HRI), 11–14.

Mohanty, Saraju P., Uma Choppali, and Elias Kougianos. 2016. "Everything You Wanted to Know About Smart Cities: The Internet of Things is the Backbone." *IEEE Consumer Electronics Magazine* 5(3): 60–70.

Ni, Jianbing, Kuan Zhang, Xiaodong Lin, and Xuemin Sherman Shen. 2018. "Securing Fog Computing for Internet of Things Applications: Challenges and Solutions." *IEEE Communications Surveys & Tutorials* 20(1): 601–628.

Norman, Donald A. 2013. *The Design of Everyday Things: Revised and Expanded Edition*. New York, NY: Basic Books.

NSW Department of Health. 2009. *Respecting Patient Privacy and Dignity in NSW Health*. Sydney, Australia: Australian Government.

Nursing and Midwifery Board of Australia. 2018. "Code of Conduct for Nurses." https://www.nursingmidwiferyboard.gov.au/Codes-Guidelines-Statements/Professional-standards.aspx.

Nursing and Midwifery Council. 2002. "Nursing & Midwifery Council Code of professional conduct." https://onlinelibrary.wiley.com/doi/pdf/10.1002/9780470774564.app1.

Olivier, Patrick, Guangyou Xu, Andrew F. Monk, and Jesse Hoey. 2009. "Ambient Kitchen: Designing Situated Services Using a High Fidelity Prototyping Environment." Proceedings of the 2nd International Conference on Pervasive Technologies Related to Assistive Environments, Corfu, Greece.

Poulsen, Adam, Eduard Fosch-Villaronga, and Roger A. Søraa. 2020. "Queering machines." *Nature Machine Intelligence* 2, 152.

Poulsen, Adam, Ivan Skaines, Suzanne McLaren, and Oliver Kisalay Burmeister. 2020. "Value Sensing Robots: The Older LGBTIQ+ Community." In *Societal Challenges in the Smart Society* edited, by Mario Arias Oliva, Jorge Pelegrín-Borondo, Kiyoshi Murata and Ana María Lara Palma, 275-287. Logroño, Spain: Universidad de La Rioja.

Poulsen, Adam, and Oliver Kisalay Burmeister. 2019. "Overcoming Carer Shortages with Care Robots: Dynamic Value Trade-Offs in Run-Time." *Australasian Journal of Information Systems* 23.

Poulsen, Adam, Oliver Kisalay Burmeister, and David Kreps. 2018. "The Ethics of Inherent Trust in Care Robots for the Elderly." In *This Changes Everything – ICT and Climate Change: What Can We Do?* edited by David Kreps, Charles Ess, Louise Leenen and Kai Kimppa, 314–328. Poznan, Poland: Springer International Publishing.

Poulsen, Adam, Oliver Kisalay Burmeister, and David Tien. 2018. "A New Design Approach and Framework for Elderly Care Robots." Australasian Conference on Information Systems, Sydney, Australia.

Putnam, Cynthia, Aaron Reiner, Emily Ryou, Morgan Caputo, Jinghui Cheng, Mace Allen, and Ravali Singamaneni. 2016. "Human-Centered Design in Practice: Roles, Definitions, and Communication." *Journal of Technical Writing and Communication* 46(4):446–470.

Sacks, Jonathan. 2009. "The Dignity of Difference: Avoiding the Clash of Civilizations." *The Review of Faith & International Affairs* 7(2): 37–42.

Sharkey, Amanda. 2014. "Robots and Human Dignity: A Consideration of the Effects of Robot Care on the Dignity of Older People." *Ethics and Information Technology* 16(1): 63–75.

Spiekermann, Sarah. 2015. *Ethical IT Innovation – A Value-based System Design Approach*. New York: Taylor & Francis.

Sprint, Gina, Diane. J. Cook, Roschelle. Fritz, and Maureen Schmitter-Edgecombe. 2016. "Using Smart Homes to Detect and Analyze Health Events." *Computer* 49(11): 29–37.

Strömberg, Hanna, Jaana Leikas, V. Ikonen, Netta Iivari, Timo Jokela, & Natasja Leurs. 2005. "User-Centred Design Guidelines for Methods and Tools." *ITEA if 02019*: Nomadic Media.

Sundaravadivel, Prabha, Elias Kougianos, Saraju P. Mohanty, and Madhavi K. Ganapathiraju. 2018. "Everything You Wanted to Know about Smart Health Care: Evaluating the Different Technologies and Components of the Internet of Things for Better Health." *IEEE Consumer Electronics Magazine* 7(1): 18–28.

Sung, Ja Young. 2011. "Towards The Human-Centered Design of Everyday Robots." Doctor of Philosophy, College of Computing, Georgia Institute of Technology.

Umbrello, Steven, and Angelo Frank De Bellis. 2018. "A Value-Sensitive Design Approach to Intelligent Agents." In *Artificial Intelligence Safety and Security*, edited by Roman Yampolskiy, 395–410. CRC Press.

United Nations General Assembly. 1948. *The Universal Declaration of Human Rights*. Paris, France.

Upton, Hugh. 2011. "Moral Theory and Theorizing in Health Care Ethics." *Ethical Theory and Moral Practice* 14(4): 431–443.

Vallor, Shannon. 2011. "Carebots and Caregivers: Sustaining the Ethical Ideal of Care in the Twenty-first Century." *Philosophy & Technology* 24(3): 251–268.

van Andel, Judith, Frans Leijten, Hans van Delden, and Ghislaine van Thiel. 2015. "What Makes a Good Home-Based Nocturnal Seizure Detector? A Value Sensitive Design." *PLOS ONE* 10(4): e0121446.

van den Hoven, Jeroen. 2017. "The Design Turn in Applied Ethics." In *Designing in Ethics*, edited by Jeroen van den Hoven, Seumas Miller and Thomas Pogge, 11–31. Cambridge: Cambridge University Press.

van Gorp, Anke. 2007. “Ethical issues in engineering design processes; regulative frameworks for safety and sustainability.” *Design Studies* 28(2): 117–131.

van Wynsberghe, Aimee. 2013a. “Designing robots for care: Care centered value-sensitive design.” *Science and Engineering Ethics* 19(2): 407–433.

van Wynsberghe, Aimee. 2013b. “A method for integrating ethics into the design of robots.” *Industrial Robot: An International Journal* 40(5): 433–440.

van Wynsberghe, Aimee. 2016a. “Service robots, care ethics, and design.” *Ethics and Information Technology* 18(4): 311–321.

van Wynsberghe, Aimee. 2016b. *Healthcare robots: Ethics, design and implementation, Emerging Technologies, Ethics and International Affairs*. London, England: Routledge.

Vandemeulebroucke, Tijs, Bernadette Dierckx de Casterle, and Chris Gastmans. 2018. “How do older adults experience and perceive socially assistive robots in aged care: a systematic review of qualitative evidence.” *Aging Ment Health* 22(2): 149–167.

Vanlaere, Linus, and Chris Gastmans. 2011. “A personalist approach to care ethics.” *Nursing Ethics* 18 (2):161–173. DOI: 10.1177/0969733010388924.

Vredenburg, Karel, Ji-Ye Mao, Paul W. Smith, and Tom Carey. 2002. “A survey of user-centered design practice.” Proceedings of the SIGCHI Conference on Human Factors in Computing Systems, Minneapolis, Minnesota, USA.

Wolf, Gary. 2011. “What is The Quantified Self?.” https://quantifiedself.com/blog/what-is-the-quantified-self/.

Xu, Boyi, Li Da Xu, Hongming Cai, Cheng Xie, Jingyuan Hu, and Fenglin Bu. 2014. “Ubiquitous Data Accessing Method in IoT-Based Information System for Emergency Medical Services.” *IEEE Transactions on Industrial Informatics* 10(2): 1578–1586.

PART 2

Socio-political Challenges of Smart Technologies

CHAPTER 6

Rules for Regulators

Jonathan Wolff

1 Introduction[1]

The theory of regulation relies on a premise – more often implicit than explicitly stated – that regulation should be relative to, or at least a response to, risk. That is to say, unless regulation is to be pointless, it should have some purpose, and that purpose should be to reduce the probability of some undesired event happening. Good regulation cannot happen except by chance unless we have first identified the risk against which we are regulating. Excellent regulation reduces risk in a proportionate, economical way by providing well-designed specific instructions to specific actors who follow those instructions with (only) the intended effects. Ideally it will be transparent, and simple, although these ideals can be hard to reach and may even be in tension with each other.

If good regulation is regulation that reduces risk in the ways I have stated then there are many ways in which it can fail. Possibly the worst regulations are those that do not reduce risk but increase it, at significant cost. This can be aggravated if not only is it ineffective, but its costs fall unfairly on those who are already vulnerable in some way. But as we will see below, there are many ways in which regulation can fail to meet its intended purposes.

My aim in this paper is to raise the question of how smart technologies, including robotics and artificial intelligence, should be regulated given some of the things we know about good regulation and regulatory failure. In a paper of this nature I cannot set out proposed regulations in any detail. My task, rather, is to explore some of the questions and considerations that need to be taken into account by those who make regulations. In the next section (Section 2) I shall set out an account of principles of good regulation, which is partially influenced by an approach to regulation that was produced by the UK Better Regulation Task Force (BRTF), founded around 1997 to improve the quality of regulation in the UK (see e.g. BRTF, 2003a; 2003b). Accordingly, in that section I sketch what I grandly call the "rules for regulators." Section 3 discusses and illustrates a series of problems regarding regulation in what we can call normal

1 I'd like to thank Penelope Rowlatt and John-Stewart Gordon for exceptionally helpful comments on an earlier draft, and Mark Hannam for many insightful discussions of these issues.

 | DOI:10.1163/9789004437876_008

circumstances, which occur when risks are fairly well understood. The problems I discuss are variations of "regulatory failure" which is when regulation simply fails to achieve its purposes of reducing risk.

If "normal circumstances" are those when risks are known and understood, not all risks of smart technologies fall into this category. Some do, such as those around data protection and the possibility that some personal information needed to drive smart technologies will fall into the hands of people able to exploit those personal details in ways that harm or exploit others. But there are also cases of smart, or other new, technologies where the risks can be known only in outline or not at all. For a comparable example from several decades ago, in discussions of nanotechnology there was speculation that nanomachines could self-replicate very rapidly outside human control and the world would be taken over by "grey goo" (Drexler 1986). Whether or not this was a genuine possibility, it was notoriously raised as a public concern by Prince Charles (Radford 2003), but brought back into perspective by the Royal Society which concluded that the age of self-replicating nanomachines was far off (Royal Society 2004). Yet at least when first brought to public attention, the risk was very vaguely formulated and its probability unknown. This is a typical situation for new technologies, where we can have a keen sense of unease, but the focus of that unease is extremely vague. Some forms of artificial intelligence and robotics are raising similar concerns now (Bostrom 2014). Hence in Section 4 I will briefly add some further considerations, drawing on other work I have done, on how to regulate in the face of radical uncertainty.

Section 5 identifies the risks said to be consequent on the adoption of smart technologies while Section 6 applies some of these lessons to the regulation of AI in robotics and information processing, although all I am able to do is raise some general considerations rather than go into details. However, it might seem that this discussion raises more questions than it answers. Regulation varies from legal system to legal system, covers different agents in different ways, can be enforced through different mechanisms, and failure can attract different penalties.[2] These are all urgent and important issues. However, they are beyond the scope of this paper. Before we can look at the details we need to have a better grasp of the general issues at stake, which is what I attempt here.

2 I'm grateful to an anonymous referee for raising these points. I intend to address them in future work.

2 Principles of Good Regulation

I mentioned that my approach to regulation has been influenced by the work of the (short-lived) Better Regulation Task Force, which was set up by the New Labour government to improve the quality of regulation in the UK. Business, as always, claimed to be "drowning in red tape" and the government resolved to streamline regulation to reduce the regulatory load. But rather than have a draconian and mechanical policy of the "cut two regulations for every new one" type that some governments have imposed, often with highly damaging results (Browne 2018), the idea was to look at issues in their individuality and complexity and see what could be achieved. Although any official body charged with dealing with regulation is likely to meet with great scorn and scepticism, the BRTF did produce some thoughtful materials.

BRTF in their policy guidance sets out "principles for better regulation" stating that regulations should be transparent, accountable, targeted, consistent and proportionate (BRTF 2003a). It is hard to argue against any of these aspirations, but these are broad values, rather than specific instructions and need to be made more concrete, which can be done in several ways. But the key inspiration for me in reading some of the BRTF material was the following remark: "The accountability burden in the higher education sector is out of proportion to the risk of financial or academic mismanagement" (BRTF 2003a, 7). Now, whatever we think of this particular example, it is clear that its argument makes the critical assumption that a key purpose of regulation (or accountability) is to reduce risk in an appropriate fashion. Once this rarely stated insight is absorbed, it becomes apparent that there are, therefore, many ways in which regulation can go wrong (to which I will return in the next section). Hence, we have the first principle of good regulation.

1. Before regulations are drafted it is essential to have a clear sense of the risk the regulation is intended to mitigate.

My suspicion is that regulation is often discussed with only a vague sense of its point, taken to be something about holding people or agencies "accountable." But free-floating accountability is an empty idea. People and agencies need to be accountable for something. And that something should be making due precautions against risk, or, in other words, not being reckless or negligent. And this, in turn, requires some form of risk assessment.

Now it may be that the risk cannot be well-specified at this stage, especially in the case of new technologies. But the limits to our understanding are part of the risk that needs to be appreciated. There can also be very reasonable disputes about the risks and their probabilities. Such disputes will need to be settled to some degree if regulation is to proceed, although it is likely to remain a contested issue. In many cases some stakeholders or observers will think that regulation overstates the risk, and others that it has not been properly appreciated. But there is no obvious way around this.

2. Regulation should address the risk identified.

This second rule may seem so obvious as not to need stating. But it is often violated, I believe. In the face of a newly appreciated risk the politically urgent thing can be to show that "something is being done," and hence putting forward hasty regulations, or regulations that require some sort of public show or display, but do not address the actual risk. Once more I will return to this in the next section.

3. Regulation should reduce the risk identified.

Again, this seems blindingly obvious, but there are cases where the regulation may well actually increase the risk, as again we will see below.

4. Regulation should not have side-effects that are more serious than the risk mitigation it achieves.

This is the familiar area of unintended consequences. Any regulation will incentivise new behaviours, or disincentivise existing ones, even though often they are outside the scope of the explicit terms of the regulation. What are those behaviours and what will be their effects? This can be a very difficult area, as many unintended consequences cannot be anticipated, or, worse, are understood but deliberately left outside the discussion. Several examples will be explored in what follows.

5. Regulation should give specific instructions to specific agents and it should be possible to verify whether those agents have carried out their instructions.

Here we have one way in which "accountability" can figure in regulation. Regulation can fail in that it is unclear who holds which responsibilities. This is the area of "risk ownership" which does receive attention in formal risk assessment by means of a standard "risk register." Although critically important in practice I will not discuss this further here.

6. Regulation should be proportionate.

In a way this is already covered by principle 4, but here I focus on the costs in terms of time and effort, rather than unwanted side-effects. This is also an area which is much discussed, and it will be touched on in what follows.

7. Regulation should be transparent.

It should be clear why the regulation is as it is. This will give those regulated more confidence that considered thought has gone into the design of regulation.

8. Regulation should not be more complex than required.

Over-complex regulation is costly in time and effort, and is less likely to be followed. Hence simplification is an obvious demand. Transparency and simplicity (the avoidance of complexity) are obvious and appealing principles. Yet it is clear that they can come into conflict. Although the example is outside the scope of regulation as normally understood, I came to understand the possible tensions between these rules when I was a member of a working party that aimed to improve my university's financial model. We were then operating with a complex model which tried to incorporate many factors, but had become extremely time-consuming for many departmental heads and managers to understand. As I understood it, in this context transparency was also a proxy for some notion of fairness and efficiency; that departments should be charged for the actual costs of the services they use. Otherwise some may be given a hidden subsidy and the model could incentivise behaviours where the cost falls elsewhere. To give an illustration of the difficulties, one area that was discussed was how departments should be charged for the library. It was suggested that as humanities students use libraries more than science students, humanities should make a larger contribution. To this it was countered that science publications are more expensive and therefore science departments should make pay more. A third view was that in the interests of simplicity, all should be charged the same, or even a fixed overhead (lump sum or percentage) to cover all central services, without trying to vary by cost or usage, which was more trouble than it was worth, even if a fixed overhead could be both unfair and inefficient. This type of discussion will be replicated again and again at all levels in many organisations, and illustrates how transparency and simplicity can push in different directions. It is a lesson learned again and again in welfare reform, for example.

But to recap, I have proposed eight principles of good regulation, albeit rules of thumb rather than precise instructions. I do not suggest that this is an exhaustive list. Other principles may also generally apply, or apply only in particular types of cases. But this provides a good starting point for thinking about examples of regulatory failure.

Some will argue that my account of regulation is misguided in that the point of regulation is not to avoid risk, but to achieve beneficial outcomes. Some think, for example, that business needs to be regulated in order to maximize economic gain. One area in which this arises is in the setting of product standards, of which there are very many. As has been persuasively argued, while standards can guard against misinformation, and reduce uncertainty, they can also build trust (Gordon and Fomin 2019). I am happy to accept this point, but in the end it may come down to the semantic question of whether the situation can equally be described as one of overcoming distrust. Nevertheless, even if it is insisted that there is a positive function to regulation beyond addressing risk, then, first, dealing with risk remains a very substantial part of regulation, and second, exactly the same considerations apply (*mutatis mutandis*) regarding the pursuit of the declared purpose.

3 Flawed Regulations

The idea that regulation can fail to meet its purpose is well-known, and everyone will have their favourite examples. There are at least six ways in which a regulation that aims to reduce a risk can fail to do so:

1. It can fail to address the risk.
2. It can increase total risk by generating new risks of the same kind.
3. It can increase risk by making us falsely believe the risk has been dealt with and thereby lead us to lower our guard.
4. Those subject to regulation may seek loopholes, leading to either no change to the risk or making it worse.
5. Although it may reduce the risk under consideration, it can lead to worse risks.
6. Regulation can be used to pursue other agendas.

The first case, of simply failing to address the risk, is the familiar "something must be done" phenomenon, or, as the BRTF put it, make a "kneejerk reaction." The BRTF give the example: "Gun control measures introduced following the tragedy at Dunblane School [1996] appear to have largely been ineffective in tackling gun crime" (BRTF 2003a, 7). This example is contested (Barnett 2017), but the general point that regulations can fail to address the problem is clear. In this case the argument would have been that criminals use guns that are unregistered or smuggled in and so tightening laws is irrelevant.

The worst case is where the measure simply does not address the risk at all. More likely, however, is that even a hasty measure will have some effect, even though very small. Consider, for example the possibility of a terrorist attack on a transport system. The authorities may take a number of measures, such as increasing the number of police and security staff, at large public expense. Yet it may be that the authorities know that the particular measures available to them do little in practice to reduce the risk, as any system has many points of vulnerability, and a security system is only as strong as its weakest element. An authority that argues that it will not increase security or surveillance on the grounds that it thinks that it will do little good in practice would be very brave. It would be akin to the medical practice of "watchful waiting" in medical services; acknowledging that symptoms exist, while admitting that there are no useful therapies available for the condition in its current phrase, but steps will need to be taken if it worsens. Not all patients accept watchful waiting and sometimes insist on useless fake remedies; similarly, the public would have to be very sophisticated to accept watchful waiting in security matters, and so steps are taken to reassure the public. I have argued before that reassuring the public is important, and should not be dismissed, even if the steps taken do not reduce risk. But they need to be understood as fear-reducing measures, rather than risk-reducing if we are to appreciate their true nature (Wolff 2006).

The second case is that of measures that could actually increase the risk. For a relatively trivial example, consider the proposal that has been discussed at several universities in the UK and the Republic of Ireland, and implemented in at least one for a time, of requiring that, in addition to the two examiners, every PhD examination has a third, senior, academic present as Chair of the examination. Why was this suggested? In a very small number of cases, candidates who did not get the result of the examination they wanted made an appeal. But as there can be no appeal against academic judgement such appeals are typically against failure of process. Appeals can be costly and time-consuming. To ensure the integrity of the process and reduce the number of appeals it was suggested that each examination should have a Chair present.

The most obvious objection was one of proportionality. In the UK every year there are tens of thousands of PhD examinations, but very few appeals. A simple cost-benefit analysis suggests that it is more efficient to allow a few appeals every year, than take the tens of thousands of hours of senior academic time that would be needed to run such a system. But a more subtle objection is that we have no evidence that the system would reduce the number of appeals. Paradoxically it could increase them. Once a Chair is appointed the process is already more complex, with more potential points of failure. Arguably, all of this time and effort could be spent to make things even worse.

Another example concerns a law introduced in the UK after the tragic death of the daughter of MP Jenny Tonge who was electrocuted after very poor electrical work was done in her house (Pook 2004). The new regulations, which came into force in 2005, required all electrical work to be undertaken or inspected by people who had achieved a level of certification (IET ND). Some electricians argued that this would actually increase risk. Good electricians didn't need regulation but the cost of certification put their prices up. Those who evaded certification would be able to offer their services more cheaply, and would more often be selected for the job. Hence, they argued, what looked like a perfectly sensible regulation would be counter-productive. I have no idea if this is true, but it does set out a possible consequence.

Examples abound of specific instances in which laws have been counter-productive. Laws designed to protect historic buildings lead to attempts to demolish some that would be very costly to maintain to the required standard, before the buildings are put on the register. The Royal Saline Works at Arc-et-Senans, France, now a UNESCO world heritage site, displays a notice saying that such a demolition was intended by previous owners, precisely to pre-empt registration and its associated costs, but luckily frustrated at the last minute.

Sadly, some very well-meaning welfare regulations can have a similar effect. When the UK government introduced security of tenure for individuals renting unfurnished accommodation, that particular rental market more or less dried up. Either properties were let furnished, but with furniture of a low standard, or, if unfurnished, only to companies (to whom the law did not apply), leading to a situation where, for example, a family business might take on a lease on behalf of a family member. These were typically wealthier people, who then had a further advantage in the rental market. Those most in need of a protected tenancy found it increasingly difficult to gain any tenancy in the first place (Shepperson 2018). Other claimed examples include water saving dual-flush toilets which, if poorly installed, are alleged to use more water because they often leak, and people do not always use them properly, as well as energy efficient buildings with low-powered heating, which can lead to people using energy-hungry portable electric heaters simply to get rooms to an acceptable heat. It is unlikely, perhaps, that there really is an overall negative effect for these technologies, but the current point is merely that measures introduced for a particular purpose can have the opposite effect, against which we must guard.

The third case is where a regulation can be believed to be effective but because it is not, it makes us complacent and thereby subject to higher risks. The example, previously mentioned, of regulation of electricians in the UK may be an example. If it is believed that an electrician must have met a safety standard in order to offer their services, we may be less vigilant in checking out their record. An extraordinary case, albeit outside the field of regulation, concerns the British businessman, Gary Bolton, convicted of selling "bomb-detectors" that were actually empty boxes and were completely ineffective. It may have been that deaths were caused because security services around the word used these fraudulent devices and thereby let bombs into areas they believed to have been secured (BBC 2013). Some bad regulations are like faulty bomb detectors believed to be reliable. The general idea is that regulation enhances trust, and thereby leads us to let down our guard and our focus. This can be a good thing when regulation is highly effective, but disastrous when it is not.

The fourth issue was that those subject to regulation may seek loopholes, leading to either no change to the risk or making it worse. This is very common, and perverse effects of regulation are very often likely. The pervasiveness of this response can be divined simply by reflecting on one's own behaviour. Anyone subject to regulation knows full well that their first reaction to any regulation they do not welcome will be to see how it is possible to carry on doing what they are doing in a way that "gets round" the regulation. These "side fixes" can make matters worse. People seek out loopholes, and when the stakes are high, such as in tax policy, it can be worthwhile for an individual or company to pay expert advisors to find ways of mitigating the effects of regulation, without breaking the law. But it is also possible to "go underground" to make their activity invisible to the regulator, thereby heightening rather than attenuating risk. There can be calls to keep regulation very light, precisely so those in an area will register what they are doing, and thereby can be kept under easier surveillance. Sometimes there are arguments that an activity should not be regulated at all, so those engaged in it will be happy to publicise their activities. This has been argued in the case of synthetic biology, for example. It is possible to set up a lab in a garage or basement that can make innovations in synthetic biology that can have far-reaching effects. In theory it may be possible to engineer new forms of virus, combining one that is easily transmitted with another with devastating effects on human beings. One obvious response is to have laws that ban, or at least heavily restrict this activity. But because it can be carried out on a small scale, with materials that, largely, are readily available as they have other uses, the probability is that the effect of any law will be to make people remain silent about what they are doing, rather than boasting or exchanging notes online as they do when the activity is permitted. In other words, when regulation is very burdensome, those who are most in need of regulation have a strong incentive to avoid it entirely (cf Kaebnick et al. 2014).

My fifth category of regulatory failure is when a regulation leads to consequences that, although different, may be worse. Consider, for example, a law in Vienna in the 18th Century that did not permit (most) men to marry unless they had reached a certain income level. The point of this, no doubt, was to ensure that women and children were well-provided for, and didn't become destitute or a burden to the state, which may seem reasonable goal in the historical circumstances (McKay 1996, 44). But the composer Franz Schubert, who was unable to marry because of his low, and irregular, income, complained that the law led to "misery or gross sensuality" (McKay 1996, 66), which in his own case included contracting syphilis which led to his tragically early death.

This is the notorious realm of unintended consequences, which can be both good or bad, but are more often noticed when they are problematic.

Finally, I want to add a somewhat different case of regulatory failure, where regulation is used to pursue a different agenda to reducing the direct risk. Here I have in mind the accusation that a regulatory body's own interests may interfere with appropriate regulation. There are at least two ways in which this may happen. The first is an application of the insight of the (rather cynical) Public Choice school of economics: that the first incentive of any employee – whether Professor, lawyer, civil servant, etc – is to ensure that they still have a job next year (Buchanan and Tullock 1962). Regulators regulate and hence need to create and enforce regulations if they are to have the greatest chance of remaining in employment. This is completely understandable, of course, but it leads to the danger of over-regulation, to make work that needs to be done. The main difficulty here is likely to be disproportionate regulation, but it is possible that unnecessary regulation can be ineffective, or even counter-productive.

The second is rather different, concerning the reputation of the regulator. To take an example from the UK, it has been suggested that some hazards are so severe that they fall outside the scope of normal range of risk cost-benefit analysis with standard values. These are sometimes called "societal concerns," which the UK Health and Safety Executive has explained in the following terms:

> Societal concerns or the risks or threats from hazards which impact on society and which, if realised, could have adverse repercussions for the institutions responsible for putting in place the provisions and arrangements for protecting people, eg Parliament or the Government of the day. This type of concern is often associated with hazards that give rise to risks which, were they to materialise, could provoke a socio-political response, eg risk of events causing widespread or large scale detriment or the occurrence of multiple fatalities in a single event. Typical examples relate to nuclear power generation, railway travel, or the genetic modification of organisms. Societal concerns due to the occurrence of multiple fatalities in a single event is known as societal risk. Societal risk is therefore a subset of societal concerns. (HSE 2001, 12)

Now, we might wonder at the choice of examples – is railway travel really in the same category as nuclear power generation and genetic modification of organisms? – but what interests me is the way the issue is framed. It concerns impacts which "could have adverse repercussions for the institutions responsible for putting in place the provisions and arrangements for protecting people." In other words, the regulator is worried about potential damage to the regulator, and therefore, it seems, is prepared to consider upping the regulatory burden on other entities in order to reduce the risk of such damage. From its own standpoint, this may seem reasonable. After all, it is important that the public trust the regulator. But first, if the regulator is seen as over-regulating, trust could be undermined, and second, the costs of high regulation to secure the regulator's reputation will be borne not by the regulator but by, ultimately, consumers.

4 Risky New Technologies

So far I have made some general points about regulation without considering the fact that new technologies often have unknown risks that put us outside traditional ways of dealing with risk by means of cost-benefit analysis. Genetically modified organisms and nuclear power are examples. I have written about this elsewhere (eg Wolff 2019), and will not dwell on the issues here in detail, although there are some specific considerations that apply to some new smart technologies.

In the general case of risky new technologies, there can be a great deal of concern, some soundly based, some not so. One problem is that hazards, and especially probabilities, are often unknown, and there is therefore no easy way of applying standard risk cost-benefit analysis. But even if we could apply standard methods, some of the risks maybe so extreme that they should be avoided, such as extinction of the human species, even if the probability is so slight the cost-benefit analysis supports taking the risk. Sometimes it is suggested that we should apply "the precautionary principle" to such cases, but, to my knowledge, there is no formulation of a precautionary principle that can play an adequate role. Rather, I propose that we should take a precautionary attitude, which means thinking through a series of steps. But the approach boils down to the following advice. If a new technology involves an appreciable risk of a highly significant, irreversible hazard, and there is no way of eliminating that risk, then we should only contemplate accepting that risk if the proposed technology solves a serious and genuine problem in a way we cannot solve without taking such a risk. Even then, we may well decide not to go ahead. But the key point is that unless a highly risky new technology solves an important problem that cannot be solved in other ways, then it hardly seems right to take such a risk. This, of course, is not a new idea, and similar thoughts can be found widely in the literature. (e.g. Jonas 1984; Hartzell-Nichols 2012; Hubert 2017).

This may seem a sane approach. Scientific curiosity or private commercial gain is not a sufficient justification for exposing the wide population to risks that are not, in some sense, necessary to take on in order to fulfil our important goals. But there is an assumption underlying this approach which makes it more problematic to apply to smart technologies. The assumption is that there is a single moment of choice where it is possible to decide whether or not to pursue the technology. In some cases, especially where a game-changing breakthrough is proposed, this is reasonable. The same applies to areas where government licencing is already a norm. But AI and robotics are not like that. They are incremental technologies, mostly proceeding by small steps, with no special moment where everything changes. This is not to diminish the truly significant changes. Lee, for example, repeatedly mentions Geoffrey Hinton et al.'s "Deep Learning" as the development that has made contemporary AI possible (Lee 2018). Whether its significance was appreciated at the time is another question, but even then it took time to develop the technologies based on the basic science, with many developments in many different times and places (and this multi-national development is another, separate, headache from the point of view of regulation). But the main point is that there is no particularly salient moment of decision, and once there are commercial developments and products on the market, the prospects of stopping the technology in its tracks are bare. When there are no clear potential stopping points, regulation, at most, can try to impose safeguards, to mitigate the most problematic effects.

But another point I have made elsewhere in connection with risky new technologies remains, and that is what we can call "regulatory drift." Where there is a new or highly specialised technology, whether by quantum leap or incremental improvement with increasingly significant effects, it is likely that there will be an initial round of intense public interest and scrutiny, and very cautious regulation. However, there are at least three reasons why regulations may change to be become less stringent over time. First, if there are no incidents, it may seem that we have over-regulated, and need to relax, which can lead to a level of complacency. Second, governments are often under pressure to "cut red tape," and look for places where, for symbolic reasons, they can show they are being business friendly. Sometimes regulations are removed entirely. Jude Browne has called this "regulatory gifting" (Browne 2018). And finally, and most pertinently to the case of high technology, it can often be difficult to find people with the detailed level of knowledge to know how to regulate the area, and have the interest in doing so (Ramanna 2015). Hence, once a new technology is approved under strict regulatory rules, conditions can settle down to a point where many of those with a continuing regulatory role have a background in the industry being regulated. This is a form of regulatory capture. Even if they no longer have a financial interest they may well have sympathies with the corporations involved in developing new products and perhaps will be less concerned about possible negative consequences than they ought to be. This type of regulatory capture, knowing or unknowing, is a real danger in any new technology, and adds a further level of risk, in that the regulators can have too much faith that the producers of the technology will be able to deal with potential problems. We have seen something similar recently in relation to the aviation industry (Steib 2019), and in the UK, building inspection (Kennedy 2018). Arguably all three mechanisms of regulatory drift – complacency, regulatory gifting, and regulatory capture – led to a high degree of semi-negligent self-regulation, with disastrous results in cases both of the 737-MAX aircraft and the cladding on Grenfell Tower. Hence, we can end up at a point where we believe regulation is in place, but in reality it does not really address the most important risks. In other words, regulatory drift is a particular example of the third case discussed in the previous section.

5 The Risks of Smart Technologies

Following the principles of regulation set out above, in thinking about the regulation of any technology the first issue must be to identify the particular risks for which regulation could be an appropriate response. The risks of different technologies will vary, of course, and this is not the place to look at technologies in detail. Earlier I mentioned AI and robotics as two related smart technologies that are thought to pose numerous risks. Although any way of dividing risks into categories can be somewhat arbitrary, here I want to divide risks of smart technologies into four broad and general categories: the individual; the economic/social; the existential; and the insidious. I will briefly outline these risks before going on to discuss, in the concluding section, what forms of regulation or mitigation could be possible.

By individual risks I mean the sort of risks that are typically associated with concerns that new technologies can lead to harms to individuals that would not have happened otherwise. Discussions concern whether the potential risks are serious enough to place limits on the technologies or to require some sort of other mitigation. Often the issues are familiar and widely discussed as they occur frequently in difference guises, and are subject to examination by academics, journalists, the broader public, and legislators, although there is no reason why there should not also be new individual risks.

There are many possible individual risks that come to mind in relation to AI and robotics (Whittlestone et al. 2019). One, obvious and familiar, source of risk concerns data: who owns it, whether it is used with consent, and how it is handled. Such issues have been widely discussed in bioethics, but data handling is especially critical in AI. After all, the generation and use of data is, as it has been put, the fuel on which AI runs (Lee 2018). Vast quantities of data are gathered and deployed, and access to that data can be very valuable. It can also be misused, or at least used for political and commercial purposes for which permission has not been granted and can be against the interests of the people from whom the data is collected. One risk, then, is simply that of privacy. A smartphone can gather data about an individual's lifestyle that could, for example, lead to an insurance or bank refusing to offer services or charging a higher price for them. This is the risk of what we could call "adverse use." The Cambridge Analytica scandal shows what can happen in a political context, where information about individuals can lead to tailored attempts to influence them with political propaganda (Cadwalladr and Graham-Harrison 2018). This is the problem of "political manipulation."

A futher risk in this category, is that of being an "unwitting experimental subject," which will apply to technologies such as internet medical services, whether diagnostic or providing services, and, perhaps, some forms of automated customer interface in consumer services. Unlike earlier technologies, such as mobile phones and microwave ovens, the immediate concern does not seem to be diseases, such as the cancer fears of microwave ovens, or at least not directly, but instead poor service delivery. Obviously, this could be very serious in the use of AI or remote services in medical care, and can also have highly detrimental effects for, say, the selling of inappropriate financial services, or the incorrect classification of people into different financial risk categories.

We must also take account of the fact that AI systems can be self-learning, and as a result process information in ways that cannot always be understood by human beings. This gives rise to two further types of risk. First, an AI system could produce a result that just seems arbitrary and unfair. For example, an algorithm could turn down an individual for a loan without anyone being able to understand why, but they would then have the further detrimental data point of being having turned down for a loan to be added to their record. Hence there is a risk of pernicious and unfair victimisation. Of even greater concern, however, is where these risks have an unfair bias. As has often been noted, algorithms could develop biases. As Benkler suggests,

> because algorithms are trained on existing data that reflect social inequalities, they risk perpetuating systemic injustice unless people

> consciously design countervailing measures. For example, AI systems to predict recidivism might incorporate differential policing of black and white communities, or those to rate the likely success of job candidates might build on a history of gender-biased promotions. (Benkler 2019)

A further risk to individuals is known as the "alignment problem" (Bostrom 2014) which refers to the issue of whether it is possible to align the values of machines with those of human beings. It is reasonable to assume that if a robot is programmed with machine learning capabilities it will come to prefer its own survival over that of human beings, thereby potentially putting the safety of human beings around it in peril. If we reduce its capacity for autonomy we then get two problems in return. First, we may lose other advantages than come with autonomy, but second, it seems we would have to programme it with a set of moral principles, which, as the case of self-driving cars is revealing, is itself fraught with difficulty (Whittlestone et al. 2019, 19).

The second broad category I mentioned was that of economic and social risks, and of course several of the risks already mentioned have an economic or social aspect to them. But thinking on a more collective scale, one of the most discussed risks in this category is that of job losses, where it has been argued that over time a very significant proportion of current jobs will be capable of being automated. What the net effects of this will be are widely debated, but there is certainly something here that needs to be put on the risk register. Taking it further, Kai-Fu Lee argues that developments in AI pioneered in the USA and China will lead to ever increasing inequalities, both within countries and between them. The rationale of this argument is very strong. Many areas in AI are, as observed above, critically dependent on data, and companies with the greatest access to data have a significant market advantage. The greater the access to data the greater the potential to dominate a market and hence reap ever-larger profits. Hence this is a natural "winner takes all" market, where coming second may well eventuate in bankruptcy or being taken over by the number one. More generally, the advance of AI will see companies, regions, social classes and countries dividing into the highly successful and the left behind. Inequality will be rife, so it is argued (Lee 2018). The combination of unemployment and gross inequality is not a recipe for social peace, and therefore Lee suggests that social disorder is another risk of AI, as is a type of general dehumanisation of society where interactions between people are mediated or replaced by machines.

One response to this worry is that AI will generate a level of wealth that will allow for high benefits for those unemployed, possibly in the form of Unconditional Basic Income. Naturally this argument makes a number of economic assumptions, but, most importantly, assumes that income alone is an adequate compensation for lack of access to the workplace. But in any case there is much to discuss here (see Wolff, forthcoming).

A different type of social and ethical risk is the concern that "increased personalisation in the digital sphere [threatens] solidarity and citizenship" (Whittlestone 2019, 2). For example, algorithms can break people down into particular categories for insurance risk purposes, thereby individualising rather than pooling risk, not only leaving some unable to find affordable insurance, but undermining the thought that "we're all in this together" which can have further adverse effects in terms of atomisation and eventual loss of trust.

A third type of risk is being called "existential" in that it suggests that ultimately the development of new technology could eventually lead to the extinction of the human race. Such a scenario was famously sketched out by Nick Bostrom, in which machines, of a higher intelligence than human beings, become self-replicating, and replace us. More generally, there could be a concern that, like Frankenstein, we create a monster we no longer can control, making socially irreversible changes (Bostrom 2014). In effect, this is an extreme version of the alignment argument mentioned above. The general consensus, however, seems to be that such fears are groundless, based on current technology. Nevertheless, it is important to keep such concerns in mind, for if there is a rapid technological advance it will be important that considerable thought has already gone into regulation. And at the same time, if robots do develop a further level of sophistication, there may come a point where we will need to debate what appropriate moral sensitivity towards robots might look like (Gordon 2018).

Finally, I have already mentioned the way in which corporate interests can, before long, capture the regulation of the new technologies, making decisions that are favourable to their own interests, and against the interests of others. This is another way, in addition to inequality, in which AI can increase the power of the already powerful. I call this fourth type of threat "insidious risk."

6 Regulating Smart Technologies

In the light of the previous sections, what can be said about the regulation of smart technologies? In accordance with the first principle of good regulation set out in Section 2, I have, in Section 5, sketched the type of risks we would need to keep in mind when regulating. The question that comes into sharp focus as a result of Sections 3 and 4, however, is whether effective regulation is even possible against many of these risks. For some, such as confidentiality and data handling, the landscape is already well-known. Here the major obstacles are practical, rather than theoretical; the combination of very fast moving and fluid technology and commercial arrangements with the fact that the businesses involved are multi-national, and international regulation can be very hard to achieve and co-ordinate. But still, there are standard protocols for handling the regulation of personal data and so forth. The "black box" problem may be more tricky. As technology has developed it may be very difficult to arrange things so that we both harness the main benefits of the technology but also avoid bias and discrimination. To some degree this may be a matter of balance, although the lurking threat of an outright ban of the technology may have the outcome of producing a variant to the technology which gives rise to fewer ethical problems.

Social and political risks, such as joblessness, inequality and social disorder, cannot be adequately dealt with in a global market by regulating technology. Such regulations would be very hard to design, but even harder to implement without extreme protectionist trade policies to prevent global competition. And those trade policies would have their own costs. Social and political risks need social policy responses. In this context, as already mentioned, basic income has been widely discussed (Graeber 2018; Van Parijs and Vanderborght, 2017; Wolff 2019), although that is only part of a possible set of policy responses, which can also include job retraining, subsidies for start-ups and a host of other imaginative schemes, including those which not yet even been devised.

The category of existential risk is very hard to handle even in theory. From the point of view of precaution, even the smallest chance of extinction means that it is not worth taking the risk. But this seems an unreasonable standard, as it would be a bar to progress in many areas, any of which could, in theory, have rapidly accumulating negative consequences if things went very wrong. Here the key aspect will be designing in break-points, so that when risks begin to look more serious the technology can be stopped in its tracks, rather than continue under its own momentum. This will differ from case to case.

Finally, the insidious risk of regulatory drift and capture needs a response not at the level of direct regulation of innovation, or in social policy, but of regulation of the regulatory (which of course threatens a regress). In some cases civil society groups form, or can be encouraged to form, in order to act as a type of conscience to the industry. Greenpeace, PETA, and GeneWatch are such examples. A comprehensive review of the broader regulatory infrastructure to overcome regulatory capture now seems essential, as the phenomenon spreads far beyond this area.

Setting out the details of potential regulation for any technology is not for a political philosopher, especially not one working alone, even though political philosophers can make constructive contributions to expert discussions. But the arguments of this paper should lead us to appreciate that one of the greatest dangers is that regulators of new technology will fall for the "something must be seen to be done" approach, and introduce regulatory measures that do not, in fact, address the risks in any substantive way. If, nevertheless, the public believe that something significant has been done, they may lower their guard, as described above, and thereby become even more vulnerable. Bad regulation can be worse than no regulation, although the point of course is to devise regulation that minimizes risks in an economically efficient fashion.

If good substantive regulation is outside current possibilities, it can sometimes be better to allow activities to go ahead with minimal regulation, but with the possibility of legal or political action if things go badly. That is, those damaged by new technologies could bring legal action against those who have harmed them. But, on the other hand, the most vulnerable will rarely have access to legal or political remedies. Such considerations suggest that when an activity risks damaging the interests of the vulnerable, regulation, even though it is far from perfect, may well be the best we can do. Nevertheless, the overriding message is that hasty regulation can be worse than useless, and a keen appreciation of the point of regulation and its possible failures needs to be kept firmly in mind as concrete measures are introduced.

References

Better Regulation Task Force (BRTF). 2003a. *Principles of Good Regulation.* https://webarchive.nationalarchives.gov.uk/20100407173247/http:/archive.cabinetoffice.gov.uk/brc/upload/assets/www.brc.gov.uk/principlesleaflet.pdf.

Better Regulation Task Force (BRTF). 2003b. *Imaginative Thinking for Better Regulation.* https://webarchive.nationalarchives.gov.uk/20100812083626/http:/archive.cabinetoffice.gov.uk/brc/upload/assets/www.brc.gov.uk/imaginativeregulation.pdf.

Barnett, David. 2017. "Firearms Act: Twenty years on, has it made a difference?" *Independent*, December 16, 2017. https://www.independent.co.uk/news/long_reads/firearms-act-twenty-years-on-has-it-made-a-difference-dunblane-port-arthur-a8110911.html.

BBC, 2013. "Businessman Gary Bolton jailed over fake bomb detectors." BBC News, August 20, 2013. https://www.bbc.co.uk/news/uk-england-23768203.

Benkler, Yochai. 2019. "Don't let industry write rules for AI." *Nature* 569: 161.

Bostrom, Nick. 2014. *Superintelligence*. Oxford: Oxford University Press.

Browne, Jude. 2018. "The Regulatory Gift: Politics, Regulation and Governance." *Regulation & Governance.*

Buchanan, James, and Gordon Tullock. 1962. *The Calculus of Consent.* Ann Arbor MI: University of Michigan Press.

Cadwalladr, Carole, and Emma Graham-Harrison. 2018. "Revealed: 50 million Facebook Profiles Harvested for Cambridge Analytica in Major Data Breach" *The Guardian,* March 17, 2018. https://www.theguardian.com/news/2018/mar/17/cambridge-analytica-facebook-influence-us-election.

Drexler, K. Eric. 1986. *Engines of Creation.* New York: Anchor Books.

Gordon, John-Stewart. 2018. "What Do We Owe to Intelligent Robots?" *AI & Society,* 1–15.

Gordon, John-Stewart, and Vladislav V. Fomin. 2019. "Ethics and Standardisation." In *Managing Corporate Standardization and Meeting Growing Demand,* edited by Kai Jakobs, 177–192. Hershey USA: IG Global.

Graeber, David. 2018. *Bullshit Jobs.* London: Allen Lane.

Health and Safety Executive. 2001. *Reducing Risk, Protecting People.* London: HMSO. http://www.hse.gov.uk/risk/theory/r2p2.pdf.

Hartzell-Nichols, Lauren. 2012. "Precaution and Solar Radiation Management." *Ethics, Policy & Environment* 15(2): 158–171.

Hubert, Anna-Maria. 2017. *Code of Conduct for Responsible Geoengineering Research.* https://www.ucalgary.ca/grgproject/files/grgproject/revised-code-of-conduct-for-geoengineering-research-2017-hubert.pdf.

IET (ND) "Frequently Asked Questions." https://electrical.theiet.org/bs-7671/building-regulations/part-p-england-and-wales/faqs/.

Jonas, Hans. 1984. *The Imperative of Responsibility.* Chicago: University of Chicago Press.

Kaebnick, Gregory E., Michael K. Gusmano, and Thomas H. Murray. 2014. "The Ethics of Synthetic Biology: Next Steps and Prior Questions." *Synthetic Future: Can We Create What We Want Out of Synthetic Biology? Special Report, Hastings Center Report* 44(6): S4–S26.

Kennedy, Seraphima. 2018. "Grenfell was a Foreseeable, Entirely Preventable, Tragedy. Here's the Proof." *The Guardian,* August 4, 2018. https://www.theguardian.com/commentisfree/2018/aug/04/grenfell-tower-prentable-tragedy-proof-warning-combustible-cladding.

Lee, Kai-Fu. 2018. *AI Superpowers.* Boston: Houghton Mifflin Harcourt.

McKay, Elizabeth Norman. 1996. *Franz Schubert: A Biography.* Oxford: Oxford University Press.

Pook, Sally. 2004. "MP's Daughter Killed in Botched Kitchen Fitting." *The Telegraph,* October 12, 2004. https://www.telegraph.co.uk/news/uknews/1473919/MPs-daughter-electrocuted-in-botched-fitted-kitchen.html.

Radford, Tim. 2003. "Brave new world or miniature menace? Why Charles fears grey goo nightmare." *The Guardian*, April 29, 2003. https://www.theguardian.com/science/2003/apr/29/nanotechnology.science.

Ramanna, Karthik. 2015. "Thin Political Markets: The Soft Underbelly of Capitalism." *California Management Review* 57(2): 5–19.

Royal Society. 2004. *Nanoscience and Nanotechnologies: Opportunities and Uncertainties*. https://royalsociety.org/~/media/Royal_Society_Content/policy/publications/2004/9693.pdf.

Shepperson, Tessa. 2018. "The Rent Act 1977 in Context." *The Landlord Law Blog*, August 8, 2018. https://www.landlordlawblog.co.uk/2018/08/08/rent-act-1977-context/.

Steib, Matt. 2019. "Report: Self-Regulation of Boeing 737 MAX May Have Led to Major Flaws in Flight Control System" *New York Magazine*, March 17, 2019. http://nymag.com/intelligencer/2019/03/report-the-regulatory-failures-of-the-boeing-737-max.html.

Van Parijs, Phillipe, and Yannick Vanderborght. 2017. *Basic Income*. Cambridge: Harvard University Press.

Whittlestone, Jess, et al. 2019. *Ethical and Societal Implications of Algorithms, Data, and Artificial Intelligence: A Roadmap for Research*. London: Nuffield Foundation.

Wolff, Jonathan. 2006. "Risk, Fear, Blame, Shame and the Regulation of Public Safety," *Economics and Philosophy* 22(3): 409–427.

Wolff, Jonathan. 2019. *Ethics and Public Policy*. London: Routledge.

CHAPTER 7

Free Speech, Public Shaming, and the Role of Social Media

Carl Fox

1 Introduction[1]

It should come as no surprise to anyone at this point that a lot of nasty and hurtful things are said on the internet. Of course, a lot of nasty and hurtful things are said offline as well, but social media is where we seem to feel comfortable turning it up to eleven. In this chapter I want to take a closer look at a particularly contentious form of hurtful speech that is flourishing in this new technological environment. Public shaming occurs when large numbers of people engage in the collective shaming of some individual or organisation perceived to have done something morally wrong. Strong moral criticism is often warranted, and we know shame to be a powerful motivational force. But however effective it may sometimes be, it is clear that many targets find it to be a deeply distressing and degrading experience. Moreover, shaming campaigns sometimes result in egregious harms being visited on ordinary people for relatively minor offences.

For these reasons, and others, we might take the view that shame is a tool we would do better not to digitise. But should we go further? Should we make the determination that public shaming ought not to be allowed and set about finding formal and informal ways of regulating online speech to prevent it? Harm reduction, though, almost always comes at some cost to liberty, and in this case it runs smack into one of the most highly-prized freedoms in the democratic tradition – freedom of speech. Our commitment to freedom of speech is why we tolerate all sorts of noxious things. To take just one example, it's why we allow Piers Morgan to have a broadcasting career.

We might, then, frame the question as follows: is public shaming protected by the right to free speech or is this a case where our old norms and principles have simply been exposed as unfit for a new era? In this chapter, I actually

1 I would like to thank Rob Lawlor and Olof Leffler for their insightful comments, and also audiences at the Smart Technology and Human Rights workshop in Stockholm University and the IDEA Centre in the University of Leeds.

 | DOI:10.1163/9789004437876_009

want to take a different approach. The standard free speech defence for objectionable speech seems to me to have no hope of successfully sheltering a permissive approach to public shaming, because public shamings are just not political in the way that such an argument would require. Instead, I will consider another possible approach which builds on a significant political dimension that *is* apparent in many public shamings. Relatively powerless individuals and groups can use social media to achieve the critical mass required to deploy shame as a meaningful sanction for wrongful behaviour that the state either can't, won't, or shouldn't punish itself. In particular, we can point to various kinds of structural injustices that the state is ill-equipped to deal with, especially when their manifestations are subtle and tangled up in other rights and liberties. So, could shaming on social media make a sufficiently important positive contribution that we should learn to live with its pernicious effects?

Different parts of the public sphere make different contributions to our discourse, and so to the quality and resilience of a democratic system of government. Journalism informs us, art moves us, satire pokes fun at us, and so on. On this line, social media would also qualify as an institution with a distinctive role to play. More specifically, by facilitating public shaming, social media would help us to realise the full value of political equality. Ultimately, I will be forced to conclude that while we have strong reasons to find new ways to buttress political equality, we should not embrace public shaming on social media. In fact, drawing on the work of Martha Nussbaum, I will offer some reasons for thinking that social media is, in fact, uniquely ill-suited to deploying shame as a sanction for bad behaviour.

I will start by explaining why traditional free speech defences cannot be extended to cover public shaming, before explaining, in Section 3, the idea that the public sphere is composed of many distinctive parts and make the case for thinking that a permissive attitude towards public shaming could be a boon for political equality. Section 4 takes a closer look at the concept of shame and argues that structural features of our interactions on social media make it particularly likely that shaming will go unacceptably wrong. Finally, I will go on to outline some principles by which we should formally and informally regulate shaming speech on social media, and briefly float another theory about the contribution that social media might make to a functioning public sphere.

Before I begin in earnest, a few words about what I will and will not be talking about in this chapter. In what follows I will use the term "social media" to mean virtual platforms that facilitate instantaneous interactions between large numbers of people, and especially between strangers. I will not address all of the practices that we find in instances of public shaming. Some of them clearly cross any sensible line for protected speech. For instance, "doxxing" is where a target's personal information, often including details such as their home address, is posted online. There is no plausible way to interpret that as anything other than action designed to intimidate and to facilitate further intimidation, if not outright violence. I am interested here in direct communications to an individual that are designed to elicit the emotion of shame. This will include harsh and often highly-offensive language which does not explicitly call for shame, but nevertheless expresses a negative judgment about that person's character. It is sometimes hard to distinguish between such language and threats. I will assume that actual threats are impermissible, but for the sake of argument I will take statements such as "You deserve to die" or "I hope you die" as crudely expressed judgments about a person's badness, as measured in the currency of liability for punishment. I should also clarify that I am interested in the online shaming of ordinary people. I will set aside the special case of public representatives as there are important reasons for ensuring that direct channels of communication remain open between citizens and their representatives which I will not be able to go into here.

2 The Standard Free Speech Defence

In his entertaining and illuminating book on public shaming, Jon Ronson (2016) discusses the case of Lindsey Stone, a care worker who in 2012 posed for a picture beside a sign in Arlington National Cemetery, a military graveyard near Washington D.C. What would later get Stone in a great deal of trouble was that the sign was asking for "Silence and Respect" while she was giving the finger and pretending to shout. As she explains to Ronson, this was part of a running joke with her friend where they would photograph each other flagrantly disobeying signage. Stone did not intend to disparage anyone buried in the cemetery, nor make light of either their service or sacrifice.

Even so, there are two ways in which she may have gone wrong here. The first is essentially negligence. It might be that her obscene gesture and general air of frivolity did, in fact, violate general norms of conduct applying to cemeteries of which she ought to have been aware, and so did show disrespect to the people buried there. The second possibility is that her attitude to authority, and perhaps even her high spirits, could have caused offence, particularly to people still coming to terms with grief. Graveyards might be places where we should be extra sensitive to the effects that our behaviour has on others, regardless of whether their reactions are justified or proportional. There does not seem to me to be anything very much wrong with making fun of a sign, even if the content of that sign is authoritative. I can easily imagine myself and many of my friends doing this kind of thing and feeling pretty pleased with ourselves. I am also inclined to think that within some common-sense limits a person might have fun and still comport herself in a way that shows appropriate respect to the dead and their families, but I accept that this is a point on which reasonable people might disagree. For the sake of argument, let's grant that Stone's behaviour was, all-things-considered, misguided. It is still hard to make a convincing case that it should be treated as a very serious wrong. Even without the ominous foreshadowing, you can predict what happened next. Here's a snapshot Ronson provides of the online response to the picture:

> 'Lindsay Stone hates the military and hates soldiers who have died in foreign wars' and, 'Die cunt' and, 'You should rot in hell' and, 'Just pure Evil' and, 'The face of a Typical Feminist. Fifty pounds overweight? Check. Sausage arms and little piglet fingers? Check. No respect for the men who sacrificed? Check' and, 'Fuck You whore. I hope I die a slow painful death.[2] U retarded cunt' and, HOPE THIS CUNT GETS RAPED AND STABBED TO DEATH' and, 'Spoke with an employee from LIFE who has told me there are Veterans on the board and that she will be fired. Awaiting info on her accomplice...' and, 'After they fire her, maybe she needs to sign up as a client. Woman needs help' and, 'Send the dumb feminist to prison' and, in a response to a small number of posters suggesting that maybe a person's future shouldn't be ruined because of a jokey photograph, 'HER FUTURE ISN'T RUINED! Stop trying to make her into a martyr. In 6 months no one except those that actually know her will remember this,' (Ronson 2016, 200–201).

As a result of public pressure, including a dedicated Facebook page, Stone did indeed lose her job. All of this is extremely difficult to defend, but it is instructive to see why a traditional free speech defence won't work here.

It starts from the idea that freedom of speech is a civil right designed to provide citizens with a weapon they can use against nascent tyrannies that overstep the limits of political authority. By speaking out, individuals can challenge the official narrative and mobilise public resistance. However, this is clearly a fragile resource. If the state has the power to decide what speech is and isn't protected by the right, then the kind of rulers we have reason to be wary of are precisely the kind of rulers who will be sorely tempted to gerrymander the confines of the right to make it progressively harder to challenge them. On that basis we might conclude that the right to free speech should be

2 I can only assume that this was a typo, and not an unexpected attack of self-awareness.

as expansive as possible, even if that comes at the cost of permitting nasty speech such as that directed at Lindsey Stone.

Although there is clearly something to this worry, it is surely overblown as a response to concerns about this case. The ability to speak truth to power does not rest on the ability to send vitriolic insults to one another.[3] Free speech as a civil right will certainly shelter discussion and criticism of Stone's actions, but this is a far cry from the ability to attack Stone for her appearance or engage in lurid revenge fantasies. A free speech advocate might reply that self-expression is hard and our emotional and instinctive reactions can often come out less polished than we might wish when we try to articulate them. Pressing the point, they might argue that offence or discomfort are a small price to pay for a society in which one can be completely assured of the right to stand before others and give voice to their thoughts.[4] The problem with this is that it only works if you agree that it is, in fact, unduly burdensome to ask people to resist their initial impulses to castigate Stone in whatever terms happen to pop into their heads. It is, of course, perfectly possible to express disagreement with Stone's actions without using demeaning or threatening language, and without contacting her directly. Once again, freedom of speech might shelter some activities that we observe in cases of public shaming, but not an absolute right to direct shaming speech at another individual, and certainly nothing like the phenomenon as a whole.

This might all seem to miss the point. Nobody disputes that we shouldn't punish people when they have done no wrong, or that we shouldn't punish people disproportionately harshly for wrongs they have committed. Let's take a case where moral criticism and shame was warranted: the 2015 decision by Martin Shkreli, then CEO of a company called Turing Pharmaceuticals, to raise the price of Daraprim – a vital medication used for controlling infections in AIDS patients – from $13.50 a pill to $750. This is literally a lifesaving drug that Shkreli wanted to sell for more than 50 times its original price. Unsurprisingly, the internet went ballistic. According to Caitlin Dewey, a reporter for the Washington Post, over the course of three days more than 50,000 Twitter users tweeted at Shkreli to let him know exactly what they thought of him and his

3 See Jerome Neu (2008) for a detailed discussion of the significance of insults.

4 See Jeremy Waldron (2012, 173–203) for a nuanced discussion of an argument that Ronald Dworkin makes against hate speech legislation which touches on similar issues. That argument turns on the thought that the legitimacy of coercive laws depends on those who are subject to them having a chance to express their opposition to them, using whatever language they wish, no matter how vile or unfocused it might be. Waldron makes a number of solid points against this argument, not least of which is that it would prove too much and permit threats, libel, and so on.

decision.[5] Presumably he expected, and was prepared to resist, some backlash, but the ferocity of the reaction had the desired effect and he soon backed down. Can Shkreli have any complaint for finding himself on the rough end of some rough language? Such a person *ought* to feel deeply ashamed. More importantly, what he attempted to do was straightforwardly wrong and needed to be stopped. That the shaming did indeed stop him looks like a pretty good justification. And shaming does work more generally. We know that human beings care a great deal about their standing in the eyes of others.[6] Geoffrey Brennan and Philip Pettit discuss the idea of an "economy of esteem," explaining it as follows:

> In short, A cares what B thinks of him, and is prepared to act in order to induce B to think better (or less badly) of him. To the extent that prevailing values are matters of common awareness, A's desire for positive esteem (and the desire to avoid disesteem) will induce A to behave in accord with those values. (2008, 176)

This suggests that public shaming can be effective in the pursuit of worthy moral goals, and is, at least sometimes and to some degree, warranted. So, an expansive right to free speech that covers shaming speech might be instrumentally justified as a way of realising other values. However, we have just seen that shaming campaigns often fix on inappropriate targets and the impact they have can be devastating. One familiar philosophical approach that we could take would be to identify necessary and sufficient conditions for a warranted public shaming campaign. However, that is not what I intend to do here. As is often the case when it comes to the enforcement of rights, it is not just a matter of working out what behaviour we should and shouldn't allow – the more relevant questions are actually about the degree to which rights are protected presumptively as opposed to retroactively. We would all be a good deal safer if nobody was allowed out after dark and everybody was electronically tagged, but almost all of us agree that that would be to get the balance between liberty

5 Dewey 2015, https://www.washingtonpost.com/news/the-intersect/wp/2015/09/23/the-successful-internet-shaming-of-pharma-bro-martin-shkreli/?utm_term=.c8e372040f20.

6 In a paper on the relevance of moral criticism to practical strategies for mitigating and adapting to climate change, Rob Lawlor (2017) discusses the case of slavery. If any practice is shameful then it is holding human beings as property, and there is reason to think that moral criticism and social pressure played a tangible role in changing minds and, eventually, laws. See, for instance, David Brion Davis (2006). If slavery were still legal today, a modern-day slaveholder ought to be ashamed of themselves and the rest of us ought to do something to stop them.

and security drastically wrong. We can call this strategy the "omelette approach" because it allows some eggs to be broken for the sake of a greater good. It is important to see that this view is not committed to saying that rights violations are permissible, but rather that we should take a relatively permissive attitude towards conditions that might be conducive to them. This is consistent with holding that there should be serious punishment after the fact for harmful behaviour. The key feature of this approach is that it does not operate under a "safety first" principle and instead views the possibility that other rights will be violated as an acceptable cost for preserving a high degree of freedom.

In this case, the question we have to ask ourselves is whether we should arrange the norms and laws governing our use of social media to prioritise free speech even though we know that it might sometimes be misused, or emphasise the various rights that shaming campaigns run the risk of violating? As above, the omelette approach is perfectly compatible with insisting on retrospective liability for threats or libellous speech, or compensation of some kind when shaming fixes on the wrong target, or simply goes too far. It simply views the possibility that other rights will be violated as an acceptable cost for preserving a high level of freedom for speakers.

But these are substantial collective punishments that we are talking about doling out. If we are going to seriously consider the idea that we should take an omelette-style approach to public shaming on social media then we would need a very good reason to do so. In the next section, I will examine an important value that might provide such a reason: political equality. Should we adopt a permissive attitude because the combination of shame and social media disperses social power, and thereby helps us to redress some of the deep and abiding imbalances of political equality that undermine our democracies?

3 Political Equality and the Role of Social Media in the Public Sphere

Although we sometimes speak of the public sphere as if it were a single, continuous space encompassing any and all communication that is not strictly private, this is inaccurate and, I believe, misleading. There are, in fact, many different parts of "the" public sphere and it is worth our while to distinguish between them. They make different contributions to the health of a democratic system of government, but in order to do so, their practitioners require special rules and permissions. We don't allow doctors to arrest people, nor do we allow police officers to write prescriptions. In much the same way we should have different standards of behaviour for journalists, pundits, satirists, artists, judges, political representatives, and so on.

Let's take journalists and satirists. Journalists actually perform a range of roles, but chief among them is the job of informing and educating the public about events and issues that are relevant to the decisions that they make in their own role as electors. Satirists may sometimes inform, but that is not a central part of what they do. Rather, satirists exaggerate, and thus expose, the vices of the powerful – and indeed of their audiences as well – to keep those people in check and ensure that the public are not blind to their faults. Of course, satire also entertains and it can help a society to blow off a little collective steam. At its core, though, it is a deeply political activity. Satirists often aim to offend their audiences and they are right to do so, because this can be an effective way of shocking people out of received wisdom or lazy preconceptions (Waldron 1987). As such, we should recognise a broad permission for satirists to push the envelope and to violate the boundaries of good taste. Journalists may sometimes produce material that they know will cause offence to sections of the population, but this is something they should only be permitted to do when it is impossible for them to discharge their professional obligations without causing offence. If journalists are to inform the public and to set, if not a common news agenda, then at least some parameters on the key issues for discussion in civil society, then they must be respected and trusted across society. If journalists took it upon themselves to regularly offend their audiences, then they would jeopardise that broad trust.[7]

Here's another example. We ensure that there are public spaces, such as Speakers' Corner in London's Hyde Park, where individuals can exercise their right to free speech and put their point of view to their fellow citizens. As Cass Sunstein (2017, 35) notes, we guarantee the right to speak as a right to have access to our fellow citizens in public places. But this is not the same thing as a right to be heard. We do not insist that anyone come to listen to them. However, we effectively enforce a right to be heard for artists when we put their work on the school curriculum, or display it in a public building to which people have no choice but to come, or simply construct it on a massive, unmissable scale.[8] Why do we do this? We do it because there are some abiding truths, such as our common humanity, that we should all reflect on from time-to-time, and because art has the capacity to make us *feel* such truths in a visceral way. The idea that I want to consider now is whether social media might make such

7 For a fleshed-out version of this argument, see Fox (2019).

8 See Sunstein (2017, 210–212) on the question of subsidising particular types of speech.

a distinctive contribution to the health of our political community and so might also merit special permissions to facilitate it.

The fundamental principle underpinning all of democratic theory is that above some threshold for competence, all adult human beings are entitled to an equal share of political decision-making power. This is the idea that whatever differences there may be between us, we must all count for the same as members of the sovereign body politic.[9] As Thomas Christiano puts it: "Individuals in a [democratic] society have equal rights to vote, to organize political parties and interest groups, and to contribute to discussions on how the society would best be organized" (1996, 16–17). Sadly, those of us who endorse the principle of political equality are confronted with the messy, and indeed shameful, reality that "[s]ocial structures often position people unequally in processes of power, resource allocation, or discursive hegemony" (Young, 2000, p. 86).[10] While clearly wrong in themselves, these kinds of discrimination also add up and interact with one another, gradually building up a social environment in which some citizens find that their opinions and interests are systematically undervalued. Dealing with the larger problem entails dealing with the (relatively) minor everyday injustices that create and sustain it. But how are we to do this?

Clearly, a significant part of the problem is down to a lack of political will. Citizens cannot always rely on their governments to guarantee their equal standing. Even if there was sufficient political will to seriously tackle structural injustice, though, many of the specific injustices to which people are subjected on the basis of their gender, race, disability, sexuality, religion, and so on, are manifested in patterns of speech and behaviour that are so familiar they can be almost invisible. This makes them difficult to deal with. One important example of this is what Miranda Fricker (2007) calls "epistemic injustice." She argues that under various descriptions, large groups of people are disadvantaged by the stereotypes that we apply to them, often unconsciously. We undermine them as knowers and discursive partners because "prejudice will tend surreptitiously to inflate or deflate the credibility afforded the speaker" (2007, 17).

There are, of course, things that the state can do to ameliorate epistemic injustice.[11] However, as Fricker (2007, 176) suggests, it is also a matter of

9 I'll set aside questions arising from the debate about the basis of equality here. See Williams (1973, 230–249) and Carter (2011) in particular.

10 Further, if we have reason to believe that the principle of political equality is being violated, then we have cause to expect that our institutions will be insufficiently representative and, for that very reason, unlikely to fully appreciate the scope and urgency of structural injustice.

11 See, for instance, Sherman and Goguen (2019).

individuals coming to possess epistemic virtues and this is not something that legislation alone will achieve. This is not an isolated problem. The law is often clumsy and slow. Looking at the gender pay gap, for example, it is apparent there too that a top-down approach is unlikely on its own to change the ingrained and, again, perhaps only partly-conscious attitudes and dispositions of hiring committees and promotions panels.[12]

Finally, aside from questions of political will or effectiveness, there might also be injustices that we would prefer the state actually refrain from policing. When it comes to discriminatory speech that falls below whatever threshold you care to set for hate speech or incitement to violence,[13] we might decide that it is inappropriate to deploy the coercive power of the state. In order to avoid the chilling effect that bearing witness to the brutal side of state power may have on legitimate political discourse, the omelette approach may tie the hands of the state, compelling it to stand by or even protect some morally objectionable speech.[14] So, there are threats to political equality that governments will not, cannot, or should not resolve through direct intervention. Where does that leave us?

John Stuart Mill offers us some assistance here when he reminds us that there are different levels of sanctions which are fitting for different kinds of wrongdoing: "We do not call anything wrong, unless we mean to imply that that a person ought to be punished in some way or other for doing it; if not by law, by the opinion of his fellow-creatures; if not by opinion, by the reproaches of his own conscience" (1979, 303–304).[15] This offers a tantalising prospect. We could, perhaps, pass the burden of addressing some wrongs to civil society and rely on one another to enforce the middle level of social sanctions, and thereby police the kinds of injustice that the state will not, cannot, or should not deal with directly. And this sense that social media might be a new and important space for pushing back against injustice is one that resonates with many users. Here's Ronson on how he felt in the early days of internet shaming:

> And then one day it hit me. Something of real consequence was happening. We were at the start of a great renaissance of public shaming… When we deployed shame, we were utilizing an immensely powerful tool. It was coercive, borderless, and increasing in speed and influence. Hierarchies

12 This is not to say that the law doesn't have an important part to play here – it certainly does. On this point see Jolls and Sunstein (2006).

13 On the difficulty of dealing with racist dogwhistles in particular, see Saul (2019).

14 Although, see Corey Brettschneider (2012) on what the state might *say* about such things.

15 For a discussion of the relationship between morality and justice in Mill's thought, see Skorupski (2006, 34–38).

> were being levelled out. The silenced were getting a voice. It was like the democratization of justice. (2016, 9)

The key point to pick out here is his thought that justice could be democratised by leaving punishment of wrongdoing to the people. Since social media can connect users across great distances, and even across national boundaries, it can allow members of minority groups to find each other (and like-minded supporters) and accumulate a critical mass to condemn discriminatory actions and practices, and assert their status as political equals. Taking a permissive attitude to shaming on social media, especially if state regulation was designed with this aim in mind, could thus endorse and encourage it as a distinctive contribution to the health of our political community. Unfortunately, I think that while this idea has a certain amount going for it, it should be rejected all the same. To see why, it's important to take a step back and get a better handle on the concept of shame, and the powerful psychological forces it can unleash.

4 Shame, Shaming, and the Perils of Social Media

Up to this point I have been trading on an intuitive understanding of "shame" and "shaming." Can we be more precise about what we mean by those terms? For Martha Nussbaum, we feel shame when we believe that we have failed in some way that really matters to us.[16] "In shame, one feels inadequate, lacking some desired type of completeness or perfection" (Nussbaum 2004, 184). The key here is the totalising nature of shame. It is not that you feel bad about doing some particular thing, or even about a series of actions, but rather the object of your disapproval is your entire self. Not surprisingly, the psychological origins of shame are complicated and Nussbaum distinguishes between two types of shame, which correspond to two different grounds for thinking that you are inadequate.

The first form of shame she calls "primitive shame" and describes it as "shame at the very fact of being imperfect and needy" (2008, 337). Nussbaum

16 For Williams (2008, 82), to feel ashamed requires, if not a real person actually watching us, then at least an imagined observer. However, he is very permissive about the form that the observer can take, and allows that it can be "abstracted and generalised and idealised" (2008, 84). It is not entirely clear to me why Williams thinks that a straightforward negative evaluation of one's own behaviour cannot generate a feeling of shame, although he is surely right that such feelings are very often tangled up in fears about what others may think of us.

draws on literature from psychology to tell a story about our deep-seated insecurities as fragile and vulnerable creatures who become aware very early on that they are profoundly dependent on favourable conditions and the beneficence and care of others in order to see their physical and psychological needs met. However much we might like to pretend that we are independent and self-sufficient individuals, reality rubs against our fantasies and time and again forces us to recognise that we have our limits. When we realise that we are not equal to some obstacle, and come to view ourselves as small, or weak, or pathetic, or so on, the feeling we experience is primitive shame.

The other form of shame that Nussbaum (2004, 211) discusses is "constructive shame." Constructive shame is much, much narrower and comes directly from a belief that one has failed to live in accordance with a suitably justified moral principle. Although the keen awareness of one's failure is similar, as is the sense of being unworthy of the respect and esteem of others, there is a significant difference between them. Constructive shame is an appropriate reaction, justified by the reasons that count in favour of accepting the relevant principle. Primitive shame, on the other hand, is an irrational response to an ineliminable feature of human existence. The two types of shame give rise to two different forms of shaming, which are rooted in the pursuit of two very different goals.

When we engage in constructive shaming the point is to prod people into change. We communicate our disapproval of their cavalier attitude to an important principle by which we think that they ought to govern their behaviour. Shame is often contrasted with guilt, because the latter is also an unpleasant emotion that comes over us when we have done something wrong. The difference is that guilt is directed towards a specific action as opposed to the whole person. Often when we fail to live up to our moral standards it is guilt that is the appropriate response. As Christian Miller (2018, 10) notes, it is patterns of action that determine a person's character, so we needn't feel bad about our *selves* every time we make a mistake, but instead direct our internal disapproval to the offending action. However, if a moral failing occurs consistently over time, as when we continue to tolerate an obvious injustice, then shame is called for. As Nussbaum says: "It is too easy to say, 'Let's not do A again.' We need to say, 'Let's not be that way any longer (greedy, materialistic, hostile to equality, etc.)'" (2004, 212).

Since constructive shaming is anchored by moral principles, we might wonder whether it is likely to be effective in contexts where there are deep moral disagreements.[17] Many of the people who engage in discriminatory speech do

17 I thank John Stewart Gordon for pushing me on this point.

so knowingly and deliberately, and quite possibly as part of a strategy to normalise and spread their political views. There is little chance they will be moved by the disapprobation of people whose very standing as equals they are calling into question. However much of a waste of time it may be to try to constructively shame a committed white supremacist, even in such situations there are still three grounds for thinking that it could be worth the effort. First, even white supremacists are responsive to reasons in a general sense. If they find out that it's raining then they'll postpone their cross burning ceremony until the weather improves. It may be difficult to reach such a person, but difficult is not the same as impossible. Second, even if the original offender is a lost cause, there may be others watching who are not. Articulating and deploying defensible moral principles might influence their thinking and development. And, third, the act of shaming those who engage in hateful speech is a public rejection of their views. If they intend to undercut somebody's status as a political equal, then this is potentially a way to solidify it.

Most of the time, though, Nussbaum thinks that shaming is not like this. Usually, when we engage in shaming she thinks that we are attempting to shore up our own self-worth by distancing ourselves from the weakness of others – weaknesses that are uncomfortably close to the ones that we perceive in ourselves: "in order to get rid of the discomfort they cause us, we project those weaknesses onto other people, saying that they, and not we, are the substandard human beings" (2008, 346).

Nussbaum takes aim at the notion that the state should employ shame punishments for breaches of criminal law, raising a number of troubling objections.[18] Her objective is to show that a policy of humiliation and degradation for wrongdoing is incompatible with the commitment of a liberal state to guarantee the equal status of all its citizens and its duty to underpin that equality by providing access to something very like the Rawlsian primary good that he calls (2003, 59) the "social bases of self-respect."[19] In general, she fears that even well-intentioned constructive shaming opens a door to our darker, more primitive side. Introducing the sensibilities of the mob to our system of punishment makes it more likely that mistakes will be made since mobs are not known for their careful deliberative procedures. More frightening, though, is the effect of the incentives associated with primitive shame. As people start to enjoy the

18 See Nussbaum (2008, 346) for a concise summary of her view.

19 Thom Brooks (2008, 327) argues that shame is unavoidable in a criminal justice system since "every person convicted of a crime carries with her some measure of shame." Nussbaum (2008, 347–348) rejects this, however, countering with a portrait of punishment that preserves the dignity of offenders.

short-term psychological high of comparing themselves favourably to the person or group being shamed, shaming becomes increasingly likely to mutate into scapegoating and, eventually, discrimination and oppression. It is no coincidence that shaming has historically been targeted at vulnerable and underrepresented groups. In the end, even though Nussbaum makes some space for constructive shaming, it's clear that she thinks that all shaming, and particularly shaming with a mob element to it, is dangerous and quite often counterproductive. Occasionally, as in the case of Martin Shkreli, it will lead to a good outcome, but the process is deeply unreliable. This may be a cost that some will be willing to bear if they believe that it will deliver more hits than misses, but we care about more than outcomes when it comes to justice.

This is bad news for the idea that public shaming can contribute to the realisation of political equality and the proposal under discussion that we should take a permissive attitude towards it on social media. We might try to draw a line in the sand here and argue that Nussbaum's worries about shame punishments lose some of their force when directed against shaming as practiced by private citizens on social media. After all, even widespread public shaming doesn't carry the imprimatur of the state, which claims to act in everyone's name. However, this response fails for two reasons.

First, we have been considering the notion that we should treat shaming on social media in a way that is analogous to the ways we treat other parts of the public sphere when they have a special and distinctive contribution to make to the smooth operation of a democratic system of government. As a community we would thus be making a statement that is at least comparable in its weight and significance to the official judgment of the state. Second, the punishments in question here would inevitably bear on the social standing of the individuals concerned and so Nussbaum's worries about political equality must surely apply. In fact, we should take this thought even further because of the potential for large-scale shaming to fix on the wrong targets. The power to successfully deploy shame as a punishment on social media may be widely dispersed, but it is still a raw form of power. In the absence of protections to harness power safely, vulnerable groups are often left the most exposed.

As we saw earlier, an economy of esteem exists because we value others' good opinions of us (Brennan and Pettit 2008), making social esteem one of the many goods that we pursue and that must be balanced against each other as we decide how to live our lives. The main problem with social media that I don't think can be overcome is that the anonymity it affords shields us from the costs of disesteem that might moderate our behaviour without significantly reducing the benefits we get from engaging in primitive shaming and from the esteem that comes our way when others like or otherwise approve of our comments. Once this point is conceded, it seems to me that we actually have *more* reason to worry that shaming on social media will go wrong in the ways we have just been considering. If this is right, then encouraging shaming speech is likely to wind up undermining, rather than supporting, political equality.

One reason why we don't shout nasty things at strangers in coffee shops when they express political sentiments with which we disagree, is that we are not prepared to be judged harshly by those around us. Online, the risk of this kind of blowback is greatly reduced because of two kinds of anonymity offered by the internet. The first is the anonymity of the crowd. When we are merely one of a large number doing the shouting, then the chances that an observer would fix their attention on us in particular is small. The second kind of anonymity is the complete anonymity of operating under an alias or assumed identity. In this case the risk of discovery is all but eliminated entirely. Few users have the skills and resources required to unmask someone who wishes to remain hidden. Even if they did, this anonymity is reinforced by the anonymity of the crowd, since it would be prohibitively costly to chase down every user with a false identity.

It would be natural to expect that as anonymity reduces the risks of disesteem, the quantity and quality of esteem on offer also fall. But it turns out that the relationship is asymmetrical. Even though we seem to be able to shrug off whatever meagre disapproval might be directed at our handles and avatars, the kudos that accrues to them still generates the little rush of endorphins that comes from feeling liked. In fact, the situation is even worse, because in order to get that rush from within a mob, one has to stand out in some way, which encourages more extreme rhetoric. It is hardly news that anonymity breeds bad behaviour, but it is perverse that the economy of esteem should also reward it.

Brennan and Pettit conclude that "the desire for esteem will play a significant role in stabilising decent communication practices and supporting the operation of other social norms on the Internet" (2008, 193). No doubt this will sometimes be true, especially for users who maintain consistent online identities in enduring and tight-knit online communities. However, these stabilising forces are patently not strong enough to counteract the anonymising and dehumanising forces unleashed in the maelstrom of a major shaming event.

These features of social media also combine with others to create further problems. The numbers who use it and the rate at which content is produced and devoured mean that the quantity of disapproval heaped on a target of shaming can be of a magnitude that our ordinary interpersonal interactions will never prepare us for. It is one thing to feel that one person or a small group of people think poorly of you. It is, I suspect, quite another when they number in the hundreds of thousands. In that kind of situation, a person might well come to believe that their community as a whole has cast judgment upon them.

Compounding this problem are the assumptions built into what Chris Bennett (2008) calls the "apology ritual." By this he means the procedure that wrongdoers must undergo in order to acknowledge the significance of failing in their capacity as a fully paid-up member of the moral community and eventually repair their relationships with the rest of that community. The apology ritual is a carefully choreographed dance, but it is one that requires an audience. It is not feasible to think that anyone who is shamed might be able to repair their reputation by convincing even a modest proportion of those doing the shaming that they are sorry and have taken the opportunity to change. In a mass social shaming, who would even have the time, let alone the emotional stamina to reply to every post, tweet, or email? General responses might be read by some, but in the absence of face-to-face contact, it will prove hard to demonstrate sincerity. With regard to another free speech issue that is playing out on social media – fake news – Neil Levy (2017) builds a strong case that it is easier to acquire beliefs than we realise, and, perhaps more importantly, harder to shift them.[20] At any rate, words are cheap and redemption takes time and effort. The internet, like the tide, waits for no-one, and the vast majority of the people whose good graces a shaming target wants to win back will simply have moved on. In criminal cases, there is powerful symbolism in being released from prison and formally re-entering society. We do not have anything analogous for wrongs that fall below the threshold for state punishment, but still play out on a national or even international stage.

Given all of this, it does not seem reasonable to me to expect that shaming on social media will generally be, in Nussbaum's terms, constructive. In the end, the most constructive approach we can take to shaming as a community may be to eschew our right to punish and restrict our focus to how best we can prod an offender's conscience, inviting them take the burden of punishment onto themselves. We would then be shifting to the bottom level of Mill's taxonomy. And perhaps this should come as no surprise. In ordinary language, when we express the judgment that some wrongdoer ought to feel shame, the most common formulation is *"You should be ashamed of yourself."*

20 "[E]ven when we succeed in consuming fake news without coming to acquire beliefs that directly reflect its content (surely the typical case), the ground level representations will play a content-reflecting role in our further cognition, in ways that may be pernicious" (Levy, 2017, 32).

Before I conclude, I'd like to take this opportunity to ask whether there are there any lessons that might be learned about how we should aim to behave on social media, and if there are any obvious principles to which we might appeal in order to guide any future attempts to regulate it.

5 Regulating Social Media

Players and fans of ice hockey enjoy a good, honest fight. But in 1971, the National Hockey League had a problem. Fights kept escalating into brawls, and the brawls kept escalating into melees. Two burly men beating the living daylights out of each other for the entertainment of a cheering crowd adds a bit of welcome spice, but melees were felt to be unseemly.[21] The NHL decided to act and instituted the so-called "Third Man In" rule which punishes the first player to intervene in the fight. Their reasoning was that the third man unbalances the situation, forcing the other team to respond and thereby establishing the dynamic that eventually spirals into bedlam. I submit that a similar approach is worth considering for shaming on social media.

Of course, it will not be easy to determine what the threshold should be in any particular instance of objectionable behaviour, such that we can precisely identify the "third man," and it may be that the threshold should be higher for more egregious behaviour to reflect the gravity of the wrong committed and increase the chances that the offender will appreciate their mistake. Even so, most of the time it will not be difficult to see that the key points have already been articulated and the invitation to feel shame has been issued. Jumping onto the pile after that point is unlikely to make the shaming any more constructive, but will add to the dynamic of escalation that encourages further and, eventually, disproportionate punishment.

This, worry points us to another principle which I believe should regulate shaming speech on social media: we should always phrase our invitations to feel shame in a way that leaves it clear that there is a way to return to the warm embrace of the moral community. Perhaps there are some wrongs that are so grievous that there is no way back, but here there are two things to say. First, most wrongs are not like this. And, second, even in the case of extreme wrongdoing, we should be wary of declaring someone's social death. The psychologist James Gilligan (1997) argues that, over a prolonged period, feelings of shame are enormously damaging to the human psyche, and can result in extreme violence.

21 Especially when captured on video.

As we saw in the Stone case, a common demand in cases of public shaming is that the perpetrator of the original offence be sacked from their job. Since shaming is already a form of punishment, this amounts to a "doubling up" of punishment. This is something that should normally be avoided. Of course, what somebody has done might constitute grounds for losing their job, and that is a decision that their employer will have to take on its merits. The shaming itself should not create an additional reason for an employer to fire someone, and we should be very wary of creating one by bringing moral pressure to bear on the employer as well.

One of the most well-known cases of public shaming was the ordeal endured by Justine Sacco in 2013. Sacco was excoriated for an attempted joke on Twitter about white privilege and the extent of the AIDS epidemic in Africa. In the aftermath of the initial shaming, the website Buzzfeed ran a story entitled "16 Tweets Justine Sacco Regrets" (Ronson 2016, 68). The point of the article was to trawl through Sacco's Twitter feed to find other tweets for which she could be shamed. Combing through somebody's past social media activity to find further cause for outrage is another instance of doubling up of punishment. And, as the Sacco example illustrates, it runs a particular risk of degenerating into primitive shaming, because it paints a damning picture of the target as irredeemably bad and so someone whose feelings and interests should not matter to us.

All of this might be summed up in a general ethos of "shaming up" rather than "shaming down," where the guiding principle is to avoid piling on someone who is vulnerable, and it should at this point be clear that a public shaming is the kind of thing which can increase someone's vulnerability in a range of different ways. How "shaming up" might be reliably achieved is another question. Government regulation may well be required, but it will also be up to us as users to develop and maintain new norms of behaviour for the online realm.

6 Conclusion

In this chapter, I have considered what I take to be the best case that can be made for the idea that we should understand the right to free speech in such a way that it supports a permissive attitude towards shaming speech even though we can predict that it will often misfire and cause unjustified harm. The idea was that public shaming on social media might be uniquely suited to redressing imbalances of political equality, and so make a distinctive contribution to the health of a democracy. Ultimately, the approach fails because of the volatility of our psychological relationship to shame, the likelihood that introducing a mob element will lead to serious mistakes, and the vulnerability of already marginalised groups to any form of raw social power.

Where does this leave social media and its place in the public sphere? We could revert to the null hypothesis. Perhaps social media is just a technology and we make a mistake when we portray it as a distinctive space. If that is true then the proper conversation to have is about the impact it has on the actual parts of the public sphere and the challenges and opportunities it presents for the contributions that we need them to make. I'm inclined to take a different view. Political equality can be served by another foundational civil right, freedom of association, and this may be where social media can really come into its own.

We live in an age of surveillance. Everybody and his tamagotchi are after our data, and people are beginning to place a premium on the ability to create private, unobserved spaces in which they can be themselves. Users are forsaking more public forms of social media, such as Facebook, for apps like Whats App, Snapchat, and Signal where they feel more confident in their ability to keep strangers, governments, and corporations out. In terms of political equality, it is particularly important that relatively powerless or underrepresented groups should have the ability to come together to find their voice, articulate their experiences, and work their way through potential solutions to the problems they face. The right to create dedicated forums for distinct groups or issues can greatly increase our chances of getting at the truth. I will park that thought here for the time being, but it strikes me as one that merits further investigation.

References

Bennett, Christopher. 2008. *The Apology Ritual: A Philosophical Theory of Punishment.* Cambridge: Cambridge University Press.

Brennan, Geoffrey, and Philip Pettit. 2008. "Esteem, Identifiability, and the Internet." In *Information Technology and Moral Philosophy*, edited by Jeroen van den Hoven and John Weckert, 175–194. Cambridge: Cambridge University Press.

Brettschneider, Corey. 2012. *When the State Speaks, What Should it Say? How Democracies Can Protect Expression and Promote Equality.* Princeton, NJ: Princeton University Press.

Brooks, Thom. 2008. "Shame on You, Shame on Me? Nussbaum on Shame Punishment." *Journal of Applied Philosophy* 25(4): 322–332.

Carter, Ian. 2011. "Respect and the Basis of Equality." *Ethics* 121(3): 538–571.

Christiano, Thomas. 1996. *The Rule of the Many: Fundamental Issues in Democratic Theory*. Boulder, CO: Westview Press.

Davis, David Brion. 2006. *Inhuman Bondage: The Rise and Fall of Slavery in the New World*. Oxford: Oxford University Press.

Dewey, Caitlin. 2015. "The Successful Internet Shaming of 'Pharma Bro' Martin Shkreli." *Washington Post*, September 23, 2015. https://www.washingtonpost.com/news/the-intersect/wp/2015/09/23/the-successful-internet-shaming-of-pharma-bro-martin-shkreli/?utm_term=.c8e372040f20.

Fox, Carl. 2019. "Journalism, Offence, and Free Speech." In *Media Ethics, Free Speech, and the Requirements of Democracy*, edited by Carl Fox and Joe Saunders. London: Routledge.

Fricker, Miranda. 2007. *Epistemic Injustice: Power and the Ethics of Knowing*. Oxford: Oxford University Press.

Gilligan, James. 1997. *Violence: Reflections on a National Epidemic*. New York, NY: Vintage.

Jolls, Christine, and Sunstein, Cass R. 2006. "The Law of Implicit Bias." *California Law Review* 94(4): 969–996.

Lawlor, Rob. 2017. "Rejecting Amanda Machin's Complacent Democracy": *Environmental Ethics* 39(2): 211–233.

Levy, Neil. 2017. "The Bad News about Fake News." *Social Epistemology Review and Reply Collective* 6(8): 20–36.

Miller, Christian B. 2018. *The Character Gap: How Good Are We?* Oxford: Oxford University Press.

Mill, John Stuart. 1979. *Utilitarianism*. Glasgow: Collins.

Neu, Jerome. 2008. *Sticks and Stones: The Philosophy of Insults*. Oxford: Oxford University Press.

Nussbaum, Martha. 2004. *Hiding from Humanity: Disgust, Shame, and the Law*. Princeton, NJ: Princeton University Press.

Nussbaum, Martha. 2008. "Hiding from Humanity: Replies to Charlton, Haldane, Archard, and Brooks." *Journal of Applied Philosophy* 25(4): 335–349.

Rawls, John. 2003. *Justice as Fairness: A Restatement*. Cambridge, MA: Harvard University Press.

Ronson, Jon. 2016. *So You've Been Publicly Shamed*. London: Picador.

Saul, Jennifer. 2019. "Immigration in the Brexit Campaign: Protean Dogwhistles and Political Manipulation." In *Media Ethics, Free Speech, and the Requirements of Democracy*, edited by Carl Fox and Joe Saunders. London: Routledge.

Sherman, Benjamin R., and Stacey Goguen, eds. 2019. *Overcoming Epistemic Injustice: Social and Psychological Perspectives*. London: Rowman & Littlefield International.

Skorupski, John. 2006. *Why Read Mill Today?* London: Routledge.

Sunstein, Cass R. 2017. *#Republic: Divided Democracy in the Age of Social Media*. Princeton, NJ: Princeton University Press.

Waldron, Jeremy. 1987. "Mill and the Value of Moral Distress." *Political Studies* 35(3): 410–423.

Waldron, Jeremy. 2012. *The Harm in Hate Speech*. Cambridge, MA: Harvard University Press.

Williams, Bernard. 1973. *Problems of the Self: Philosophical Papers 1956–1972*. Cambridge: Cambridge University Press.

Williams, Bernard. 2008. *Shame and Necessity*. Berkeley, CA: University of California Press.

Young, Iris Marion. 2000. *Inclusion and Democracy*. Oxford: Oxford University Press.

CHAPTER 8

Smart Technologies and Fundamental Rights

Global Governance of AI: Pressure on Political Legitimacy

Stephen Rainey and Aníbal Monasterio Astobiza

1 Legitimate Political Decision-Making

Big Data and Artificial Intelligence (AI) have the potential to transform our societies in a variety of ways (Mayer-Schönberger and Cukier 2013). This includes using data and automated processing as an efficient means of contributing to social goods. But liberal democracies especially should be cautious about the temptation to use Big Data and AI techniques. Bring politics under the greater control of a *technological ethos* could have unexpected effects concerning legitimacy. The problem can be seen as follows: Data science -the blend of Big Data and AI- cannot inform political actions without reducing politics to technocratic solutions (Morozov 2013). The very same technologies that could bring us decentralized and efficient solutions could also bring about a kind of totalitarianism in the form of service delivery without any normative grounding or any direct participation by citizens themselves. This requires some further description.

Firstly, technological advances create inequality in their own right – for example in terms of the digital divide (Chinn and Fairlie 2007). An elitism that separates those who are technological savvy from those who are less sophisticated is a possible negative outcome of technologism in general. But secondly, and most important, AI powered by machine learning algorithms could drive political transactions on citizens' behalf. Where political legitimacy rests on voluntary cooperation in civil society to produce effective government, this ought to be concerning.

Second, the development of AI can not only erode the public space for political debate (Woolley 2018), it can endanger the very idea of civility and subsequently political legitimacy. The whole point of civility is to have a consensus, not just on an issue of debate, but on the way to conduct and govern our disagreements (Bejan 2017). With the advent of Internet-based social platforms, Facebook, Twitter etc. which are powered by sophisticated algorithms coming from various AI techniques, citizens and individuals are exposed to a

 | DOI:10.1163/9789004437876_010

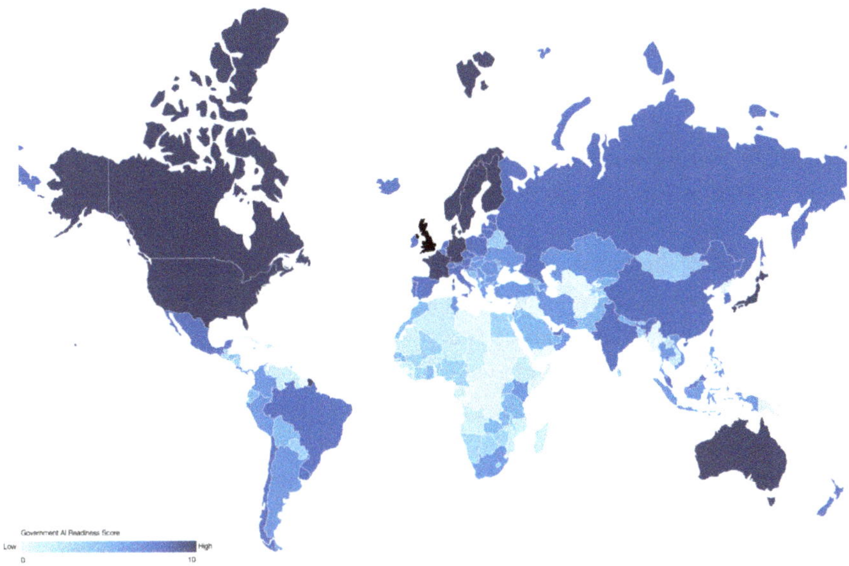

FIGURE 8.1 AI readiness index. Degree of government preparedness for AI being implemented in government functions.
OXFORD INSIGHTS

great deal of information and certain dynamics that are a danger to democracy because they cause political turbulence.

Internet-based social networks, such as Facebook, give us the opportunity to surmise what others think through the posts they upload on the platform. We meet with those people who think similar to us, so this produces selection effects and confirmation biases that lead us to only accept information that reinforces what we already believe (Bozdag 2013).

In this *infosphere* there is room for unintended outcomes such as "filter bubbles" that mobilize people collectively around certain ways of seeing reality that may be at odds with reality itself (Pariser 2012). These social networks use algorithms that build in a very sophisticated way the informational sphere to which we are exposed by selecting what we see and who we see it from etc.

These idiosyncrasies of digital technology generate effects of social polarisation and extremism in politics and public life that ultimately destroy civility. All over the world we've seen concerns for civility in public and political discourse. The election of Donald Trump as president of the United States, the resurgence of right-wing nativism in Europe etc., are clear examples. Openly, Donald Trump seems to despise women, undocumented migrants, Muslims and political rivals and the rising populism of the left and right maintain an intolerant, degenerating discourse across many regions of the globe. That these eventualities have likely been manipulated and exploited by companies such as Cambridge Analytica ought to give pause to those who would be otherwise optimistic about the datafication of the political realm.

A traditional conception of civility tells us that we have to establish rules and norms for an inclusive and egalitarian coexistence, because, ultimately, civility is about establishing the rules of social belonging. In the current digital age we are witnessing how products and services powered by AI deform the public space of political discourse and debate putting in jeopardy the very notion of civility. What do we mean here by "civility"?

Civility is about tolerance and consideration towards opposing points of view, and stewardship over the processes of consensus and dissensus. Civility is a matter of peaceful resolutions of political disagreements (Thomas, 2018, 102). And because political legitimacy rests on voluntary cooperation in civil society to produce effective government, without civility we have no political legitimacy.

In essence, political legitimacy is constrained by the view of output-oriented governance and the input-oriented dimension of political legitimacy (Scharpf 1999). These two facets of political legitimacy are reflected in the sharp contrast between, on the one hand, Lincoln's "government for the people" and "by the people" and, on the other hand, "responsible" or "responsive" government (Mair 2009). A technology-driven form of governance is sided more towards this latter form of governance in which justice and distributional issues are less important than efficiency gains or more responsive government.

Democracy in general is about influencing the ways in which representation is conducted. Citizen voices are mediated via representatives who sit in a parliament, for example, and so this parliament becomes a proxy for general societal discourse. The problem is that if we automate government with the use of algorithms and AI we delegate our political power to virtual entities and therefore we can't be sure they really represent us. The generalising force of electing members to a parliament is based in the expressed preferences of aggregates of individual citizens. In a Big Data and AI context the generalising force is more fine-grained: it operates on preferences, subsystems within individuals, and aggregates according to these. Rather than subjects taking centre stage, predicates become key. This is what we call *virtualising citizens*, administering the state according to virtualised imperatives based in statistical projections from these virtual citizens' apparent preferences. Too much trust in numbers is an oversimplification of something as complex as politics made by individuals of flesh and blood. All these issues echo a sociological maxim that "not everything that can be counted counts, and not everything that counts can be counted" (Lohr 2015, 10).

Institutions shaped by a data-driven approach such as this can only care about statistics. Imagine police chiefs who are obsessed only with crime statistics, not with what happens on the streets: if they were to change everything that needs to be changed to fix the numbers, this would put the cart before the horse. Thus begins the wonderful book *The Tyranny of Metrics* (Muller 2018) showing the pervasive consequences of what is known as Goodhart's law: "When a measure becomes a target, it ceases to be a good measure." And we can project that an automated government could only care about statistics, not about citizens and politics. It would not have access to the latter two notions in being based in projections from aggregations of apparent preference.

So far this may sound too general. Is anyone proposing a completely automated government? Perhaps not. But policy development increasingly aims for a basis in evidence. If policymakers are to avoid policy-based evidence-making, they require a lot of confidence in the evidence they can use. Political legitimacy in representative democracies is founded on, at least, government by consent. This means that "we the people" agree to an arrangement wherein elected officials enact political programmes according to proposals agreed to via elections. The realities of this often leave much to be desired, and innovations are sought to improve accountability.

One such innovation is the proposal to include AI in the development of policy. The promise is that data on "the people" will be collected, analysed, and used more efficiently, ensuring better representation. Where algorithmic analysis of vast datasets, comprising citizen behaviours and preferences, can lead to better policy insight, better politics and more just societies can emerge. This is the promise, but in fact this is better described as an aspiration or ideal.

We now proceed in analysing the issues associated with this kind of eventuality, diagnosing the problem in terms of citizens' representation, and suggest what kind of role AI ought to have, and how should it be deployed in governance.

1.1 *The Citizen and the State*

AI-augmented government could transform the way political administrations deliver services to the general public. But we need a sober reflection on what are the opportunities for AI in government and what threat it poses.

For decades, researchers in AI have sought to automate many tasks once the province of humans (Nilsson 2009; Gideon 1948/2014). AI software can play games, recognize voices and faces, learn and also make informed decisions within a multiplicity of domains. AI is behind technologies such as computer vision, speech recognition systems, machine learning, robotics, which are very powerful. And they can make government smarter. For instance, AI applications could potentially make public transport in cities and in transit more efficient reducing traffic jams, free workers from mechanical and routine tasks, scan millions of documents and identified relevant information from them, but more importantly, cut costs and reduce resource constraints.

Despite all these applications and benefits of AI in government, if not implemented carefully they can pose many challenges or exacerbate risks not only in service delivery, but also in privacy and ethics. We think that one of the most important challenges is the potential and revolutionary transformation of a new relationship and mode of communication between the citizen and the state.

Federal, state or local government provides services to citizens, but citizens' experience interacting with each of those forms of government can be very frustrating. The main problem is how to engage with elected representatives. If governance becomes more automated due to applications of AI, this problem of how citizens engage with governments and the often frustrating experience in return could grow and multiply.

If it is to be asserted that C20th governance infrastructure and logistics are growing obsolete, and the use of 21st technology governments all across the globe can improve citizens services, then there are many areas in which AI can be applied. Big Data and AI were mentioned above. However, our confidence in institutions and democratic representation is situated at a very low historical level both in Europe and more widely in the World (EVS 2011; WVS Wave 6 2010–2014.) Is automated government by means of AI is the right way to regain confidence in democratic institutions? Without high trust in the status quo, how can major revisions of practice be assessed by the citizens to whom they are to be addressed?

Various media frequently report the inherent bias in machine learning systems and other AI applications (e.g. racial bias, gender bias). Another type of bias, and one much less widely discussed, comes from the statistical projection of, and inferences made about, citizen's preferences based on large-scale data or databases. This introduces very general, and acutely problematic, biases based in general projections from aggregated examples. Such projections can be used for the profiling of citizens' behaviour, or even policy and decision-making. This is what we called *virtualising citizens*; a spurious projection of citizens' behaviour carried out by an automatic system of AI. Racial and gender biases, among others, are not less important than this kind of virtualisation, but this type of bias is more general in the sense that state automation involves generating digital profiles of every person regardless of race, gender, or any particular characteristic. The application of AI to the state involves the virtualization of people's preferences, essentially administering the state according to virtualised imperatives based in statistical projections from virtual citizens.

If our politicians can be subject to lobbyists, imagine the way in which an automatic system can be trained with corrupted data so that reflects in its construction the hidden interests of politicians, private companies, lawyers etc. And we can imagine a lot of real examples where AI tools and systems can be applied with dramatic consequences if implemented without any strategic foresight. For example, to refer to a real case, the use of risk-scoring system or the applications of AI in the criminal justice system. Thanks to the work of investigative journalism made by Propublica (Angwin 2016) we know that black people were more likely to receive harsher sentences than white people for the same crimes.

Many experts and lay people think that AI systems are logical and rational, but multiple studies show that machines are a reflection of human activity and acquire human-like biases (Caliskan, Bryson and Narayanan 2017). In other words, AI systems and machines echo human prejudices. When we apply AI to government we should not suppose that we are applying a neutral and innocent tool.

2 Perceived Problems in Political Legitimacy

AI could dramatically improve our lives at various levels, but we need to have a discussion about global governance for AI because of its perceived problems with political legitimacy when apply to government itself. It is true that government, from the perspective of diurnal activities and operations, is not a question of leaders and individuals but a process of rules, mechanisms and procedures. In other words, the governance of citizens and societies does not depend on individual actions but on systematic conditions in which our actions take place. AI technology might well be suitable for analysing this context marked by rule-based procedures where there is interdependence between factors.

Taking the example from Daniel Innerarity (2018) imagine the following thought experiment devised by Robert Geyer and Samir Rihani (2010, 188):

1) What would happen if the governors of the Bank of England were replaced by a room full of monkeys?
2) What would happen if Great Britain were to copy Norway's educational system exactly?
3) What would happen if a super-medicine were invented that suppressed all the symptoms of the common cold (or of our students' hangovers)?

If we had to answer quickly to these questions, immediate intuition derived from our "System 1" style of thinking would lead to the following answers:

1) The British economy would collapse.
2) Educational results would improve, since Norway's educational system is far better ranked than the United Kingdom's.
3) It would be a marvellous advance for personal health, since the patient would feel much better.

Nonetheless, as Innerarity points to, if we think carefully on our intuitive and immediate responses and start looking at things from the perspective of complexity, answers would look very, very different.

1) The government of monkeys would make manifest exactly to what point we are governed more by systems than by people, with checks, balances and counterbalances, and so the monkeys would do less harm than might be supposed.
2) The transfer of an educational system to another country would not be as successful as all that. There is of course much to be learned from the best practices of others, but the success of a system as complex as education depends a great deal on factors that are not automatically transplantable.
3) Being healthy is not the same thing as feeling well, and removing bothersome symptoms is equivalent to depriving oneself of signals and learning mechanisms that are precisely at the service of our health, understood as something more valuable than a mere absence of ill-being at any given moment.

This thought experiment is very interesting because it highlights the inadequacy of a too-individualistic reduction of complex systems to simplistic components. In terms of governance, this type of thinking may suggest that governance or government is a heroic pursuit made by single individuals, the leader or president, instead of complex processes of interdependence between many different factors and actors. With the reductive type of thinking we undervalue the underlying systematic properties of social complexity.

If we take seriously the systemic thinking and complexity of our socio-technological systems as exemplified by the conduct of governance or governments, it may seem easy to delegate control of this complicated machinery to artificial systems (computers or software) that compute and process the interdependence of the various factors of this social complexity more efficiently. Governments are expected to do relatively well irrespective of bad governors due to the fact that, at least in liberal democratic societies, the prolonged experience of democracy has provided a constellation of structures, processes and rules (e.g. institutions, constitutions etc.) that provide democracies with a resilience based on a collective intelligence. A government in this sense seems to be a collective intelligence system *par excellence*. Why not codify these ineliminable, resilience-enabling subsystems in technology? AI could be the instantiation or embodiment of the perfect collective intelligence.

But one central risk in seeing such technology as a genuine solution to the collective problems arising from governance is that of political legitimacy. Government does not arise from individual actions (those actions of leaders or single politicians) but from the aggregation of many individual actions mediated by norms, rules and procedures that give rise to a complex whole. This complex whole cannot be understood from the study of one of its parts but we have to approach it from the complexity of networks: the interdependence of multiple individual actions, *correctly enacted in legitimate contexts of action*. To understand this complexity we can fall into the trap that artificial computational systems that we design can better understand this complexity among the elements.

One of the most important factors responsible for a deficit in political legitimacy is the idea that representative democracy, as a collective choice based on peoples' preferences, is in retreat. Expressions of dissatisfaction in governance are global. The claim is made that politicians no longer represent people; They are deluded by many Siren calls from lobbyists, supranational pressures, short-termism. All of this provokes distrust in politicians and politics in general. And for reasons like these, many people could be tempted to succumb to the idea of *technological solutionism*. The freeing of governance from human partiality seems, on paper, a good idea.

3 Technological Promises for Political Legitimacy

To many people the solution or the way to fix the perceived problems of political legitimacy could be technological solutionism. In the face of poor governance, these people might wonder: Can Artificial Intelligence improve democracy? Can we use Artificial Intelligence to transform our governments? Technological solutionism might then be conceptualized in terms of two approaches: The application of strong, and of weak AI to government. We sketch each approach below.

– Strong version of AI applied to government

Dreams of an automated state that provides solutions and the best services to citizens are not new. In the fall of 1971, Stafford Beer landed in Santiago (Chile) to start an ambitious project: to assist the Chilean government in the design of a system that would help manage the economy by monitoring its behaviour in real time.

Beer was an internationally renowned operational research consultant who had experience in applied cybernetics, the science of feedback and self-regulation, based on Norbert Wiener's work at MIT in the 1940's (Wiener 1948). Beer had successfully applied these cybernetics ideas in the steel industry and when the Chilean government approached him, he saw the opportunity to test these ideas at a more large-scale.

Beer's conception of an effective government or to use our preferred term, automated state, was inspired by the metaphor of the nervous system that had inspired so many before him and would inspire so many after. The ability of the nervous system to respond to changes in the environment and to adapt is what governments need to imitate to survive and thrive (Beer 1974, 43).

The project materialized in the form of a telex network connecting different areas and industries of the Chilean economy with one central computer that processed the data received. An operations room (the brain of the system) was designed to be the centre of reception of all the information to make the most optimal decisions.

The cybersyn project as was called, a portmanteau of the words "cybernetics" and "synergy," was never completed as the 1973 military coup ended the dreams of Beer and his team. However, this unfinished project and the experience of designing this type of system highlights many of the problems we confront today with the development of digital technologies that they are increasingly entrenched in policy, governance, and politics in general.

Critics of the project saw its hidden authoritarian nature and its potential to become an all-controlling panopticon. At this time, the notion of "algorithmic regulation" or "algorithmic government," as it relates to the future of government, continue to respond to the very same aspirations. One of those aspirations is to increase the efficiency and adaptability of the decision-making processes.

Today this dream is still present in researchers working in computer science, data science and/or applied theoretical physics. For example, Cesar A. Hidalgo, director of the Collective Learning Group at the MIT Media Lab and associate professor at MIT, firmly believes that democracy can be updated or improved using technologies. In the future we will be able to automate various government functions. Politicians will be augmented by algorithms that work with teams of experts to *really* serve the people.

But his vision goes beyond that. His project, he calls "augmented democracy," it is intended to replace politicians because they are "filled with compromises" (Anzilotti 2018). And because of that, democracy has a "bad user interface." To correct this, many authors have thought of a direct and participatory democracy. In such a system every citizen votes on every issue. But Hidalgo says that there's a problem with that, a "cognitive bandwidth problem."

Our politicians, senators and/or representatives have to study very high numbers, on average, of documents, bills, etc. so it is very likely that they will not be able to asses all of them properly. It's their job to do so nonetheless. But for ordinary citizens, who work outside of politics, this can represent a very high cognitive burden. According to Hidalgo, the solution is to delegate our vote to a digital avatar. It is an algorithm that learns from you, from your preferences, that you train, that feeds on your data and predicts your choice. And as our thinking and preferences change and evolve, the algorithm and our avatars also change and evolve.

Technologies that allow people to get directly to power, bypassing representatives, creating a political participation which takes place through an AI agent. The only problem for Hidalgo and those who think like him, is if we are willing to implement something like this: Do we have a society where we really want to amplify the voices of all people?

For Hidalgo and his followers, an augmented democracy with AI would be a true democracy, but maybe we are still not prepared for it.

– Weak version of AI applied to government

The image above shows the degree of government preparedness for AI being implemented in government functions and tasks given the current state of the art in technology. According to this report from which the image is taken and written by Richard Stirling, Hannah Miller and Emma Martinho-Truswell; AI will revolutionise public service delivery or in other words governments. The report asked the following question: How well-placed are the national governments in the OECD to take advantage of the benefits of automation in their operations?

Taking into account different factor and measures (e.g. innovation as measured by Global Innovation Index, digital public services as measured by UN eGovernment Survey, etc.) the report provides a timely insight into the capacity of governments to capitalise on the innovative potential of AI.

Actually, AI can be applied to the public sector in the following areas:

1. Automatic evaluation of political and legal documents (reducing human effort and being less time consuming) making tedious tasks a thing of the past
2. Better decision making in education and planning (welfare payments, immigration decisions)
3. Using AI to redesign work processes
4. Predicting criminality, wildfires or poaching

5. Calculating where to allocate limited resources or prioritize where to invest
6. Predicting outbreaks or spreading of diseases
7. Identifying fraud
8. "Listening" to social media for public feedback (e.g. chatbots)
9. Understanding traffic patterns for planning
10. Understanding human mobility for planning
11. Strengthening cybersecurity
12. Drone surveillance (e.g. identifying violent individuals, monitor illegal border crossings)

This weak version of the application of AI to government contrasts with the strong version, but we believe that in no way can technology replace the foundations of political legitimacy based on communication, discourse as a *modus operandi* among citizens and values such as civility.

3.a) *Big Data and AI*

Big Data and AI have captured people's imagination and are shaping multiple economic and political aspects of our societies. The promises of these technologies are that they will solve all our problems, from the cure of cancer, crime, education, design of our cities, employment, to name just a few. Perhaps the boldest aspiration or promise is that Big Data and AI will be able to solve the problems of politics and cure the ailments of democracy.

The tech industry has built an overly optimistic rhetoric that exceeds the methodological capabilities of existing technology. To be sure, there's nothing new in Big Data and AI. They only represent the new incarnations of a type of logic present in industrial capitalism and currently in neoliberalism (Foucault 2010). Big Data and AI are just two quantification techniques that are part of a long history of using statistics applied to business for profit, control, exploitation of resources etc. The term Big Data was born in business. During the 2000's "Big Data" as a term enters the business discourse defined as the 4 Vs: volume, velocity, variety and veracity. Big Data is a neologism but collecting data for statistical purposes and to measure and manage populations dates back centuries (Muller 2018).

Today, there is no absolute consensus on the proper definition of Big Data but we can generally say that it is the belief that combining data from multiple sources could lead to better decisions. AI is not very easy to define either. On the one hand, it is a science and engineering that tries to automate processes, but, on the other hand, it is an aspiration to create intelligence in artificial systems. The concept of AI first came into contemporary use during 1950s (McCarthy, Minsky, Rochester, and Shannon, 1955).

Both the hype surrounding Big Data and AI can often serve to distract from the reality they inhabit. Company profits (e.g. Google) and research grants are at stake in this field, after all, and so overstatement is tempting for those involved. It is worth noticing that these very human socio-economic pressures are like those felt by politicians. From the beginning, it was predicted that with large amounts of data and new algorithms the machines would become more intelligent than human beings. This exaggeration continues to this day with a kind of magical thinking about AI (and Big Data) that dominates public discourse. This includes the fear that such technology could dominate and enslave us (Bostrom 2014). During the first decades of the birth of this new discipline, most AI research was conducted in academic labs. But progressively with the development of "expert systems," its application began to be seen in commercial and industrial settings. Our intention is not to offer a complete history of Big Data and AI but to observe how from its first steps now is being discussed its application in government.

4 Big Problems

4.1 *Virtualisation of Citizens – "You" Are Not Your Data*

The generalising of group preferences, values, and ideals in a political system is familiarly carried out in elections of members to a parliament. Being a citizen includes having a participatory right in this kind of exercise. As we have suggested, however, the virtualisation of citizens includes the displacement of the level of analysis from the citizen – a person, with collections of values, preferences, judgement, etc. – to subsystems within individuals, and aggregates according to these. Moreover, a kind of double abstraction is involved as the parameters of this subsystems are to be inferred from Big Datasets, using algorithmic means. The administration of a state according to virtualised imperatives based in statistical projections from these virtualised citizens' apparent preferences ought to be seen as a problem.

One arena from which the data on preferences might thought to be drawn would include social media. This has already featured in political discourse in the election of Donald Trump, and the UK's turgid Brexit debacle. In each of these examples (which are surely unlikely to be isolated cases) spurious projections of citizens' behaviour were carried out by algorithmic means, and then instrumentalised to steer opinion toward a pre-defined goal. This ought to be enough to demonstrate the non-neutrality of social media data, but the relation between citizens and their online behaviour nevertheless is worth exploring a little more.

In general, the mediation in social media posts of daily activities tends to suggest various ways in which "we" collectively endure, flourish, get along, and encounter strife. The advent of ubiquitous communications technology, and platforms to share rich personal data, has honed this to the point where a mythic "natural collectivity" is posited, and mediated digitally (Couldry 2014). The supposedly casual (and ubiquitous) nature of the data volunteered, throughout *a day in the life*, grounds this notion that natural, unvarnished behaviour is being collected. Specifically where Big Data is concerned, this issue becomes acute. As noted, the volumes of data make them multiply interpretable, as well as apparently suggestive of answers to questions unasked.

This is taken to represent a problem for sociology, for one thing, as social knowledge is at stake and rendered uncertain in this interpretive data "fog" (Scott 1999). Others see an opportunity, and self-consciously reject the idea of interpreting the data to get "an answer" to a question. Instead, these Big Data enthusiasts opt to generate from the data itself "proxies" to stand for what ought really to matter. Proxies can be related to one another, their interactions mapped, quantified, and evaluated. For instance, using location data from mobile phones might be used as a proxy for mapping relationships among people.

Proximities between phones could indicate regularities of association that indicate types of relationship; colleague, friend, partner. From these kinds of abstractions, it is hoped that inferences can be drawn to account for reality, without having to engage directly with that reality, opening a door to more intensive and wide-ranging data-driven research. This "datafied" (Mayer-Schönberger and Cukier 2013, 93) account is held as an improvement upon prior practices of observation, interpretation, evaluation, or critique of phenomenon. It is an improvement as it provides an objective view over a wide scope of information in a way that human observers could not. This has been reported as the "end of theory" (Anderson 2008) and a step up from more traditionally recognisable scientific methods.

Questions remain (boyd and Crawford 2012), however, and they can be seen to centre on the construction of the objects of study – how is this construction approached, with what material, and why? In a context of understanding the "we" in terms of online political behaviour – gaining insight on political views based on expression and comment in online fora – it is noted that "an understanding of the political and social contexts in which digital political phenomena unfold" is required as well as "taking into account the subjective viewpoints and motivations of those involved" (Gerbaudo 2016). The data, in other words, may stand as evidence, but some theory is required if it is to be understood appropriately, meaningfully, and in a way that will permit conclusions.

Echoing Bourdieu's nod to Kant, "theory without empirical research is empty, empirical research without theory is blind"(Bourdieu 1988, 775).

This can be taken to be a more general point relating to Big Data analysis in general, which is a set of procedures that are not self-explanatory, and do not present a context of discovery that can be easily assessed for bias (Bollier and Firestone 2010). The lesson from the sociological experience is that datafication of subjects, using proxies to stand for salient features of social reality, can become an end in itself and serve to obscure sociological insight. One might always argue that datafication of subjects is aimed at gaining particular knowledge about those subjects, so that datafication is a tool, not an aim. But the "end of theory" context just described suggests more is hoped to lie latent in data *per se*, in providing not only a basis for particular knowledge, but also the very questions to be pursued.

Where too much faith is put in the data and the operations upon which it is analysed, vital information, to do with context for example, can be overlooked. This potentially overlooked information can be central to giving meaning to the analysis, and so to the legitimacy of the sociological endeavour. What ought to be asked is something like: "Can relevant contextual data be discerned from the objective data before me? Can this objective data really be detached from this, or another, sense of its context?"

Essentially, there is no means of characterising an action "once and for all," in a mythically objective sense. This is irrespective of whether data on such action is gathered in a causal and comprehensive sense. Meaningful social action is enacted relative to a rationale, or in a way that connects to rationality, such that it isn't capturable once-and-for-all by a procedure. This can be illustrated by challenges to "rational choice theory," for example (Tversky and Kahneman 1986).

Bourdieu's "Habitus," or the internalised context of action that "frames" (Entman 1993) that action, perhaps tacitly, is required to understand a given action as such. This context can be discerned via interpreting action and so coming to an account of how it means what it might mean for the subject in the light of the context in which they act. The alleged objectivities of, "theories, conceptual schemes, questionnaires, data sets, statistical techniques, [...] define the visible and the invisible, the thinkable and the unthinkable; and like all social categories, they hide as much as they reveal and can reveal only by hiding" (Bourdieu 1988, 778).

The data can't neatly distinguish between the perception of observer-relative and observer-independent phenomena, for instance. It can't explain how these inform collective intentionality, and other such essentially perspectival dimensions of intersubjective association. Yet these are parts of what account for rational, social interaction (Searle 2006). The foregoing all serves to suggest that Big Data is not the "end of theory" as has been suggested (Anderson 2008), especially when it comes to political legitimacy.

4.2 *Virtualisation of State Interests – Governance Becomes Service Delivery*

Whereas it might be taken for granted that democratic governance is governance "by the people for the people," the datafication of citizens underwrites a transformation in this. Where the interests and perspectives of citizens are displaced by data that would propose to represent such things more clearly than before, states must respond to the apparent evidence they have before them. The administration of a state responds to virtual citizens if it bases its action upon such data.

Political exclusion of groups is a widespread issue, seeing marginalised groups underrepresented or actively ignored in important respects. The issue in such cases of exclusion often comes down to how power relations divert influence toward the interests of particular groups. However, at the extreme end, the datafication of a citizenry through a reliance upon Big Data representations of their interests, removes any particular citizenry's interests, replacing them with supposedly objective ones. The citizen, unframed, displaced from any intelligible habitus, becomes an objective contribution to a general dataset. The notion that a political system could operate objectively on data such as these is a deeply problematic one.

Political legitimacy is not something that ought to be thought of as a binary value that, once achieved, can be stabilised and proceduralised. Jürgen Habermas conceives of democratic legitimacy as a *modus operandi* for a state, and one that must be fostered as a dialogue rather than maintained as a mechanism. This is because citizens ought to expect only the force of argument to be wielded in civic living. Political legitimacy in this Habermasian picture maintains that citizens ought only to be exposed to the "unforced force of the better argument" (Habermas 1996, 306). On this basis, plurality and change within states is accounted for, rather than the static model implicit in the hopes for an automated state.

> To be sure, the liberalisation of values, an increasing willingness to include strangers, and a corresponding transformation of collective identities can at best be stimulated through legal-administrative means. Nevertheless, there is a circular, either mutually reinforcing or mutually inhibiting interaction between political processes and constitutional norms, on the one side, and the networking of shared political and

> cultural attitudes and convictions, on the other side. Old loyalties fade, new loyalties develop, traditions change and nations, like all other comparable referents, are not natural givens either. (Habermas 2011)

Institutional reality, political reality as it is encountered by individuals in a social setting, provides a sphere of valid reasons for social actors – it offers a mode in which reasons can come. Institutional directives, for example, can compel human actions as they offer reasons that can form the basis of action. The legitimacy of such reasons comes from just this point. Reasons are required for action *qua* action and institutional reality provides a scheme from which are derivable desire-transcendent reasons (Searle 2001), and a scheme through which identity can be placed or articulated (Brandom 2008), and so a scheme from which action can result.

Habermas' conception leaves plenty of room for uncertainty, for messiness, for change. This is what we ought to expect from a political system, as human circumstances themselves are variable, despite there being some fundamental needs such as nourishment, and shelter. Political processes ought to be modelled upon dialogical encounters, not the delivery of services. In expecting their flexibility to be replaceable with an algorithmic and objective scheme of needs-detection and ameliorative solution delivery, political legitimacy is replaced with technocratic convenience. This would be to posit a singular norm for public governance, as described here by Bohman with reference to John Rawls and public reason,

> [...]public reason is singular if it represents itself as a single norm of public deliberation; in light of this norm, agents come to agree upon some decision for the same publicly accessible reasons. As Rawls puts it: 'There are many non-public reasons and but one public reason.' (Bohman 1986, 83)

The guiding values for a state in the context of a deliberative, or dialogical model of politics include reason, accountability, just representation. But for an automated state, the sole value might be reason as efficiency – in data collection, processing, needs-identification, service delivery. The risk here is the erasure of the citizenry for the sake of the automated state's efficiency.

5 A Legitimate Role for Technology in Governance

A weak version of the application of AI to governance looks like the only potentially legitimate approach. A strong application looks like one that would obliterate citizens. But even where a weak application is considered, it must be carefully analysed in terms of motives, data, and processes at least.

The outcomes of algorithmic processes are political objects. The very idea that a state ought to automate at all, in what areas, for what reasons, are each dimensions of political concern. Do those in power wish to shrink the state for their own reasons? To enrich state justice by seeking greater detail for policy-responsiveness? To identify perceived allies or foes within a social group? All of this requires socio-political scrutiny. The data itself too, besides the motives for its use, is far from unproblematic. As we have been emphasising, it is not necessarily data that is proffered for some purpose. It may be derived, or inferred, from a large set of social media expression intended for quite another purpose (or no particular purpose at all). Data-driven outcomes are not ends in themselves, somehow Platonic answers to political problems, or pristine solutions to disorderliness. They ought to be treated as such, and scrutinised as any other political object like some manifesto, pamphlet, or speech.

Big Data sources, especially those connected with social media use, are not in fact neutral repositories of high quality data. They don't represent a people in a uniform sense. They are intentional acts, as in "acts done for a reason" (Davidson 2001). While their content may appear straightforwardly objective, the reasoning for it, its intentional content, is not readily reproducible simply by reading it off as data. Social media posts are *highly curated snippets* of various types of expression, of disclosure, of play, and other such human activity. If they represent anything as a bloc, it is the variety of this kind of activity.

If datafication is to have a role in emerging governance approaches it must take care to bear in mind the following conditions:

- Citizens are not identical with their data
- Data exploitation for the purposes of governance must feature as part of generally discursive governance measures, not in their place
- The unforced force of the better argument ought not to be overshadowed by a desire for coercion by objective data
- Modes of datafication must be engineered and tested in the light of citizens' interests
- They ought to be deployed only proportionally to their usefulness
- They ought to be abandoned where they add nothing to, or detract from, citizen based procedures
- Data sources must be recognised as interpretable in a variety of ways
- A collection of data is always the outcome of some kind of curation, tacit or explicit

These sorts of conditions will help to mitigate the risks of virtualising citizenries by acknowledging that citizen self-representation within a political system is more legitimate than apparent pattern detection within a dataset. Political problems may simply not be reducible or soluble. "Political problems" might as well be thought of as "politics" and gotten on with in a context of an adequate discourse.

6 Conclusion

Policymakers are understandably interested in Big Data analytics and "agent simulation," to "optimise" decision-making and policy implementation, among other things (Milano, O'Sullivan, and Gavanelli 2014). One upshot from this is that policy implementation becomes predicated on data processing, and simulation, not on direct disclosures from a citizenry. Thus a central part of policy becomes a technical exercise, based potentially in an AI black box. There are attendant presuppositions about data and how it relates to citizens, i.e. as a faithful representation, rather than a tendentious interpretation.

A critique of the thinking behind these interests is required, and timely. There is already an AI role in governance, it is already a part of political action. "Evidence-based policymaking" relies upon information provided by companies who deal in exactly the sort of abstracted-from-citizens "data" mentioned here (Einav and Levin 2014). The point seems uncontroversial - that there is no "raw" data (Gitelman 2013), and that tacitly "cooked" data playing a role in the political realm requires scrutiny. The data, not the citizens, steer policy implementation in this sense, and implementation is where accountability ought to really bite. How are these data "accountable"?

Data aggregated from the sorts of sources (e.g. companies just mentioned) represent no particular citizen or group of citizens. These data represent values and interests curated from broader sources. So any "optimised" policy implementation process they might contribute to is predicated on no-one in particular's interests or values, but a gestalt of "value and interest" that emerges from the (maybe automated) interpretation of a large dataset. That is a fundamental shift in the nature of policy implementation that represents a shift from politics as we know it, to service delivery. Ease is no justification for such a shift, as it has no mandate: A tyranny – rule by fear – would likely be easier to administer than a democracy, but we don't accept administrative ease as a justification for changing central political concepts. We ought not to shy from serious critique of technological interventions in political legitimacy, nor from remembering that technology is our tool to use, not an inevitable condition to which we must adapt.

References

Anderson, Chris. 2008. "The End of Theory: The Data Deluge Makes the Scientific Method Obsolete." *Wired* 16(7).

Anzilotti, Eillie. 2018 "This Plan for an AI-Based Direct Democracy Outsources Votes to a Predictive Algorithm." *Fast Company*, March 4, 2018. https://www.fastcompany.com/40557688/this-plan-for-an-ai-based-direct-democracy-outsources-votes-to-a-predictive-algorithm.

Angwin, Julia, Jeff Larson, Surya Mattu, and Lauren Kirchner. 2016"Machine Bias." *ProPublica*, May 23, 2016. https://www.propublica.org/article/machine-bias-risk-assessments-in-criminal-sentencing.

Beer, Stafford. 1974. *Designing Freedom*. Toronto: CBC Publications.

Bejan, Teresa. 2017. *Mere Civility: Disagreement and the Limits of Toleration*. Cambridge: Harvard University Press.

Bohman, James. 1986. *Public Deliberation: Pluralism, Complexity, and Democracy*. Cambridge: MIT Press.

Bollier, David, and Charles M. Firestone. 2010. *The Promise and Peril of Big Data*. Queenstown: The Aspen Institute.

Bostrom, Nick. 2014. *Superintelligence: Paths, Dangers and Strategies*. Oxford: Oxford University Press.

Bourdieu, Pierre. 1988. "Vive La Crise!" *Theory and Society* 17(5): 773–787.

Boyd, Danah, and Kate Crawford. 2012. "Critical Questions for Big Data." *Information, Communication & Society* 15(5): 662–679.

Bozdag, Engin. 2013. "Bias in Algorithmic Filtering and Personalization." *Ethics and Information Technology* 15(3): 209–227.

Brandom, Robert. 2008. *Between Saying and Doing: Towards an Analytic Pragmatism*. Cambridge: Oxford University Press.

Caliskan, Aylin, Joanna Bryson, and Arvind Narayanan. 2017. "Semantics Derived Automatically from Language Corpora Contain Human-like Biases." *Science* 356: 183–186.

Chinn, Menzie, and Robert Fairlie. 2007. "The Determinants of the Global Digital Divide: A Cross-country Analysis of Computer and Internet Penetration." *Oxford Economic Papers* 59(1): 16–44.

Couldry, Nick. 2014. "Inaugural: A Necessary Disenchantment: Myth, Agency and Injustice in a Digital World." *The Sociological Review* 62(4): 880–897.

Davidson, Donald. 2001. *Essays on Actions and Events: Philosophical Essays*. Oxford: Oxford University Press.

Einav, Liran, and Jonathan Levin. 2014. "Economics in the Age of Big Data." *Science* 346(6210): 1243089.

Entman, Robert M. 1993. "Framing: Toward Clarification of a Fractured Paradigm." *Journal of Communication* 43(4): 51–58.

EVS. 2011. *European Values Study 1981–2008*, Longitudinal Data File. GESIS Data Archive, Cologne, ZA4804 Data File Version 2.0.0.

Foucault, Michel. 2010. *The Birth of Biopolitics: Lectures at the Collège de France, 1978–1979*. Translated by G. Burchell. New York: Picador.

Gerbaudo, Paolo. 2016. "From Data Analytics to Data Hermeneutics. Online Political Discussions, Digital Methods and the Continuing Relevance of Interpretive Approaches." *Digital Culture & Society* 2(2): 95–112.

Geyer, Robert, and Samir Rihani. 2010. *Complexity and Public Policy. A New Approach to 21st Century Politics, Policy and Society*. London: Routledge

Gitelman, Lisa, ed. 2013. *Raw Data' Is an Oxymoron*. Cambridge: The MIT Press.

Gideon, Sigfried. 1948 [2014]. *Mechanization Takes Command: A Contribution to Anonymous History*. Minneapolis. Minnesota University Press.

Government Office for Science. 2016. "Artificial Intelligence: An Overview for Policy-Makers – GOV.UK." https://www.gov.uk/government/publications/artificial-intelligence-an-overview-for-policy-makers.

Habermas, Jürgen. 1996. *Between Facts and Norms: Contributions to a Discourse Theory of Law and Democracy*. Cambridge: MIT Press.

Habermas, Jürgen. 2011. *The Crisis of the European Union: A Response*. Cambridge: Polity Press.

Innerarity, Daniel. 2018. "The Government of Intelligent Systems." In *Smart: Paths to Sustainability*. ACCIONA. https://sustainabilitybook.files.wordpress.com/2017/12/smart_acciona_en.pdf.

Lohr, Steve. 2015. *Data-Ism: The Revolution Transforming Decision Making, Consumer Behaviour, and Almost Everything Else*. New York: Harper Collins.

Mair, Peter. 2009. "Representative versus Responsible Government." *MPIfG Working Paper* 09/8. Cologne: Max Planck Institute for the Study of Societies.

Mayer-Schönberger, Viktor, and Kenneth Cukier. 2013. *Big Data: A Revolution That Will Transform How We Live, Work, and Think*. Houghton Mifflin Harcourt.

McCarthy, John, Marvin L. Minsky, Nathaniel Rochester, and Claude E. Shannon. 1955. "Proposal for the Dartmouth Summer Research Project on Artificial Intelligence." Retrieved from https://www.aaai.org/ojs/index.php/aimagazine/article/view/1904.

Milano, Michela, Barry O'Sullivan, and Marco Gavanelli. 2014. "Sustainable Policy Making: A Strategic Challenge for Artificial Intelligence2." *AI Magazine* 35(3): 22–35.

Morozov, Evgeny. 2013. *To Save Everything, Click Here: The Folly of Technological Solutionism*. New York: Public Affairs.

Muller, Jerry. 2018. *The Tyranny of Metrics*. New Jersey: Princeton University Press.

Nilsson, Nils. 2009. *The Quest for Artificial Intelligence: A History of Ideas and Achievements*. Cambridge: Cambridge University Press.

Pariser, Eli. 2012. *The Filter Bubble: What the Internet is Hiding From You*. London: Viking.

Scharpf, Fritz W. 1999. *Governing in Europe: Effective and Democratic?* Oxford: Oxford University Press.

Scott, James C. 1999. "Seeing Like a State: How Certain Schemes to Improve the Human Condition Have Failed." *Utopian Studies* 10(2): 310–312.

Searle, John. 2001. *Rationality in Action*. Cambridge: MIT Press.

Searle, John. 2006. "Social Ontology: Some Basic Principles." *Anthropological Theory* 6(1): 12–29.

The Economist. 2016. "Re-Educating Rita." *The Economist Special Report: Artificial Intelligence*, June 25, 2016. https://www.economist.com/news/special-report/21700760-artificial-intelligence-will-have-implications-policymakers-education-welfare-and.

Thomas, Keith. 2018. *In Pursuit of Civility: Manners and Civilization in Early Modern England*. New Heaven: Yale University Press.

Tversky, Amos, and Daniel Kahneman. 1986. "Rational Choice and the Framing of Decisions." *Journal of Business* 59(4.2), S251–S278.

Wiener, Norbert. 1948. *Cybernetics: Control and Communication in the Animal and the Machine*. New York: John Wiley.

Woolley, Samuel. 2018. *Computational Propaganda: Political Parties, Politicians, and Political Manipulation on Social Media*. Oxford: Oxford University Press.

World Value Survey Association. 2010–2014. *World Values Survey Wave 6 v.20150418*. http://www.worldvaluessurvey.org/WVSDocumentationWV6.jsp.

PART 3

AI and Law

CHAPTER 9

The Rule of Law and the Protection of Fundamental Human Rights in an Era of Automation

Tanel Kerikmäe and Katrin Nyman Metcalf

1 Introduction

Society has changed more rapidly in the last decades than in previous centuries. Even if it is tempting for any era to consider any changes as more important than those that have happened earlier, there is evidence that new technologies spread more rapidly in recent times[1] and affect more parts of our daily life. Internet permits communication and interactivity which is almost instantaneous, just as in ancient tribal societies (Dutt and Kerikmäe 2014, 246; Howes 2001, 41). The increasing possibility to undertake various transactions without direct human contact, via different electronic means, is a change that not only affects social relationships but also governance (Hood and Margetts 2007, 202). From this viewpoint, it also affects the protection of fundamental rights and may influence the rule of law. Saying that it affects these matters does not include any judgement about whether it weakens or strengthens the protection, but it does underline the need to think about what is affected, how, and – if there is an effect – how to ensure that new ways of doing things do not weaken the protection of rights or the right to a fair, efficient and rule-based governance.

This chapter examines the protection of rule of law and fundamental human rights in an era of automation, looking at existing e-governance systems and plans for reforms by innovative uses of existing technology as well as by introducing more advanced technologies like artificial intelligence (hereinafter – AI) to a greater extent. We discuss why technology should be considered also from a legal perspective and what changes, if any, technology should entail for the law and protection of rights. We mention the example of Estonia in many contexts, not just – or even primarily! -because we are Estonian, but as Estonia is advanced in using e-governance and automation of various kinds. It is easy to fall into the trap of thinking that if technology is complex, the laws and rules relating to it must also be complex – that changes in society will always mean

1 See for example https://www.visualcapitalist.com/rising-speed-technological-adoption/.

 | DOI:10.1163/9789004437876_011

major changes to the law and the system of governance. Our starting point is instead that rule of law as well as the protection of fundamental rights are values that can and should remain in an era of automation, without the need for far-reaching changes. For this to be possible, it is nevertheless important to understand what it is that changes, how and why, and how it is perceived by people. With this knowledge, one can distil what is new and needs reflection in the legal system in order to better protect rights and strengthen the rule of law nature of governance. Although machine-learning is a less developed technology than AI, for the purpose of this chapter, we use the umbrella terms of automation and AI, referring to several inter-connected innovations used in governance and business that can have an impact on human rights.

2 e-Governance for Protecting Rights

Estonia is among the leading countries in the world in the field of e-governance – even *the* leading country for many issues. In Estonia, we can handle all communication with authorities electronically and sign all official documents with a widely used digital signature (Madise and Vinkel 2014, 61). Data is electronic by default, just as for example the valid, formal version of legislation is the electronic one.[2] Government went paperless in the year 2000, and 28% of Estonian voters chose to cast their ballot online in the latest elections.[3] The benefits of e-governance for a cheaper, more efficient and faster governance are much talked about and delegations from countries all over the world come to Estonia to study e-governance examples. It is interesting to notice that rich, democratic European countries are among those that have the most problems with comprehensive and rapid reforms, of the kind made in Estonia more than 20 years ago. Developing countries and countries in transition to democracy are more successful at taking decisive steps – making use of the same approach that Estonia used: when you have little to lose, you can leap-frog the developments others have gone through and go straight to something new.

The use of e-governance tends to be presented as if there is a need to weigh whether benefits of efficiency and speed make up for data protection risks. What is less well understood is that e-governance solutions and use of electronic data can, if properly used, provide greater data protection than traditional, paper-based data. In Estonia, one of the important features of the interoperable data system, the X-road, is that individuals (citizens or residents

2 https://www.riigiteataja.ee/abiLeht.html?id=1.

3 https://rk2019.valimised.ee/en/participation/participation.html.

of Estonia) can easily see what data the authorities hold on them, by accessing one web-site (www.eesti.ee) from which all public services and databases can be reached (Nyman Metcalf 2014). Whenever an authority accesses the data of an individual, this leaves a footprint, which the person can easily see. It says which authority looked at the data at what time and the person is entitled to ask the authority to explain why they did this. Inside the authority, it is also seen which person in the authority that accessed the data, as any data access is only possible after the authorised person has identified themselves – with the same ID-card and codes that is used in Estonia for all data access and access to public services (or the digital signature, valid also in private transactions). The authorisations for access to data within an authority are not given per authority but specify which persons can access what type of data. As the data access is electronic and the person must identify him- or herself for each data access, there is no risk of unauthorised individuals gaining access to data. Thanks to the footprint, there is furthermore no risk of authorised persons abusing their access rights out of curiosity, as there is a major chance that they will be called on to explain the purpose and proportionality of the data use (Rull, Täks and Norta 2014). Not only can individuals see who accesses their data, but there are controls conducted both within authorities and by the data protection inspectorate, who will react if unusual or unexplained data access appears to have happened (Nyman Metcalf 2017).

The protection of human rights is thus strengthened by use of electronic tools, but such aspects will not be appreciated by the population at large if they are not presented and promoted. In many countries, a majority of people still feel that electronic data is more vulnerable than paper-based data (Nyman Metcalf 2014b). To some extent, this may be a generational issue but there are also other reasons. The new things that can be done thanks to electronic data and automatic data processing, like face recognition or generally the combination of enormous amounts of data that at least in practice would be impossible to do manually, do entail new risks. In addition, as people provide so much data to social networks, commercial firms and to each other, there are understandably many instances of personal information being abused or at least too widely known. Most people do not in their appreciation of risks make a distinction between such issues and the public sector use of data, where – in democratic rule of law states – there are much stronger protection systems in place and fewer problems. No one likes to hear that the main problems are entirely self-created: you cannot expect privacy if you put your life on Instagram or Facebook! However, it cannot be ignored that this leads to people having an uneasy feeling about electronic data in general that may lead to scepticism also regarding e-governance. We cannot say that people should just know better and understand the distinctions. Not only is such an arrogant approach to people's fears and hesitations wrong in a democratic society that cares about its people, but it would also be counterproductive for the development of e-governance. If people do not trust services, they will not use them and in such a case those who develop services have less of an incentive to do so – leading to less availability of e-services so even those who would be willing to try such services have no special incentive to learn how to do so. A state cannot justify spending money on transitioning to a new form of governance if this is just a niche project, used by an elite, while the traditional ways to keep data or provide services are kept in just the same way as before. This means that efficiency gains of e-governance are not felt, nor can improved protection of rights be fully explored and exercise a real effect.

In Estonia, the level of trust in e-government keeps growing – the degree of digital transformation is intense and, e.g., the number of digital signatures collected in February 2018 hit 10 million, beating the previous year result of 6 million. Every look at person's data is registered in the system, and a misuse of private information constitutes a criminal offense. The trust towards the state processing of personal data in Estonia did not drop even during a security crisis back in 2017, when a group of Czech scholars identified vulnerabilities, which resulted in the need to renew the protocols of almost all the id cards issued. Nevertheless, the public admired the high level of transparency in how the officials have been handling the accident, thus the i-vote technology was employed by 31,7% of the voters during the same year elections.

Apart from seamless public services delivery, the digital era brought e-participation – a tool that empowers the civic dialogue and facilitates collecting feedback on planned or completed government actions. Accompanied by publishing open data sets, e-participation makes authorities much more vulnerable to public observation and criticisms, compared to internet-free bureaucracy. Information shared on government portals is sometimes regarded as providing a "false sense of transparency and accountability" (Gajendra, Xi and Wang 2012), because it is the agencies themselves who decide which data to share or hide from the public, but the e-platforms (forums, petitions) can make people's voices heard.

Estonia provides a good showcase for what e-governance means in practice. In addition to the mentioned benefit to the protection of rights, what may be noticed is that Estonia does not have a lot of specialised e-governance legislation. Indeed, in addition to e-identity legislation and regulations on the X-road, there is little that could be said to be for e-governance only. The focus remains on *governance*, with the electronic means just a tool to achieve the aim of governance measures. Mainly, the new way to do things can be applied to situations through a different interpretation of law with perhaps some adjustments (Brownsword and Goodwin 2012, 19–21).

3 Proactive Government: We Know What You Want

Automation permits proactive government. This is a new concept of government and governance, where individuals do not need to apply for things – or even contact authorities at all – unless they want to do something different, meaning not what is expected and mostly done in a certain situation. The government process sets in motion at the birth of a child, which is registered, and the child is given an identificator. From that moment, child benefit is determined, the child is assigned a doctor, it is registered for a future place in kindergarten and school, and so on. If the parents do not want a place in kindergarten for example, as they arrange alternative child care, they can inform about this in a simple, electronic manner, but if they have no special requests, the decisions will be taken without application. Special needs due to medical conditions and similar will be known by the system, as they are entered into the relevant databases by authorised professionals (medical personnel) so these can also be taken into consideration when decisions are taken. Such a system is possible due to automation and the fact that databases contain relevant information, as the decisions will be taken automatically based on the data entered into the system without a need to re-enter or apply specifically. The data in question to the most extent already exists in governmental systems, so the new aspect is not one of collecting more data but rather of how this data is handled.

In Estonia, there is a legal possibility to presume the will of persons. There is a Government Regulation from 2017 on "Principles for Managing Services and Governing Information" (adopted 3 June 2017, in force 7 July 2017).[4] The Regulation stipulates (in Article 2, extract) that (*3*) *Proactive services are the direct public services provided by an authority on its own initiative in accordance with the presumed will of persons and based on the data in the databases belonging to the state information system. Proactive services are provided automatically or with the consent of a person.* Proactive services are based on life events, rather than on applications. In the Regulation it states (Article 2, extract): (*4*) *Event services are the direct public services provided jointly by several authorities so*

4 English translation available at https://www.riigiteataja.ee/en/eli/507072017004/consolide, adopted under subsection 27 (3) of the Government of the Republic Act and subsection 6(2) of the Archives Act.

that a person would be able to perform all the obligations and exercise all the rights conferred on the person due to an event or situation. An event service compiles several services (hereinafter component service) related to the same event into a single service for the user.

The use of proactive services is mentioned in Article 7: *(8) If the information needed for the provision of a direct public service exists in the databases of the state information system, the authority shall devise a proactive service, where appropriate, in cooperation with the authorities administering the databases.* Some services are already proactive, but there is a plan to increase the number and variety of such life events with automatic consequences.

As mentioned, Estonian public administration is based on citizen centric electronic services in all areas, accessed from one web portal and in the same manner for all services, which makes them easy to use. An important enabler for seamless and efficient public services is the interoperability of databases. The system – called X-road – means that databases can communicate directly, meaning that authorities who are in need of data that exists in any public database will access this database rather than create their own – a practical application of the once-only principle, which means that a person never has to provide data more than once to the authorities. If the data already exists in a database, it has to be used from that database. It is prohibited by law to create a database with information that already exists, and persons should not be asked to provide the same information more than once (Public Information Act).[5] This system is more secure, as it eliminates the need for officials to have to ask for data and the need to send data between authorities in some form. Such situations of asking and sending are points of risk for data protection as well as for the accuracy of data. Interoperability of databases makes it possible to use data from the source, avoiding the need for duplication but also for gigantic, centralised databases – both risks for data protection as well as data accuracy and quality. The Administrative Procedure Act[6] contains requirements of efficiency, supported by the electronic system. The system provides a basis for moving on, relatively easily, to more proactivity.

It is well-known that for data protection reasons, personal data can only be used if it is in accordance with reason accepted in data protection legislation, like in the European Union (EU) General Data Protection Regulation (GDPR). In the debate, the discussion tends to be around consent and the question whether the consent people give to use of their data through ticking boxes or similar is proper informed consent. For the public sector, consent is most often

5 https://www.riigiteataja.ee/en/eli/514112013001/consolide.

6 https://www.riigiteataja.ee/en/eli/ee/Riigikogu/act/523012018001/consolide.

not the direct ground on which data is provided. It is possible also in this context to reach an explanation of consent, through a more philosophical discussion on the social contract as a basis for living in communities, but in practice such philosophical discussions are not held but instead the other possible bases for data processing are referred to, like the need connected to carrying out public tasks. The use of data needs to be for a purpose, proportional and so on. Greater automation of decisions does not add any need to move away from these requirements – if anything, eventually not only will people not have to provide data more than once, they may have to provide less data as the efficient use of existing data allows to "know what is needed" based on fewer elements.

4 Automation of Society: Anything New?

Global corporations were the first ones to find out the fascinating potential of machine learning and entrust the robots with decision-making power. Automation of decisions can provide important efficiency gains. Not only do computers take decisions based on large amounts of data more rapidly than humans, they are also less likely to be prejudiced, to get tired at the end of the day or to suffer from other human weaknesses. Thus, introducing this innovation into public administration was a matter of time, and today a large number of countries already apply machine-learning technologies in a broad spectrum of public sector services (from sustainable urban systems to criminal justice). In Estonia, the decisions on unemployment benefit are examples of automated decisions.[7] Anyhow, it's crucial to realise that even the "smartest" machines of today do not understand the ethical grounds of the rules they follow but base their decisions solely on data calculations. The latter can include both objective data (budgetary constraints, weather forecasts) and human-produced data (archives of court sentences), which are prone to discriminatory bias and other errors. The underestimation of this risk has been most clearly demonstrated in the US, where the predictive justice algorithm has developed a racial bias towards black offenders, due to inheriting the corrupt logic from the court decisions it has been built on (Hannah-Moffat 2018). Hence, in order to prevent the automated decision-making from becoming a frightening form of social control, the legislators have to set clear rules, based on the rule of law principles, strong ethical values and fundamental human rights.

7 https://digi.geenius.ee/rubriik/uudis/tootukassas-hakkab-sulle-robot-huvitist-maarama-kui-sa-tootuks-jaad/.

Our present-day era of great and ever-increasing automation is thus on the one hand not something entirely new for the legal system, but on the other hand it still in many ways challenges the legislators as well as the implementors of laws. Automation increases by degrees rather than incrementally, so in many ways existing principles are still valid. If a factory becomes almost totally automated rather than having some machines among many (human) workers, this may mean important changes in practical supervision, in the organisation of work, in the job descriptions of the remaining humans and so on. For the legal system, there is however little that needs to change. Responsibility and liability can still be determined, there may be licensing and insurance requirements to meet before the operations can start and so on. Autonomous vehicles moving among people in public spaces would look like something that leads to more new issues and indeed, in this case legal amendments may be required because existing laws will normally presume a human in charge. Precedents exist for autopilots in airplanes as well as for driverless trains, which are not recent inventions. Rules of responsibility will show a combination of legal and practical rules, combined with technology. With the more unpredictable nature of road traffic, for which there is no efficient general control system like air traffic control or central rail traffic control, new challenges exist. Fundamentally and philosophically the questions are not unknown, however. The very idea of compulsory vehicle traffic insurance was designed as motor vehicles are regarded as capable of causing great harm, for which victims should have a right to be compensated, even in situations where the person responsible for the harm does not have the means to pay and without the persons who suffer having to carry the entire burden of claiming the compensation themselves. This principle still fits, although the question of who is responsible in given cases is more complex. Both human and AI drivers can face a disastrous moral dilemma, when a car will, e.g., either inevitably crash into a group of pedestrians, or to sacrifice the vehicle and the lives of the passengers. At the same time, the policy-makers should assess the potential value of automated drivers not being at risk of driving under influence of drugs or alcohol, which regularly puts to death thousands of victims across the globe.

Machines form part of our society and need to be accommodated by the legal system. This ranges from the mentioned compulsory vehicle traffic insurance to rules on internet banking, from broadcasting legislation to safety standards for conveyor belts: machines need to be considered in law-making in order to ensure that risks they pose, that the different forms of interactions with them, do not weaken the protection of human rights or the possibility to enforce such rights. There are many amusing examples of how rule-makers have underestimated the pace of technical change, like the English law that a man needs to walk before every motor vehicle with a red flag to warn others[8] or the alleged 1899 statement that as everything that can be invented has already been invented so the patent office may be closed down.[9]

In early 2018, the global community was devastated by the revelations about the Cambridge Analytica algorithm, applied prior to the US presidential elections in order to collect psychological patterns and target voters with highly personalised political advertisements. Automation is also used for content moderation on the biggest social media platforms, but the detailed criteria for deleting "clearly illegal consent" remain hidden from the public, which makes civic activists talk about a new censorship era – there is no guarantee neither that well-meaning governments have the ability to really influence the tech giants, nor that the companies can protect against ill-meaning governments using algorithms. The various scandals that have been unearthed have raised a plethora of questions, like how governments should regulate the digital innovations so that the balance is maintained between public and private interests, and whether algorithms – due to their lack of transparency – pose a threat to privacy, democracy and rule of law as we know it? Some argue that if a product is free, then you are the product, implying that the users can only get a service in exchange for sensitive data, which is usually mentioned in the voluntary signed respective agreements. Anyhow, in our opinion, even the most expedient and high-quality services delivered by private actors cannot be regarded as a justification to underestimate the threats imposed by possible leaks or misuse of data by malicious users. The pivotal issue here is the *de facto* monopolies of GAFA (Google, Amazon, Facebook and Apple) in the digital space, which enable these corporations not only to simplify our lives and entertain us, but also to have an indeed incomparable impact on our news feeds, which have replaced the press as the Fourth Estate in liberal democracies. Today big businesses usually advocate for more flexible regulations of possible conflicts through ethical codes and amendments to their corporate policies, while their opponents urge policy-makers to prevent a mind-control apocalypse and to be more agile in designing new restrictive legislative frameworks.

Fortunately, although the basic question of what should come first – technology or law – remains a "chicken or egg" question, with no simple answer, humans have learnt to accommodate technology in the legal system in a

---

8 1865 England Locomotive Act. See https://archive.org/stream/statutesunitedk30britgoog#page/n246/mode/2up.

9 Attributed to Charles H. Duell, Commissioner of the US Patent Office. The quote is widely repeated in books or on-line, even if there is no actual evidence it was said or any credible original source. It is quoted as 1889 or 1899 in different places.

more flexible manner. This is expressed through principles such as technology neutrality, purpose-based regulation and specific regulatory agencies with specialised competence. The responsibility of a state to protect against human rights abuse by business enterprises domiciled in their territory has been stipulated in the UN Guiding Principles on Business and Human Rights,[10] which is entirely fitting for further development in international and regional agreements on AI challenges for our societies. The changes that need to be accommodated by the legal system to protect rights may have more to do with the importance of certain firms than only with the technology.

5 Artificial Intelligence: The Robots Take Over?

Where the situation is more fundamentally different is where AI is involved – the ability of the machine to go beyond the commands given to it. It is possible to automate many decisions also without AI if they are easily computable based on existing data and what is known about normal behaviour and similar. AI can however go further and itself decide also on issues where discretion and deliberation are needed. This is something that mostly happens step-by-step, so it does not have to suddenly present big new legal problems. Automatic data processing has led to computers "taking decisions" for a long time already, and in practice – at least for the layman – it may be difficult to understand when this is just a sophisticated execution of a command and when the computer decided something for itself. Clearly, the latter can entail new issues for our legal systems that are anthropocentric – based on humans. One of the central principles of decision-making in public sector is *non-delegation doctrine*, which conveys the impossibility for any branch of power to delegate its functions to either of the other, or to private entities. Thus, if a police officer, judge or a national security service official authorises an AI assistant to deal with discretional decision-making (parole, licence withdrawal etc.), could that be regarded as a non-delegation rule breach? Firstly, it has to be clarified, that the AI assistance in public sector could be used in two ways: either for collecting necessary set of data (weather forecasts, passenger mobility etc.), or decision-making *per se* (deciding whether a person is more likely to commit a felony again or which parent could better take care of a kid). In our view, full delegation of power is not the case in these examples, as the AI is used by judiciary and other agencies only as a tool which expands the horizon of human

10 https://www.ohchr.org/documents/publications/GuidingprinciplesBusinesshr_eN.pdf.

cognition and helps to grasp the whole picture in order to put on the scale all possible arguments *pro et contra*.

The discussion on the human-centred nature of legal systems has been held for example in the context of outer space law, where in the early days of such legislation (where much less was known about outer space than today) there was a serious discussion on how to – legally – interact with aliens if they should be encountered but also on what rights humans would have in other parts of the universe. Most commentators eventually agreed that inhabitants of the Earth can only legislate for other inhabitants of the Earth and consequently outer space law is the law on activities of Earth inhabitants – humans – in outer space.

Current research on legal and regulatory governance of AI systems tend to focus on whether existing legal systems can handle the consequences of further AI implementation. The main issue here is achieving a better understanding of the distribution of responsibilities and liabilities between humans and their use of AI systems, considering the plethora of interacting technology (e.g. chatbots) and the various players in the field, like developers, cloud providers, hardware manufacturers, telecommunications and network operators, and application and platform service providers. One of the approaches has been the "hard-wiring" of norms and values into socio-technical systems. This includes issues like compliance with security or data protection norms – including so called privacy by design, concerned with the implementation of AI techniques for differential privacy – and algorithmic governance through formal analysis and representation of regulatory principles, allocating rights, distributing liability, and managing legal identity.

Others have pointed out that law and privacy cannot be fully hardcoded (Koops and Leenes 2014) and may produce "legal by design" (Lippe, Katz and Jackson 2015) rather than "legal protection by design" (Hildebrandt 2017). According to Pagallo (2013) one can image a strict liability set-up being applicable for putting AI systems that may cause damage or harm onto the market, with the possibility to refute liability in case of a certification or evidence of security or safety by design. Such rules will "automatically" force people into compliance but may also restrict the technology. On the other hand, adopting ideas of legal protection by design allows to build transparency and contestability into AI infrastructures, as required, for instance, by the GDPR. The discussion around the legal situation for AI will need to look at whether conformity with normative values can be better addressed through "legal technology" applications that filter out human agency, such as blockchain technology, smart contracts, agreement technologies, and cryptocurrencies. It is in the nature of algorithms that they "find their own way" and thus are not transparent. The more intelligent AI gets, the more this will be noticeable. Therefore, as it has been recently pointed out by many authors (Brownsword 2016; Hildebrandt 2017), further research is required into the relationship and interaction between management of liabilities by legal technology and legal protection by design.

One central aim of regulation should be on introducing the responsibility to police and monitor systems and react against risks. AI is faced with demands concerning transparency, meaningful explanations of applications used, data protection, non-discrimination, presumption of innocence and privacy (Hildebrandt 2011). The question is, how such requirements – or any requirements – can be enforced. Countries consider whether it is possible to use existing regulatory structures (or a modification of these); if national insurance funds can be used; what possibilities there are of coordinating, aggregating and storing statistics of negative events, which allows private insurance companies to step in, scale up and develop relevant and agile products; whether humans must be able to (manually) regulate and control autonomous intelligent technologies or whether a regulatory technology or supervisory technology may be necessary. The area is very fast-moving both in the academic research area and among legislators. Good ideas abound but accepted solutions are still rare.

6 Humans: Co-existing or Dominating?

We touched upon the anthropocentric *nature* of the legal system related to outer space law and the question if "earthly" outer space law covers also aliens. The anthropocentric nature can be challenged in other ways as well, as the inhabitants of the Earth are not only humans. In recent decades, the idea of rights of animals not just as objects to be protected by animal cruelty legislation but as legal subjects in their own right, has caused a great deal of interest. The debate goes even further, with a river in New Zealand given status as a legal subject. Such discussions tend to be based on ideological ideas of emphasising the importance of protection of animals or nature, respecting traditional rights and customs and similar, rather than on a real expectation that these subjects will be able to appear in courts or other authorities to claim their rights. Indeed, the inability of exotic subjects to exercise their rights is an important reason why giving AI – autonomous agents or robots – legal subjectivity should be treated very cautiously, at least at the present level of development. It may appear as an easy solution to the pressing question of who is responsible for the actions of an autonomous machine, as it appears to solve the issue of whether the manufacturer (of software or hardware), the owner, the operator or someone else should be responsible if the machine causes damage – with the added complication of how it will even be possible to find out who these different subjects are. However, unless there is a way to extract responsibility from the machine itself, legal subjectivity does not actually solve anything.

These are important practical considerations for the ongoing discussions, all over the world, on how to ensure that the increasing presence of autonomous agents in society does not weaken important features of the legal system, like ensuring compensation for harm and instilling a responsibility for actions. As concerns the philosophical question of what changes with autonomous machines in our midst, there are several interesting points to make, starting from the same anthropocentric background. As opposed to space aliens or animals, the potential new subject of law is now something *created* by humans. AI indeed means that the machine has gone a step further and no longer needs the commands of its creator. Potentially, machines may start creating new machines without the involvement of humans, but the starting point is still a human creation. If we look at the situation of space aliens, the legal question would be – if and when we meet them – one of coexistence of two systems with the same fundamental rights to decide over themselves, with no priority of the rights of anyone over those of others. The legal matters will be those of how to ensure peaceful coexistence. The situation with animals (or nature, like rivers) is also one of peaceful coexistence where different creatures should be accommodated. With AI, the question may eventually also be one of peaceful coexistence, but at the moment it is also one of whether to allow these creatures to exist at all. Perhaps, for some time still (although we cannot be sure of this), humans have the option to actually limit the advent of these creations and thus also limit the rights given to them. We do not wish to suggest that this is a good idea, as such limits may seriously hinder innovation, especially as technological innovation tends to be such that one technology does not just do one thing, so it is not as if we can determine what is good or bad innovation.

AI can however go further and itself decide also on issues where discretion and deliberation are needed. Theoretically, this is not a problem from a rule of law perspective. It is indeed a principle of good administration that there should be discretion in administrative decision-making. One difference between authoritarian and democratic states is how the persons in position of authority take decisions. Interestingly, this difference and the desired outcome for a democratic, rule of law state, pulls in two apparently opposing directions. It is important that decisions are based on law, proportional and any limitation of rights is necessary in a democratic society. In authoritarian states it often happens that decisions are taken randomly, without a basis in law, due to factors that are irrelevant in the given case (like sentencing a known dissident to prison even for a minor traffic offence, even when regular traffic cases may be handled properly). However, another feature of authoritarian societies is that persons in authorities – apart from at the top levels – have no right to weigh alternatives and take circumstances into account, but have to take a set decision, prescribed from their hierarchy. This was the case for the Soviet administrative system, where employees were afraid to use their discretion but had to refer to higher officials even about details. Thus, the ideal for a rule of law state is something where there is a secure basis in law, but where those who take decisions are sufficiently prepared, educated and confident to make their own conclusions of what is suitable in any given case. If AI does this, there appears to be no reason why it – robots – could not take administrative decisions.

The technical challenge of creating understandable and value-based automated systems has already been addressed by the experts from the emerging field of Computational Ethics (aka Machine Ethics or Roboethics). Research findings demonstrate that scientists have already started composing models which enable the machines to plainly imitate human cognitive processes, adapt to a dynamic environment, and be capable of perceptual learning, e.g. the LIDA ethical agent, developed by Franklin et al. (Baars and Franklin 2009). Of course, it's impossible to completely "reverse engineer" the complexity of the human mind, but the pure possibility to teach robots moral yet impartial decision-making could be an advantageous assistance for policy-makers.

The supposed non-biased nature of AI has been discussed a lot in the debate. It may appear as if a robot is less likely than humans to be prejudiced against people of a certain ethnicity, sex or disability. However, some examples have been mentioned where this has not been the case, but AI has grouped people based on similar characteristics as what humans use, like racial features, and treated some categories worse than others. This may be a consequence of the way how AI is created, which is based on human input of data. Even if the idea of AI is that it "liberates" itself from human commands – it still was fed the data which is its lifeblood by humans, with all their prejudices. Maybe the clever robots will cast off such prejudice eventually. However, perhaps they will develop some of their own? The untransparent nature of the all-powerful algorithms means that it may be difficult for humans to understand such prejudices. Just like humans will not point explicitly to some characteristic as the reason for not hiring a person, but hide the prejudice in a web of apparently unrelated matters, it may take years before it is noticeable that the robot who decides on allocation of disability benefit always gives less to applicants with blue eyes who are between 32 and 46 years of age, for example.

7 Aims Rather Than Means

New technologies present many new questions. However, as we have shown in different contexts, law should not try to detail all possible answers to all these questions. The law does not need to provide detailed provisions but does need to give the conceptual tools for solving something. It should focus on outcomes and adapt to the circumstances and environment it is in: *"A new concept is justifiable when it enlarges our understanding of the phenomena under investigation"* (Todorov 2006, 15). If a rule is well written, the facts on which we apply it may change, but the activity we perform is the same and the tool given by the law is still useful. We can dig snow or plant a rosebush with the same spade, as long as it is a spade that is not too specific for one or the other activity. This is the essence of technology neutrality: you use tools to do something and it is not how you do it but the result you want to obtain that is the key.

The main novelty in creating rule of law in a digital world may in fact not be linked directly with the technology: It may be found in the way technology has changed the role of different actors. Private firms have importance of a greater magnitude than what we have ever seen before. When the importance of technologies increases, the requirements posed on operators of these technologies may have to be amended to support for example e-governance, where access to the communications network becomes essential for participation. The security of the network, of data and of the infrastructure increase as the importance of the use of the communications channel increases. Cyber security has to be considered as no state can expect its citizens to move their official interactions online if the environment is not secure. The emphasis in recent EU and national rules on security of network and information systems is no coincidence.[11]

The EU moved to what was long called "the new approach" in the 1980s,[12] moving from detailed requirements on technology to instead regulating what

11 Directive (EU) 2016/1148 of the European Parliament and of the Council of 6 July 2016 concerning measures for a high common level of security of network and information systems across the Union.

12 Especially for telecommunications and especially in the European Union. European Commission (1999) *Towards a new framework for Electronic Communications infrastructure and associated services.* COM (1999) 539. It was incorporated into the 2002 so-called

the aim of that technology should be (Kamecke and Körber 2008). For example, instead of describing the technology to be used in car engines to reduce harmful emissions, it was decreed what level of certain particles that could be permitted, which allowed the manufacturers to determine the best way to meet that goal. Various features like compulsory impact assessments for types of technology or for specific purposes (like environmental impact) serve this same purpose: to ensure certain aims are met but without details on how to do it. Technology neutrality should be taken into consideration at different stages of rule-making, in impact assessments and application.[13]

The absence of a supranational, democratic system for regulation in the increasingly inter-connected cyber world has pushed for other ways to regulate, leading to networks of actors as Morgan and Yeung (2007, 320–321) expressed it already more than a decade ago. This is what is often called multistakeholderism,[14] which has become a popular catch-phrase as well as an important phenomenon not least in the internet community. Multistakeholder fora include expert communities, activists and administrative officers. An excellent example of this collaboration result is the Toronto declaration on machine learning challenges for human rights, a joint effort from the domains of human rights advocacy and academia. The document is open for signature to public and private sectors and contains practical recommendations on how to balance technological advances with fundamental human rights (mostly the aspects of non-discrimination).[15] Morgan and Yeung explain how such untraditional groupings lead to complex and non-hierarchical patterns of influence and a hybrid range of techniques. This should ideally lead also to a more flexible regulation, which can withstand the changes to technology.

---

telecommunications regulatory package, reformed in 2009. It may however not always be well understood.

13 *Commission Staff Working Document: Impact Assessment, Accompanying the document Commission Recommendation on consistent non-discrimination obligations and costing methodologies to promote competition and enhance the broadband investment environment,* SWD(2013) 329 final. It can entail considerations like: *"The participants did not have an official single view on the issues raised in the public consultation. They mainly raised the need to further assess to which extent price differentials stem from national specificities or from the inconsistent application of costing methodologies and emphasized that any proposed costing methodology should respect the principle of technological neutrality and avoid promoting a particular network roll-out"* (p. 21).

14 Like the Freedom Online Coalition, https://www.freedomonlinecoalition.com.

15 Toronto declaration, https://www.accessnow.org/cms/assets/uploads/2018/08/The-Toronto-Declaration_ENG_08-2018.pdf.

8 Concluding Remarks: Do We Have Any Idea of What the Future Holds?

It is very pertinent what the European Court on Human Rights said in the *Marper* case in 2008,[16] that we must imagine things that we cannot even imagine today (as paraphrased). Technology develops so rapidly that it is impossible to regulate it in any detail. It may appear tempting to think that the big changes to society have happened and now we are in a phase of improvement and incremental change, but this does call into mind the idea in 1899 of closing the patent office… The balance between regulation and freedom to innovate may never be optimal, but that should not be seen as a huge problem: very few situations in which one has to balance different reasons lead to a perfect result. Indeed, just as digital signatures are not 100% secure but much more secure than traditional ones, the automated system of governance may be much more efficient and user-friendly than the traditional one, even while falling short of being perfect.

The human aspect of the situation must not be forgotten. It is quite possible that a system that is best at securing *human rights* is still not the most *humane* system. Protection of rights and the possibility to exercise freedoms are very important elements of having a good existence, but the complexity of human life and human nature means that there is much that cannot be expressed in terms of rights and freedoms, matters than cannot easily be expressed in conventions and laws. Humans already now do not always prefer to interact with other humans, rather than machines, but despite personal differences, there are still some things where most humans feel more comfortable with other humans. We tend to fall in love more easily with persons than machines and in situations of grief, comforting words or gestures of a human person mean more than those of a machine – even if in both situations it would be possible to rationally explain why this is not correct. A robot could be made to be the perfect partner and could say exactly what is suitable for the nature of grief. It is of course possible that the preference for the human is only because of the insufficiency of the machines until now, but already existing preferences in other situations tend to show that appreciation and emotions generally are indeed irrational and deeply "human." People still prefer books to e-readers,

16 Case of S. and Marper v. the UK, applications no. 30562/04 and 30566/04, Grand Chamber, 4 December 2008, https://rm.coe.int/168067d216. The actual statement was (paragraph 71): *Indeed, bearing in mind the rapid pace of developments in the field of genetics and information technology, the Court cannot discount the possibility that in the future the private-life interests bound up with genetic information may be adversely affected in novel ways or in a manner which cannot be anticipated with precision today.*

even if the latter would appear to have many benefits. Turning back to administration, until and if people will be happy to interact with a robot, it may be a duty of democratic societies to provide the human touch, for no other reason than that this is what people like. Liberty, privacy and security have to be ensured through a mutual effort of businesses, governments, NGOs, academia and other stakeholders.

We should not forget to see technology as an opportunity, but an opportunity not for itself but for what benefits it offers. When this is well understood, it can be determined what the risks are and how they can be avoided. This approach reduces two different risks. First, that so many threats are seen or imagined, that development of the new technology will not be allowed to proceed. Secondly, that technologies are developed that do not have any practical use or benefit. There is long experience in rule-making of environmental (and to some extent social) impact assessments. Such impact assessments should be made for AI and it should be an impact assessment, meaning benefits as well as risk seen as a package. Existing legislation on use of robots like health and safety legislation relating to industrial robots should not be forgotten as a source of inspiration, as in many instances introduction of AI is a change in degree rather than fundamentally (robots get gradually more intelligent, but they are already among us).

References

Baars, J., and Stan Franklin. 2009. "Consciousness is Computational: The LIDA Model of Global Workspace Theory." *International Journal of Machine Consciousness* 1(1): 23–32.

Brownsword, Roger, and Morag Goodwin. 2012. *Law and the Technologies of the Twenty-First Century*, Cambridge University Press.

Brownsword, Roger. 2016. "Technological Management and the Rule of Law." *Law, Innovation and Technology* 8(1): 100–140.

Dutt, Pawan, and Tanel Kerikmäe. 2014. "Concepts and Problems Associated with eDemocracy." In *Regulating eTechnologies in the European Union,* edited by Tanel Kerikmäe, 285–324. Cham: Springer.

Gajendra, Sharma, Bao Xi, and Qiang Wang. 2012. "E-government: Public Participation and Ethical Issues." *Journal of e-Governance* 35(4): 195–204.

Hannah-Moffat, Kelly. 2018. "Algorithmic Risk Governance: Big Data Analytics, Race and Information Activism in Criminal Justice Debates." *Theoretical Criminology*: 1–18.

Hildebrandt, Mireille. 2017. "Saved by Design? The Case of Legal Protection by Design." *NanoEthics* 11(3): 307–311.

Hildebrandt, Mireille. 2011. "Legal Protection by Design. Objections and Refutations." *Legisprudence* 5(2), 223–248.

Hood, Christopher C., and Helen Z. Margetts. 2007. *The Tools of Government in the Digital Age.* Palgrave Macmillan.

Howes, David. 2001. "e-Legislation. Law-making in the Digital Age." *McGill Law Journal* 47: 39–57.

Kamecke, Ulrich, and Torsten Körber. 2008. "Technological Neutrality in the EC Regulatory Framework for Electronic Communications: A Good Principle Widely Misunderstood." *European Competition Law Review* 5: 330–337.

Koops, Bert-Jaaps, and Ronald Leenes. 2014. "Privacy Regulation cannot be Hardcoded. A Critical Comment on the 'Privacy by Design' Provision in Data-protection Law." *International Review of Law, Computers and Society* 28(2): 159–171.

Lippe, Paul, Daniel Martin Katz, and Dan Jackson. 2015. "Legal by Design: A New Paradigm for Handling Complexity in Banking Regulation and Elsewhere in Law." *Oregon Law Review* 93(4).

Madise, Ülle, and Priit Vinkel. 2014. "Internet Voting in Estonia: From Constitutional Debate to Evaluation of Experience over Six Elections." In *Regulating eTechnologies in the European Union,* edited by Tanel Kerikmäe, 53–72. Cham: Springer.

Morgan, Bronwen, and Karen Yeung. 2007. *An Introduction to Law and Regulation.* New York: Cambridge University Press.

Nyman Metcalf, Katrin. 2014a. "The Future of Universality of Rights." In *Protecting Human Rights in the EU*, edited by Tanel Kerikmäe, 21–35. Heidelberg: Springer.

Nyman Metcalf, Katrin. 2014b. "E-Governance in Law and by Law." In *Regulating eTechnologies in the European Union,* edited by Tanel Kerikmäe, 33–53. Cham: Springer.

Nyman Metcalf, Katrin. 2017. "Drafting e-Governance: A New Reality for Legislative Drafting." *International Journal of Legislative Drafting and Law Reform* 5(1).

Pagallo, Ugo. 2013. *The Law of Robots.* Dordrecht: Springer.

Rull, A., E. Täks and A. Norta. 2014. "Towards Software-Agent Enhanced Privacy Protection." In *Regulating eTechnologies in the European Union,* edited by Tanel Kerikmäe, 73–94. Cham: Springer.

Todorov, João Claudio. 2006. "The Metacontingency as a Conceptual Tool." *Behavior and Social Issues* 15(1): 92–94.

CHAPTER 10

AI in the Context of Prevailing Privacy Concepts: In Search of a New Approach

Julija Kiršienė and Vygantas Malinauskas

1 Introduction

What does privacy mean in the current legal dimension and framework? At present, privacy as a concept encompasses a myriad of areas, among which freedom of thought; solitude in one's living environment and protection from invasions into one's home; control over personal information and ability to prevent disclosure of facts about oneself; freedom from surveillance, persecution, interrogations, and searches; the right of control over one's body; the protection of one's reputation; and some life decisions, like diaries, sexual behavior, reproduction, rearing of children, home, etc. can be distinguished.

We often expect to have privacy even when in public. For example, we consider the goods or services we buy, the books we read, the people we meet, the associations we join, and so on, as private matters. As Judge of US Supreme Court L. Brandeis pointed out, privacy is "the right most valued by civilized men" (*Olmstead v. United States*, 1928). So, privacy is of the utmost importance for a civilized, democratic, wealthy, free state, as well as for social wellbeing, as it promotes respect, love, friendship, and trust, likewise furthering individual autonomy and liberty. "Privacy is an issue of power; it affects how people behave, their choices, and their actions" (Solove 2002, 1143). Without privacy, social life is dehumanized and people are degraded.

On the other hand, as a matter of fact, interests of pivotal importance like public security, efficient law enforcement, freedom of speech and media, innovation, business efficiency, etc. are protected at the expense of privacy. For example, the need to disclose or prevent a crime can justify the use of surveillance techniques, public morality can justify a certain control of a person's sexual life, and freedom of expression can justify invasion into a person's personal life space and justify disclosure of intimate or confidential information (Foster 2003). In other words, in a situation in which competing interests intersect, often the right to privacy, figuratively speaking, loses. For example, in

 | DOI:10.1163/9789004437876_012

Court of Human Rights jurisprudence there is a provision that the right to information (media) outperforms the right of privacy in some cases of invasion into publicly known persons' private living spaces (*Lingens v Austria* 1986).

It is not hard to imagine why artificial intelligence (AI) raises unprecedented privacy concerns. For example, robots of all sizes and shapes can exceed the human capacity to observe; by sensing, recording, and processing the world around them, they have perfect memories, are tireless, and cannot be embarrassed. For example, household robots help us with vacuum cleaning, lawn mowing, window cleaning, or even more delicate and sensitive tasks like playing with children, taking care of pets, or preparing food or drinks. Since many household robots come equipped with sensors like cameras and microphones, it is obvious that they can be transformed into spying or other malicious tools to invade one's privacy and intimacy. Moreover, recent studies show that these products are vulnerable to technological attacks. So, if as "Bill Gates predicts, robots soon reach the prevalence and utility that personal computers possess today," this could have profound implications for household privacy (Calo 2011, 194). The problem seems even more serious taking into consideration the development of nanorobots, due to their small size, which makes them invisible to human eyes.

Furthermore, daily sellers of goods and services, like Google, Facebook, or Amazon, as well as public authorities collect huge amounts of information about us. Indeed, we passively acquiesce to this fact in exchange for the convenience of buying, choosing, and consuming. If we add the "algorithmic eyes" of various state authorities that observe us daily to this picture, looks similar to George Orwell's Big Brother. We could ask whether it is still important for human being to read books anonymously. But, as a matter of fact, private reading evaporated at least in virtual space with the internet and the e-book. We believe that we choose what we read and buy only what we decide for ourselves. However, we are reading selected news provided by social portals, because it is decided by algorithms based on our online identity features, and it is naive to expect that we still have some free choice online. In cyberspace, a person is only a set of information. Personal identity is perceived through patterns that are formed through the analysis of this information. What happens when a huge database of personal data is connected to an intelligence capable of analyzing such data? For AI, it is already feasible to sort and profile, using that data for decision making with regards to individuals and groups. However, there is a high risk that AIs will eventually start to marginalize and discriminate against certain groups in society. This has already been noted by the European

Parliament in its resolution of 17 March 2017 on big data. Thus, in terms of technological development, it is pivotal to adapt to the changing realities of this AI age, which causes numerous complex privacy issues, disrupting our traditional social practices of high personal value. New opportunities created by technological progress and the insufficiency of existing legal and regulatory frameworks make it necessary to search for alternative approaches to ensure an adequate protection of personal privacy in cases where others have access to private information.

In this publication, we argue that concepts of privacy, formed before the dawn of the AI era, do not provide sufficient legal instruments for the adequate protection of people from adverse consequences that may be caused by AI technologies. One of the assumptions underlying a conceptually new approach to privacy in the age of information technologies is the fact that there is less and less confidential information the disclosure of which can be still controlled by a person. These circumstances encourage us to shift the focus from legal restriction on the disclosure of private information to legal restriction on the use of personal information. In other words, in the age of AI, the protection of privacy should focus not on the protection of private information possessed by person, but on what kind of private information of a person can be used for what purposes and in what manner. For this purpose, the notion of *secondary private information*, which denotes private information derived by the means of AI technologies from publicly available sources is presented and elaborated. We argue that for the purposes of more adequate privacy protection in the age of AI, there is a need to differentiate between primary and secondary private information. Also, we propose that legal norms should establish the admissibility and conditions of the de facto use of secondary private information against interests of individuals.

2 Common Denominators of the Prevailing Privacy Concepts

As a conception of privacy can guide us toward shaping privacy policy and laws, first, it is profitable to make a brief survey of the prevailing privacy concept. As Solove argues, "philosophers, legal theorists, and jurists have lamented the great difficulty in reaching a satisfying conception of privacy" (Solove 2002, 1088), and because of its complex, various, distinct, even chaotic meanings, it is often difficult to address it usefully.

2.1 *Prevailing Privacy Concepts*

In this section, we address these questions: How has privacy metamorphosed due to changing social and political attitudes, living conditions, and technologies? Is privacy something like a bubble zone, which surrounds us in physical and mental senses as a person? What is the impact of this on the resolution of various privacy-related legal issues? How can conceptual analysis of the development of a privacy paradigm help us to understand the privacy concept better in the contemporary context?

2.1.1 The Right to Be Let Alone

A profound beginning toward developing a conception of privacy presented in article by Warren and Brandeis in 1890 is – "the right to be let alone" (Warren and Brandeis 1890, 193). This concept of privacy can also be regarded as one of the fundamental ideas of US privacy law (Solove 2002, 1105), the foundations of which can be found in the US Bill of Rights.[1] Under this concept, the privacy of a person primarily includes the privilege of each person to shape his or her own life as he or she wants to live without giving anyone the right to interfere with his or her free choices (except the legal basis provided by law) or, in other words, the right to plan his or her daily life according to his or her wishes.

2.1.2 Limited Access to the Self

A more progressive strain of the right to be let alone concept could be the concept of limited access to the self. Among the leading scientists who have developed or are currently developing this concept are E. Godkin (Solove 2002, 1103), S. Bok (Bok 1989), H. Gross (Gross 1967), E.D. Haag (Haag 2017, 149–169), D.M. O'Brien(O'Brien 1979) and others. This concept encompasses the idea that privacy as a need, and that a right is created by person's relationship to society. Privacy, as an aspect of social structure, is a condition in which a person can feel safe from the unacceptable and unwanted influence of others by attention (psychological aspect), by contact (physical aspect) or access to personal data.

This concept extends the scope of the right to be let alone concept in terms of including in it the relation to society and notions of the freedom from state and from public media interference. In summary, the basic idea of this concept can be described as the right to decide for oneself about the level of sharing with others.

Secrecy, control of personal information, personhood, and intimacy are often perceived as a subsets of this privacy concept, but they can also be distinguished as separate concepts of privacy.

1 "The right of the people to be secure in their persons, houses, papers, and effects, against unreasonable searches and seizures, shall not be violated, and no warrants shall issue, but upon probable cause, supported by oath or affirmation, and particularly describing the place to be searched, and the persons or things to be seized."

2.1.3 Secrecy of Personal Information

Secrecy of personal information is narrower than limited access to the self-concept. Privacy in this regard means the guarantee of full confidentiality of personal information as a way to limit access to the self. According to Posner (Solove 2002, 1106), privacy is a form of self-interested economic behavior, because one seeks to conceal some true but harmful facts about oneself for one's own gain, or because others might use it to one's disadvantage. The disclosure of such information without the consent of the person is considered to be an essential breach of privacy. This conception encompasses more than avoiding disclosure; it also involves the guarantee that personal information will be used for the purposes one desires. On the other hand, the secrecy as privacy concept can be also understood in the opposite sense: when any private information, in any shape or volume is disclosed, it becomes no longer private (and thus no longer protected). This is highly relevant in terms of using modern technologies.

2.1.4 Control over Personal Information

So, in addition to the traditional perception of privacy as a right to be let alone, control over personal information is one of the essential (still quite narrow) aspects of privacy and a subset of the limited access conception. According to Westin, privacy is the right of the individual, the group, or the institution to decide when, how, and to what extent private information can be communicated to others (Solove 2002, 1114). At this point, this is one of the most widely used privacy concepts in the legal framework. In Solove's view (Solove 2002, 1112), this concept is derived from the ideas of prominent philosopher and law theorist John Locke, who states that a person has the right to own property and the fruits of his work. So personal information is considered as a form of ownership of information. However, information, unlike material objects, can be easily transmitted and possessed simultaneously and it cannot be taken or eradicated from people's minds. Not only the psychological assumptions justify the need to protect a person's privacy, but also the economic ones. People constantly have information, including facts about themselves and the contents of their communication. Sometimes this information is valuable to others. According Richard A. Posner, the pioneer and most prominent representative of economic law analysis, privacy embraces at least two distinct interests. One is the interest in being let alone, and the other is concealment of information. Certain private information about individuals can be valuable, for example, to a tax administrator, creditor, competitor, partner, or fiancé. From an economic point of view, personal information that allows a person to carry out certain socially beneficial activities, such as commercial secrets, knowhow, goodwill, or reputation should be defended as property rights (Posner 1978, 397). However, it is less obvious why this information may be needed by a friend, colleague, employer, or neighbor. Of course, it makes it possible to get to know a neighbor, friend, or colleague better, and this, of course, affects relationships between individuals. Obviously, information on whether a candidate for a presidential election has been convicted can influence voters' decisions, so there is usually no question that such information should be public. On the other hand, it is not so clear whether a person has the right to know that his fiancé was convicted. As for now, the problem also lies in how the information is collected, processed, aggregated, and used when it is out of control, especially if it is used to make important decisions about one's life. This concept was complemented by the privacy aspect of *personhood*, according to which the essential aspect is the integrity of the right to be a person.

2.1.5 Personhood

Although the control over personal information concept, as mentioned earlier, is dominant in most current legal systems, its main drawback is the lack of attention to a person's private life, which cannot be reduced to a form of personal information. The concept of privacy as personhood focuses on the protection of the integrity of the personality. The right of every person to become, sustain, and develop can be considered a cornerstone of privacy. According to H. Gross, one's ability to perceive the uniqueness of one's being among other members of society, as well as the ability to perceive physical and psychological environmental factors, determines the perception of the privacy of each person, and thus it creates the structure of society we have today (Gross 1967, 35–40). It is noteworthy that continental European and American approaches to this concept differ, as America is much more oriented toward the value of liberty ,and by contrast in Europe, the values of dignity and honor are indeed at stake (Whitman 2003, 1160–1171).

In terms of liberty, private life is about lifestyle, relationships, rearing children, etc. So, privacy is about the individual liberty to choose and respect persons as self-creating enterprises (choosers) (Benn 2017, 1–26). Every person has the right to define his or her own existence, so liberty to choose is central for personal autonomy. For that, a focus on limiting state interventions in decision making and anti-totalitarianism must be maintained (Solove 2002, 1118–1120, 1095–1099).

As was mentioned, this concept also encompasses individuality, dignity, and autonomy as the cornerstones of individual privacy. Privacy-related social habits are vulnerable, and violation of personal information may have a negative impact on the balance of human dignity. So, the privacy right is one of a "much wider class of legal protections for interpersonal respect" (Whitman 2003, 1164). The concept of dignity gives rise to norms of respect for everybody – prisoners, women, minorities, etc. (Whitman 2003, 1160–1171), the ability for everybody to feel civilized and worthy of respect.

The relation of privacy to personhood is captured in *personality law*, elaborated in the doctrine of the German Constitutional Court (Eberle 1997, 979). The aim of personality law is to protect the "right to the free development of personality" (Eberle 1997, 979), which is closely related to the very concept of human dignity and autonomy. Personality law encompasses two components: "freedom of action and guarantee of a personal sphere" (Eberle 1997, 979). Both components are integral parts of a personhood (Eberle 1997, 992). However the protection of the personal sphere has special significance. According to the German Constitutional Court, basic law "guarantees individual citizens an inviolable area of personal freedom in which one can freely form one's life" (Eberle 1997, 993).[2] The Constitutional Court further elaborated the concept of *inner space* [*Innenraum*], in which individuals can "develop freely and self-responsibly their personalities, an *Inner Space* which they themselves possess and in which they can retreat, banning all entrance to the outer world, in which one can enjoy tranquility and a right to solitude" (Eberle 1997, 994).[3] The protection of the inner Space or the intimate sphere of each person is of paramount importance to the free development of personality. Any breach of privacy affecting the most intimate sphere of personal inner space inevitably infringes personal autonomy and makes the person vulnerable to outer control or other unwanted consequences. In this sense, privacy can be understood as an essential aspect of personhood.

2.1.6 Intimacy

Finally, it is necessary to mention intimacy as the latest attempt to conceptualize privacy. Privacy protection is important for the creation and preservation of personal relationships, because the maintenance and development of social relationships are impossible without intimacy (Gerstein 1978). It is common for people to hide some part of their relationships (often this part is called the intimate relationship, and privacy belongs to the nature of such relationships). According to S.R. Gerstein, privacy and intimacy are inseparable, and they must be

2 *Microcensus*, 27 BVerfGE 1 (1969).
3 *Microcensus*, 27 BVerfGE 1 (1969).

considered together for the sake of privacy protection (Gerstein 1978, 77). As Farber argues, the virtue of privacy today expands beyond simple rational autonomy: it is one of the fundamental factors influencing the development of human relationships (Farber 1993, 510, 514–515). Love, care, friendship, and trust are unthinkable without intimacy. However, the word intimacy is very vague, too, and it is often used as different word for privacy.

Thus, right to privacy is a Western cultural discourse about human self-determination and freedom. The basic concepts of privacy have always reflected the fundamental trends of the political-economic environment and the legal system of a certain period. In Table 10.1, the prevailing privacy concepts are summarized.

All these conceptions are either overly abstract and broad or too narrow and reductive. Can privacy be described as : the right of everything (Purtova 2018), because it seems responsible for any aspect of social and individual reality, like body, family, home, relationships, businesses, other institutions, and the state?

Briefly, ideas about privacy concepts can be summarized into three main clusters: (1) the spatial realm, mostly in terms of territorial and physical limited access, including shielding from unwanted invasions by observation, objects, or signals; (2) the informational realm, mostly in terms of use, disclosure, and control of personal information; (3) the expressive realm, in terms of one's ability to express oneself and to make choices by speech or action without interference. From the point of view of the key assumptions for protecting privacy, sustained personhood, preservation of human dignity and intimate relationships are quintessential for the full existence of an individual (autonomous, creative, responsible, etc.). Accordingly, the person's psychological wellbeing and health can be considered the connecting link in the framework of any privacy protection system.

3 Common Denominator: Personal Information

Despite the fact that the scope of the right to privacy remains the subject of discussion and there are different concepts of the right to privacy, it may be observed that these concepts share a common denominator, i.e., all of them are related to the information that is at the disposal of a person, the disclosure of which may have potentially adverse effects on a person's opportunities to nurture his or her life plans, and which may negatively affect a person's social relationships or social status. A person's interest in freely nurturing and pursuing his or her life plans is one of the fundamental interests of a person that derives from his or her personal dignity. Modern free and democratic society

TABLE 10.1 Main privacy concepts

Privacy concept	Short description	Key words
The right to be let alone	The privilege of each person to shape his or her life according to his or her wishes	Person, free choice, only limited by law against interference of others
Limited access to the self	The right to decide for oneself about the level of sharing with others; the condition in which a person can feel safe from unacceptable and unwanted influence of others by attention, by contact, or by access to personal data	Society, freedom from state, freedom from media interference, physical aspect, psychological aspect, personal data
Secrecy	The guarantee of full confidentiality of personal information; however, disclosed information is no longer private	Confidentiality of personal information, avoiding disclosure of personal information
Control over personal information	The right of the individual or institution to decide when, how, and to what extent private information can be communicated to others	Ownership of personal information, value of private information, economic considerations
Personhood	A person's private life cannot be reduced to personal information; every person has the right to become, sustain, and develop as he or she wishes	Human dignity, individuality, lifestyle, relationship, individual autonomy, respect, antitotalitarism
Intimacy	Privacy protection is important for the creation and preservation of personal relationships; privacy and intimacy are inseparable, and they must be protected together	Human relationships, trust, love, care, friendship

could not exist if these interests were ignored and neglected. The identification of the interest on which the right to privacy is based at the same time allows one to identify the nature of problems for the right to privacy that may be posed by AI technologies.

Different definitions of the right to privacy associate the interest of privacy with different values such as freedom from external interference (the right to be let alone) (Warren and Brandeis 1890, 193), the ability to decide on one's relation with the rest of the society (limited access to the self),[4] the ability to control private information (Solove 2002, 1114) and to preserve dignity and social respect (Gross 1967, 34), the preservation of the intimacy of personal relations (Gerstein 1978), personal autonomy (Levin and Abril 2009, 1015), etc. The infringement of privacy directly affects the above-mentioned values that underpin the interest of privacy. However, the infringement of privacy also has an indirect effect, i.e., it adversely affects a person's opportunities to nurture his or her life plans. A direct consequence of the disclosure of private information about a person's private life may be a violation of person's dignity and honor. Still, indirectly, it may adversely affect a person's abilities to maintain certain social relationships and status, which in turn may be a prerequisite for achieving personal goals in life. For instance, in moral terms, the disclosure of controversial aspects of personal life may prevent one from belonging to social circles or associations that follow a certain ethos (Levin and Abril 2009, 1015). It can therefore be reasonably stated that the protection of privacy must encompass not only the values based on which different concepts underpin the right to privacy, but also the concern to avoid undesirable consequences that are a direct or indirect consequence of the infringement of values protected by the right to privacy. These two aspects of the infringement of the right to privacy – a direct threat to values that underpin the right to privacy and the possibility of giving rise to negative consequences for a person – are important interests of a person, which should be taken into consideration when analyzing the challenges posed by AI to the right to privacy.

4 New Challenges to Privacy from AI

Developing AI technologies are gaining more and more opportunities to cause negative consequences for personal values protected by the right to privacy. AI can function as a virtual Sherlock Holmes, which, even before a person starts to say something, can tell about the person's marital status, political views,

4 See 1.2.2.

nationality, religion, income, place of residence, eating habits, addictions, financial situation, phobias, health status, etc. With the help of AI technologies, information, that is classified by the EU General Data Protection Regulation (GDPR) as special data[5] can be gathered without a person's knowledge from publicly available sources. AI can fully eliminate human privacy and make human efforts to protect private information futile. In a similar way, Sherlock Holmes's clients tried in vain to conceal some facts of their private life from the detective. However, AI can potentially disclose even more information about a person than the most intellectual or analytic detective can, which only exacerbates the problem.

Another problematic aspect of AI is that it can cause negative consequences for a person, which can be caused by the disclosure of private information, even without directly disclosing information about a person's private life. AI algorithms can assign a person to a certain category of persons without directly infringing his or her privacy. For example, a dating site algorithm can recognize a person's inclination towards certain undesirable habits and, in this way, determine people who could get recommendations to meet him or her. It should be pointed out that in this case, the AI algorithm does not make a person's habits public, but it leads to reduced opportunities to meet people who would be perfect for each other in all other respects. In addition, persons are deprived of the opportunity to decide for themselves whether bad habits are indeed a decisive factor in respect of the overall set of factors for deciding on the maintenance of future relations. Thus, the AI algorithm reduces a person's opportunities in ways that the person sought to avoid by not disclosing private information about him- or herself. In the case in question, the algorithm does not directly infringe the person's right to privacy. As a result of the application of this algorithm, private information about the person is not disclosed publicly. However, the AI algorithm, by formally preserving the status quo of a person's privacy, bypasses it, leading to undesirable consequences for the person, and, at the same time, for the goods that are protected by the right to privacy. This aspect is particularly important when considering the scope of the right to privacy in the context of AI technologies. In summary, it should be noted that when solving challenges posed by AI to a person's right to privacy, both of these aspects should be taken into account, i.e., the direct infringement of a person's private sphere and the causing of undesirable consequences that

5 EU Regulation 2016/679 of the European Parliament and of the Council of 27 April 2016 on the protection of natural persons with regard to the processing of personal data and on the free movement of such data, and repealing Directive 95/46/EC (General Data Protection Regulation). OJ L 119, 4.5.2016, p. 1–88. Art. 9.

correlates with the infringement of the right to privacy. The above-described situation makes it necessary to search for new legal safeguards of privacy and legal instruments ensuring satisfactory levels of protection. The need for these new legal instruments becomes obvious when examining the limitations of prevailing concepts of privacy in the context of the existing situation, where information technologies have the tendency to penetrate a person's private life unnoticed. These limitations become apparent when taking a closer look at the prevailing concepts of the right to privacy, i.e., control over personal information and protection of personal dignity (Levin and Abril 2009, 1009).

One of the prevailing traditional approaches to the right to privacy is associated with a person's ability to control personal information (Levin and Abril 2009, 1008). In the age of AI technologies, this concept is no longer sufficient, since a person has less control over his or her personal information because of the omnipresent eye of AI-powered data-mining technologies. The problem is exacerbated even more by the fact that AI technology allows companies to collect private information without intrusion into private space or secretly reading confidential personal files. AI technology may be employed to gather private information about a person through the analysis of publicly available data. For instance, AI can identify a person's religious and political views or his or her life plans for the nearest future, e.g., an intent to have children, taking into consideration patterns of personal behavior in virtual space. The analysis and gathering of information are done without a person even being aware of the fact of the private information gathering. Moreover, because of the black box effect, employing AI technologies for data processing makes it difficult to determine what private information and assumptions were extracted from publicly available sources and were used to make final conclusions.

4.1 *Hypothetical Situation*

The ability of AI to gather private information from publicly available sources allows us to imagine the following hypothetical situation: a woman looking for a job does not disclose private information about her plans to have a baby in the nearest future. The woman, who is in all respects eligible to take up the position she is applying for, decides not to disclose private information, known only to her, about the possible pregnancy, knowing that it may reduce her chances of getting the job. However, the woman's request is rejected, since AI provides a conclusion to the employer that there is a strong chance that the applicant will leave the job after a while. AI makes such a conclusion after having analyzed patterns of the applicant's behavior in virtual space. Based on pattern recognition technology, AI identifies that there is a great possibility that the woman is going to have a baby in the near future. This conclusion made by AI is only intermediate. It remains in the black box in the form of algorithms understandable only to AI. Neither the applicant nor the employer who used the AI to evaluate applicants are aware of the fact that the recommendation of AI to reject the applicant's claim is based on the information that belongs to the applicant's private sphere and which she decided not to disclose to anyone. The only thing that comes to the knowledge of the employer is the fact that the AI system, after evaluating the totality of available data, reached the conclusion that there is a strong probability that if employed, the woman will not stay long in her post. As the employer is looking for permanent and loyal staff, the conclusion provided to him or her by the AI is a sufficient basis for rejecting the application. Neither the woman nor the employer knows the real reasons why the AI system reached such a conclusion. Neither has any ground to suspect that the virtual Sherlock Holmes reached its conclusion using private information on the candidate that it discovered through analysis. The employer is quite satisfied with such a situation. Formally, private information is not disclosed, and the employer is not responsible for the disclosure of private information. However, the woman suffers consequences that are analogous to those that would occur if information about her personal life became publicly available.

An analogous situation may also be envisaged in the case of an application for a visa, loan, or insurance policy. A similar situation can be imagined in which a person pursues a professional or academic career, expects an early release from prison,[6] etc. In all these cases, AI may cause consequences that would normally result from an unwanted disclosure of personal information. Moreover, the way the AI works makes it very difficult to verify whether indirectly disclosed personal information was the basis for its conclusions. AI machines with learning capability can easily hide in the black box the fact that personal information derived from publicly available sources was taken into consideration by the machine when producing a conclusion. All this makes the concept of privacy, which is understood as a person's right to control private information about him- or herself (Levin and Abril 2009, 1008), less effective. In the age of AI, a person is no longer able to control and preserve large amounts of private information, which is understood as information "that identifies an individual or information about or relating to a person that most individuals in a given society at a given time do not want widely known about themselves"(Parent 1992, as cited Levin and Abril 2009, 1006). Therefore, to

6 In the UK, human rights groups are concerned about the use of algorithms in prisons. "Police warned about using algorithms to decide who's locked up," https://www.bbc.com/news/uk-politics-41996422.

preserve the relevance of the right to privacy in the age of information technologies, it is necessary to rely on alternative concepts of privacy. The concept of privacy as the protection of a person's dignity could be regarded as one such concept (Levin and Abril 2009, 1012).

4.2 *Pattern Recognition Algorithms and Prevailing Privacy Concepts*

The concept of the right to privacy as the right to the protection of dignity in its turn relies on the protection of values that are the basis of personal inviolability (Warren and Brandeis 1890, 205–207 as cited in Levin and Abril 2009, 1012). Such values protected by personal inviolability include respect for human dignity, personal integrity, and the independence of a person (Bloustein 1964, 962, 971 as cited in Levin and Abril 2009, 1012). In this respect, the infringement of personal privacy is like uncovering of the nakedness of a person and exposing him or her to public discussion, hypercritical evaluation, and critique, and, in this way, making a person vulnerable (Bloustein 1964, 962, 971 as cited in Levin and Abril 2009; 2013). Making a person socially naked by spreading information that belongs to his or her private sphere may cause huge negative social consequences for a person and create obstacles to the implementation of his or her personal life plans. However, both the concept of the right to privacy, which promotes the value of a person's right to personal inviolability, and the previously discussed concept of privacy as the right to control personal information relate privacy to the disclosure of certain information to other members of society. And yet, as has already been shown in the analysis of AI capabilities, AI can lead to a person's social vulnerability and prevent the implementation of a person's life plans without directly disclosing and making public specific private information to third parties. By using pattern-recognition algorithms, AI can circumvent the protection of personal privacy at the same time as apparently preserving confidentiality. As specific information about a person remains confidential, it becomes complicated, and, in some cases, even impossible to establish the infringement of the right to privacy in the traditional sense. As a result, the prevailing concept of privacy as the protection of personal dignity is not sufficient when dealing with problems related to the protection of personal privacy that are caused by rapidly developing AI technologies. The problem of the concept of privacy as control over personal information, as well as the concept of privacy as personal dignity, in protecting the privacy of a person in the context of AI is that the discussed concepts relate the protection of privacy to the protection of private information that is at the disposal of a person from unauthorized disclosure to third parties. However, the limitation of these concepts lies in the fact that they do not protect from the reconstruction of private information through the analysis of publicly available sources. Also, these concepts of privacy, formed before the dawn of AI era, do not provide sufficient legal instruments for an adequate protection of a person from adverse consequences that may be caused by confidential AI technologies. All this leads to a search for new legal approaches and instruments providing protection of personal privacy in the era of information technologies.

5 New Approach to Privacy in the Age of Information Technologies

One of the assumptions underlying a conceptually new approach to privacy in the age of information technologies is the fact that there is less and less confidential information the disclosure of which can be still controlled by a person. Another assumption is that with the rapid development of information technologies, the ability to reconstruct private information from publicly available sources is increasing. Thus, new opportunities created by technological progress and the insufficiency of existing legal and regulatory frameworks make it necessary to search for alternative approaches to ensure an adequate protection of privacy in cases where a person's privacy is infringed by using private information.

These circumstances encourage us to shift the focus from legal restriction on the disclosure of private information to legal restriction on the use of personal information. In other words, in the age of AI, the protection of privacy should focus not on the protection of private information possessed by a person, but on what kind of private information on a person can be used for what purposes and in what manner.

Such direction of legal regulation could help to achieve not so much the protection of formal private information (which is difficult, if at all possible, to protect in AI age), but the protection of the values that underpin the right to privacy. In other words, the concept of the protection of the right to privacy should include not only the aim to protect private information, but also the aims to protect people from negative consequences, if such information has been disclosed or used, by maintaining confidentiality.

5.1 *Primary and Secondary Private Information*

For the purposes of this proposed approach to legal regulation of privacy in the AI age, it is also appropriate to differentiate between primary and secondary private information. Primary private information constitutes private information that is at the disposal of a person. This is information that has been traditionally included in the notion of private information. Meanwhile, secondary private information is private information that has been reconstructed from publicly available sources by using the capabilities of AI technologies. Accordingly, it is possible to talk about the primary and secondary privacy of a person. Primary privacy is the right of a person to be protected from the disclosure of information that belongs to a person's private sphere. Secondary privacy means the right of a person to be protected from adverse impacts on his or her interests due to the use of indirectly disclosed or reconstructed private information of a person by third persons. The distinction between primary and secondary private information may be conceptually justified by the fact that secondary private information, which is obtained through the analysis and reconstruction of publicly available sources, is characterized by the probability dimension. Secondary information about a person's private life based on the conclusion made by AI that, for example, the husband is unfaithful to his wife, relies on probability (which may be higher or lower). That, in fact, is what makes it different from primary private information, the content of which constitutes of the fact of the husband's infidelity.

The different nature of primary and secondary private information, as well as the different possibilities for obtaining this information, justify and substantiate the need for the application of different legal regulations and different legal safeguards. Today's realities call for new legal safeguards of privacy capable of protecting individuals from negative consequences that may be caused by the use of secondary private information.

5.2 *Safeguards for Secondary Private Information*

It should be noted that some steps have already been made in this direction. For example, in 2017, the Rathenau Institute issued a report on "Human rights in the robot age," commissioned by the Parliamentary Assembly of the Council of Europe. The report addresses the potentially negative impact of robotics on a number of issues related to human rights, including respect for private life, human dignity, ownership, safety and liability, freedom of expression, prohibition of discrimination, access to justice, and access to a fair trial. One of the novel human rights recommended by this report is the right not to be measured, analyzed, or coached (related to possible misuses of AI, data gathering and the Internet of Things).

Also, some measures, at least partially protecting individuals from the use of secondary private information, have been enshrined in the GDPR,[7] the purpose

7 EU Regulation 2016/679 of the European Parliament and of the Council of 27 April 2016 on the protection of natural persons with regard to the processing of personal data and on

of which is to provide the data subjects with better control of their own personal data. The definition of personal data under article 4(1) of the GDPR[8] is mostly the same as in the former Data Protection Directive 95/96,[9] i.e., any information relating to an identified or identifiable natural person.

The regulation arguably establishes the right to clarification in the case of automated decision-making. The preamble to the regulation states that "[t]he data subject should have the right not to be subject to a decision, which may include a measure, evaluating personal aspects relating to him or her which is based solely on automated processing and which produces legal effects concerning him or her or similarly significantly affects him or her, such as automatic refusal of an online credit application or e-recruiting practices without any human intervention."[10] The regulation also stipulates that in cases of automated data processing, there should be "suitable safeguards, which should include specific information to the data subject and the right to obtain human intervention, to express his or her point of view, to obtain an explanation of the decision reached after such assessment and to challenge the decision."[11] In addition, Article 13 of the Regulation establishes the right of the data subject to receive, in cases of automated data processing, "meaningful information about the logic involved, as well as the significance and the envisaged consequences of such processing for the data subject."[12] Meanwhile, Article 22 of the regulation establishes the right of the data subject "not to be subject to a decision based solely on automated processing, including profiling, which produces legal effects concerning him or her or similarly significantly affects him or her."[13]

Even though the provisions of this regulation establish certain legal safeguards, they do not totally solve the problem of the use of secondary private information. This is because the regulation establishes the subject's right to require that no decisions are made concerning him or her relying solely on automated processing; however, the regulation does not specify the right of the data subject to require that secondary private information obtained after carrying out the analysis of publicly available data, including the analysis of data that have been publicly disclosed without the consent of the data subject, are not used at all against his or her interests when making a decision. Moreover,

the free movement of such data, and repealing Directive 95/46/EC (General Data Protection Regulation). OJ L 119, 4.5.2016, p. 1–88.

8 Regulation 679/16;

9 Regulation 95/96 Article 2(a);

10 Ibid. Recital 71.

11 Ibid.

12 Ibid. Art. 13, para. 2, (f).

13 Ibid. Art. 22, para. 1.

Article 22(2) of the regulation includes a derogation allowing the use of profiling and automated processing in case these are "necessary for entering into, or performance of, a contract between the data subject and a data controller."[14] Therefore, the provisions of the regulation do not solve the problem that arises when secondary private information obtained through the use of AI technologies is used against the interests of a person in making a decision to enter into a contract or for the policy purposes.

Furthermore, in terms of the practical application of the GDPR, there are concerns that in the hyperconnected online world, it will literally not be feasible to obey the requirements of the GDPR, because despite good intentions to provide the data subjects with the most complete protection, "the intensive compliance regime of GDPR will become 'the law of everything,' well-meant but impossible to maintain" (Purtova 2018, 40). Scholars argue that the broad material and personal scope of European data protection law combined with the high intensity of compliance under the GDPR make data protection apply to nearly anyone processing nearly any information at nearly any time, with the threat of serious sanctions omnipresent. Although the material scope of data protection in the GDPR was meant to be broad, because of the progress in data processing technologies and the amount of data available for analysis (Big data, Internet of Things, AI), the binary distinction between identifiable and non-identifiable information is bound to become meaningless[15] (Purtova 2018, 48). So, it seems that soon enough, any information like weather, waste, children's drawings, video, exam scripts, etc. can plausibly constitute personal data "regardless of the format, medium, or form, which could be 'alphabetical, numerical, graphical, photographical or acoustic,' 'kept on paper [or] stored in a computer memory' as a binary code" (Purtova 2018, 49).

The problem of secondary information was also addressed by "What if [...],"[16] recently created by Google and Fairness 360 Kit.[17] It contains software tools created by IBM that allow the identification of bias and profiling of AI systems based on unauthorized grounds, such as ethnicity, religion, sex, political views, etc.. Both tools can at least partly protect from the bias of automated categorization of individuals, and they make it possible to adjust AI preferences; however, they do not remove the problem of the use of secondary private information against the interests of a person.

14 Ibid. Art. 2, para 2, (a).

15 Many legal (often cited Ohm) and technical scholars (e.g., Narayanan) agree that at this rate, a meaningful distinction between identifiable and non-identifiable information, is not sustainable much longer.

16 https://pair-code.github.io/what-if-tool/.

17 https://www.bbc.com/news/technology-45561955.

6 Concluding Remarks

The discourse on prevailing privacy concepts associates the interest of privacy with different values, such as freedom from external interference (the right to be let alone), the ability to decide on one's relationships with the rest of the society (limited access to the self), the ability to control private information and to preserve dignity and social respect, the preservation of the intimacy of personal relations, personal autonomy, etc. It may be observed that these concepts share a common denominator, i.e., all of them are related to the information that is at the disposal of a person. However, in the age of AI technologies, there is less and less confidential information the disclosure of which can be still controlled by a person, because the rapid development of information technologies is increasing their ability to reconstruct private information from publicly available sources. The problem of the concept of privacy as control over personal information, as well as the concept of privacy as personal dignity in protecting the privacy of a person in the context of AI, is that the discussed concepts relate to the protection of private information that is at the disposal of a person from unauthorized disclosure of such information to third parties. These concepts of privacy, formed before the dawn of AI era, do not provide sufficient legal instruments for an adequate protection of a person from the adverse consequences that may be caused by confidential AI technologies.

For the purposes of new approach to legal regulation of privacy in AI age, it is appropriate to differentiate between primary and secondary private information. Primary private information constitutes private information that is at the disposal of a person. This is information that has been traditionally included in the notion of private information. Meanwhile, secondary private information is considered as private information that has been reconstructed from publicly available sources using the capabilities of AI technologies. To ensure an adequate protection of privacy in cases where a person's privacy is infringed by using secondary private information, it is crucial not to protect the private information of a person, but to protect a person from the use of such private information in cases where such information may make a person vulnerable. Such direction of legal regulation could help to achieve not so much the protection of formal private information (which is more difficult, if at all possible, to protect in AI age), but the protection of the values that underpin the right to privacy. In accordance with the proposed approach, legal acts should establish provisions regulating the admissibility and conditions of the use of secondary private information against the interests of a person, and they should establish safeguards protecting individuals from the de facto use of secondary private information against their protected interests.

References

Benn, Stanley I. 2017 "Privacy, Freedom, and Respect for Persons." In *Privacy and Personality*, edited by Russell L. Ciochon. New York: Routledge.

Bloustein, Edward J. 1964. "Privacy as an Aspect of Human Dignity: An Answer to Dean Prosser." *39 N.Y.U. Law Review:* 962–1007.

Bok, Sissela. 1989. *Secrets: On the Ethics of Concealment and Revelation.* Vintage Books.

Calo, M. Ryan. 2011. "Robots and Privacy." In *Robot Ethics: The Ethical and Social Implications of Robotics*, edited by Patrick Lin, George Bekey, and Keith Abney, 178–201. Cambridge: MIT Press.

Eberle, Edward. 1997. "Human Dignity, Privacy, and Personality in German and American Constitutional Law." *Law Faculty Scholarship*, January 1, 1997. https://docs.rwu.edu/law_fac_fs/68.

Farber, Daniel A. 1993. "Book Review: Privacy, Intimacy, and Isolation by Julie C. Inness." *Constitutional Commentary* 10(2): 510–519.

Foster, Steve. 2003. "Human Rights and Civil Liberties." Pearson Education.

Gerstein, Robert S. 1978. "Intimacy and Privacy." *Ethics* 89(1): 76–81.

Gross, Hyman. 1967. "The Concept of Privacy." 42 *NYU Law Review*: 34–54.

Levin, Avner, and Patricia Sánchez Abril. 2009. "Two Notions of Privacy Online." *Vanderbilt Journal of Entertainment & Technology Law* 11: 1001–1051.

Lingens v Austria. 1986. 8 EHRR 407.

O'Brien, David M. 1979. *Privacy, Law and Public Policy.* New York: Praeger Publisher.

Olmstead v. United States. 1928. 277 U.S. 438, 478.

Posner, Richard A. 1978 "The Right of Privacy". Sibley Lecture Series. 22. https://digitalcommons.law.uga.edu/lectures_pre_arch_lectures_sibley/22/.

Purtova, Nadezhda. 2018. "The Law of Everything. Broad Concept of Personal Data and Future of EU Data Protection Law." *Law, Innovation and Technology* 10(1): 40–81. https://doi.org/10.1080/17579961.2018.1452176.

Solove, Daniel J. 2002. "Conceptualizing Privacy." *California Law Review* 90, 1087.

Van Den Haag, Ernest. 2017. "On Privacy." In *Privacy and Personality*, edited by Russell L. Ciochon. New York: Routledge.

Warren, Samuel D., and Louis D. Brandeis. 1890. "The Right to Privacy." Harvard Law Review 4(5): 193–220.

Whitman, James Q. 2003. "The two Western Cultures of Privacy: Dignity versus Liberty." *Yale Law Journal* 113(6): 1160–1171.

CHAPTER 11

Artificial Intelligence as a Subject of Criminal Law: A Corporate Liability Model Perspective

Edita Gruodytė and Paulius Čerka

Corporation- an ingenious device for obtaining individual profit without individual responsibility

Slapper and Tombs (1999, 22)

1 Introduction[1]

"Criminal law generally is retrospective and individual-oriented and not suitable to direct the development of technologies" (Beck 2016, 141). The content of criminal liability depends on the norms of social behavior prevailing at that time (including moral and ethical norms; Drakšas 2008, 11). Criminal law is currently based on the protection of human rights and freedoms as well as society and state interests (Wessels 2003, 30) and consists of three main components: the human being, the society, and the state (Leroux et al. 2012, 51). In essence, a criminal offense is a violation of legal good and duty. It could be argued that "the illegal content of a criminal offense is revealed through the harmfulness of the consequences of the activity (violation of a protected object and encroachment upon it) and harmfulness of actions (the nature and manner of the committed actions)" (Wessels 2003, 30). One important criteria that distinguishes criminal liability from other branches of law is that the aim in criminal law is not compensation of damage, but punishment of the perpetrator by state authorities (Leroux et al. 2012, 51), since it could be argued that, for certain egregious offenses, there should be also condemnation of a person on behalf

1 This research was funded by the European Social Fund according to the activity "Improvement of Researchers" qualification by implementing world-class R&D projects of Measure No. 09.3.3-LMT-K-712.

 DOI:10.1163/9789004437876_013

of the state. If artificial intelligence (AI) committed a criminal offense, could AI be treated as a criminal? This question is especially relevant given recent societal transformations and the "changing nature of society's perception of, and relationship to, crime" (Keiler and Roef 2019, 37). The previous conception of the welfare state shifted the risk to society as a consequence of "modern industrialized and globalized societies" confronted with new risks such as modern technologies (Keiler and Roef 2019, 38).

The rapid development and modernization of technologies may improve human lives, yet new technologies also provide uncertainty and require the prediction of risks and solution to risks that may arise in the future. A key element in ensuring the responsible use of technology for advancing human interests are human rights, which perform a crucial task in "preventing the wrong" (Koops et al. 2013, 15). The rapid development of AI could present risks to human rights in the near future, thus the development of an optimal regulation model should be established in order to protect human rights. Criminal law may present an instrument that could be used in certain cases dealing with AI risks. The main goal of this article was to determine whether AI could be treated as a subject of criminal law using a corporate liability model perspective. We used the definitions of AI, the rationale and arguments establishing criminal liability for corporations, and the historical development of corporate criminal liability to analyze the question of legal capacity of AI. Throughout our analysis, we considered arguments of mens rea found in corporations and the main theories of legal entity and their rationale for grounding AI criminal liability.

2 Limits of the Research

Criminal liability and corporate criminal liability models are discussed as possible alternatives that could be applied to AI liability. The possibility of legal person criminal liability was not used as an argument here for granting AI legal personhood, but rather it served as a defense to the objection that if such personhood would be granted to AI, then it does not contradict to the concept of criminal liability of AI. Legal personality itself does not relate to real personhood, since legal personality is a metaphysical or ethical notion of personhood. "When we say that an actor has legal personality, we mean that a legal system addresses its rules to the actor, both to give the actor rights and to subject it to obligations. Legal personality is not necessarily correlated with a metaphysical or ethical notion of personhood" (Bryson, Diamantis, and Grant 2017, 277). The personal liability of an operator or creator of AI due to negligence and arguments against or for granting legal personality are not analyzed here. The concept of legal personhood, originating in the legal systems of ancient Rome, is far older than the concept of criminal liability of a person (Horsman and Korsten 2016, 277–285). Therefore, it is obvious that the validity of considering AI as a legal person or full moral agent cannot be answered by idea of criminal liability of AI, because the idea of legal personhood preceded that of criminal liability.

3 Artificial Intelligence as a Subject of Criminal Law

For the purposes of this article, we adopt the definition of AI as a technologies "that display intelligent behavior by analyzing their environment and taking actions–with some degree of autonomy–to achieve specific goals" (European Commission, 2018, 2). Only autonomous AI–the ones designed to learn to adapt their behavior in accordance with the interaction of environment–are considered in this article (Figure 11.1).

AI autonomy could be defined as "the capacity to operate in a real-world environment without any form of external control, once the machine is activated and at least in some areas of operation, for extended periods of time" (Bekey 2005, 1). Therefore, a programmer or owner of AI may not be criminally

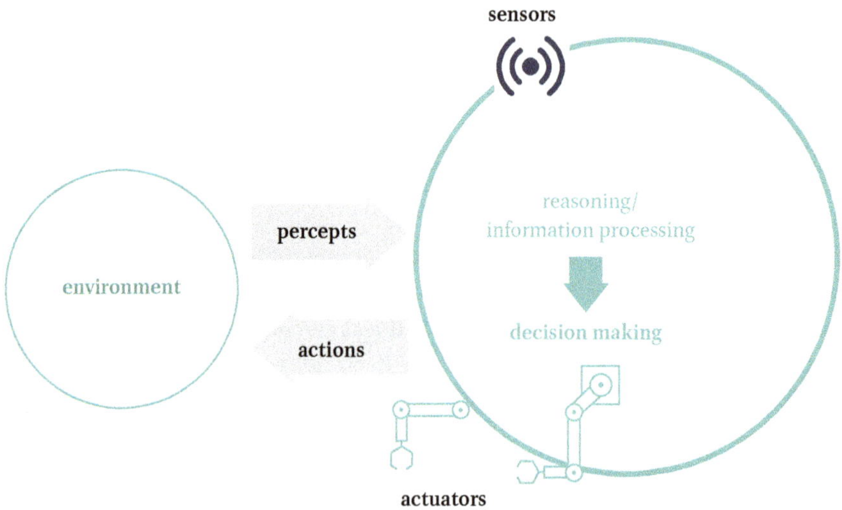

FIGURE 11.1 The illustration of an AI system
TAKEN FROM EC HIGH LEVEL EXPERT GROUP (2018, 2)

liable for the actions of AI, since these persons cannot foresee or fully control the decisions of autonomous AI. One of the two main elements of traditional criminal law grounding is the guilty mind (mens rea),[2] but since neither the owner or the programmer cannot control or foresee the decisions of AI, mens rea cannot be established in AI criminal cases and therefore owners or programmers cannot be held liable. Therefore, it is clear that "a new relationship between human and machine is developing that needs to be guided by law and regulation" (Holder et al. 2016, 576). Furthermore, current AI technology "does not recognize the reality, but the version of reality formed by the data collecting and processing mechanism, depending on the content and quality of both the algorithm and the entered data" (Stevanovic and Pavlovic 2018, 36), indicating that fully autonomous AI currently exists only in theory.

The issue of granting legal personality to various objects, including AI, is becoming increasingly relevant in the public domain. For example, on April 20, 2015, representatives of the organization Nonhuman Rights Project announced that, for the first time in history, a Manhattan Court had recognized two chimpanzees (Hercules and Leo) as legal persons but this ruling was later reversed by the New York Supreme Court in Stanley (Supreme Court Of The State Of New York New York County, 2015), that pronounced that chimpanzees are not legal persons. In another example, Hindu idols have long been judicially recognized in some jurisdictions, such as India, as legal persons, founded upon religious customs (Duff 1927, 42–48). And in 2017, the humanoid robot Sophia was given Arabian citizenship (Krainska 2018).

At present, the characteristic of being human is closely connected to legal capacity, yet there is no rule that would explicitly forbid granting non-humans legal capacity. From a technoethical perspective "fundamental rights ... may apply not only to humans but also to legal persons and, perhaps in some future, to machines or man-machine combinations that are capable of moral reasoning" (Koops et al. 2013, 16). The problem of AI not only gaining, but granting moral rights is a major issue, which "has become of utmost importance because of current and anticipated technological developments in the fields of computer science" (Gordon 2018, 2) that has been explored in many scientific analyses and arguments for and against the issue (Atabekov and Yastrebov 2018; Solaiman 2017; Naučius 2018) without consensus. There are no doubts that the development of AI and the consequences of its use should be regulated by legal acts, as serious damage including deaths of human beings could result from the use of AI (King et al. 2018, 14). In fact, the use of AI could

2 *Mens rea* refers to criminal intent. The literal translation from Latin is "guilty mind."

lead to intentional killings if AI is used as means in war (Marchant 2015, 2879–2910), which indicates necessity of AI regulated by criminal law.

Positions of scientists are controversial on the issue of AI as a subject of criminal law. A recent open letter published by experts in the field of AI expressed that "the legal status for a robot can't derive from the Legal Entity model, since it implies the existence of human persons behind the legal person to represent and direct it" (Open Letter 2018). In contrast, a different group of experts stated that "the criminal liability of legal person has constituted an innovative break-through in criminal law and the models used to support such an advance can provide us with invaluable clues to unveil a plausible dogmatic framework for the criminal responsibility of artificial entities" (Freitas, Andrade, and Novais 2014, 152).

At this stage of technological development, it is rather difficult to imagine AI having a culpable state of mind (notwithstanding the opposite arguments; Hallevy 2010, 200), and it therefore initially seems that criminal liability is impossible for AI. However, different more pragmatic position and arguments posit that "even though considerations about consciousness might contradict the view that AI systems are moral agents, they will not make it impossible for such systems to be legal actors and as such own property, commit crimes and be accountable in legally enforceable ways" (Risse 2018, 4).

The environment for such development of AI is quite favorable, taking into account the development of corporate criminal liability, which complicates the identification of concrete liable natural persons behind wrong decisions enacted in huge international corporations because of their specific corporate culture in the context of increasing economic powers and development of new technologies (Mishra and Tatiya 2009, 70–71). Such situations require implementation of an additional principle that states that "an individual perpetrator must not be a prerequisite for corporate liability" (OECD 2015, 27), or that a conviction of a corporation should not depend on a successful conviction of a guilty natural person (OECD 2015, 27). These new principles of corporate liability are suitable for establishment of AI criminal liability (Beck 2016, 141; Risse 2018, 4–5).

Using a formal approach, a legal person is defined as some fictional status provided to certain entities by decision of a certain legal system. As a consequence, a legal person, sometimes in the form of a corporation, acquires certain rights and obligations (Bryson, Diamantis, and Grant 2017, 277). Granting legal personhood allows for contracts for selling and acquiring property or acquisition of labor, and grants certain legal responsibility such as the ability to sue and be sued. Because corporations can be granted legal personhood, they oftentimes appear "to act as a liability shield, protecting … shareholders from personal liability" (Sellwood 2017, 862). This "liability shield" could be successfully used in criminal cases involving AI, as Sellwood (2017, 868–871) demonstrates with an example of autonomous vehicles treated as the legal entity and their owners take the role of a shareholder and a manager. Scholars consider that "the status of legal entity today is related to *Homo sapiens* and this is clearly stated in the most important legal acts, both international and national" (Stevanovic and Pavlovic 2018, 37).

Illegal enrichment of persons using defective AI could also draw parallels with enrichment of shareholders of liable corporation. Just as the creators of AI may not be held negligent for defectiveness of AI because of the autonomy of the technology, shareholders cannot foresee all actions which corporations can take and are likewise unlikely to be held liable. Both AI and legal entity (LE) are artificial objects, conceptualized by human beings because of technological and economic growth and globalization (Buell 2006, 476). As a consequence for a criminal act, AI could be punished similarly to punishment in a corporate criminal liability model, where there is a closing or halting of a company's operations. Under this model, stakeholders (clients, stockholders, and employees) suffer losses by not benefiting from the business. Likewise, "putting AI out of use for a determinate period of time or deletion" (Hallevy 2010, 196–197) would provide analogous negative financial consequences to the owners/users of AI.

One important difference between AI and legal entity is that human beings are the main actors behind corporations and liability is still based on human agency (Lima 2018, 687–688), while AI is driven by algorithms that dictate actions. If algorithms did not drive the actions of AI, and instead autonomous learning without interference of manufacturer, software programmer or user intelligence was the main actor behind AI, then AI "could be conceived of existing … independently and without the involvement of humans and it could reason" (Lima 2018, 688). This argument at least partially could be rebutted with establishment and latest development of corporate criminal liability.

Historical overview of corporate criminal liability indicates that the world went from questioning whether corporate criminal liability should exist at all to the phase of how it should be defined and structured (Coffee 1998, 9). This evolution of thought resembles the first stages of AI liability and personality issues. Traditionally, corporations have been treated as fictitious, artificial persons under the law (Kim 2000, 785). "Since corporations had 'no soul to damn,' business entities were initially deemed incapable of forming the malus animus required for criminal intent" (Robson 2010, 115), just as AI is currently treated (Solaiman 2017, 176). Up till the end of 19th century, enterprises, because of their specific goals, functions, and duration limits, played a rather modest role in society. Similarly, the use of AI is continually increasing and will undoubtfully be developed into an artificial mind in the near future (Lima 2018, 696).

Corporate criminal liability was established to increase the economic power of corporations. As corporations grew in their image and importance in society, they appeared both unaccountable and uncontrollable with traditional means (Robson 2010, 113). In the same way, AI currently appears unaccountable and uncontrollable (Gordon 2018; Bryson, Diamantis, and Grant 2017). "Changing perceptions of appropriate allocation of risk also fueled legislation that expanded criminal liability" (Robson 2010, 118) of corporations, including change of attitudes (Robson 2010, 119). Corporations were first held liable for nonfeasance, but liability was soon extended to affirmative acts of malfeasance (Robson 2010, 114).

During much of the 20th century, many countries held the French attitude that corporations could not be criminally liable because it was "irreconcilable with the guilt principle" (Orland and Cachera 1995–1996, 115–116) or because it contradicted traditional concept of criminal law (Coffee 1999, 10). These same arguments are used when discussing AI criminality (Lima 2018; Stevanovic and Pavlovic 2018, 43; Charney 2015, 71). However because of the economic and technological growth and globalization at the end of 20th century, corporations obtained a certain character which was distinct from their composers or employees (Kim 2000, 803–804). When "large global corporations can be treated as actors similar to sovereigns" (Paraschiv 2013, 101) the need to get more control over huge companies arose. The same processes could be seen today as evidenced by the urgency for AI legal regulation on political agendas of various actors including EU institutions, Member States (MS), and producers of AI (National and International Strategies). In certain cases, AI liability is more convenient than conviction of natural persons, as an excessive burden of legal liability could lead to a programmer hiding their identity, or the hinderance of further technology development in the official markets, moving all the programming work to unofficial markets.

AI plays and will continue to play an important role in society. Granting AI liability would require introduction of a separate artificial legal person similar to other legal entities that still house natural persons behind another entity. The arguments for the introduction of legal personhood to AI and technical obstacles, at this stage of technological development, must consider if and how AI corresponds to one of two traditional obligatory constituents of criminal law: mens rea and actus reus (Hallevy 2010, 187). However "attributing the internal element of offenses to AI entities is the real challenge in most cases" (Hallevy 2010, 187) and requires deeper additional analysis.

4 Mens Rea Factor for Legal Entity and AI

Mens rea, "understood as a state of mind, a psychological disposition that relates to actus reus of the offence in question" (Keiler and Roef 2019, 112), can be categorized as direct or indirect intent, and recklessness or negligent. Thus, a person's guilt relies "on free will and ensuring ability to determine one's behavior in line with its likely or possible consequences" (Dyson and Vogel 2018, 50). Psychological issues that are specific only to natural persons as culpability are directly related with notion of morality and "can be only imposed upon moral agents" (Cavanagh 2011, 429).

To establish a guilty mind (culpability) criminal law requires demonstration of guilt (directly related to the ability of a natural person being accountable/having legal capacity), a concrete category of fault (intentional or negligent), and awareness of unlawfulness (Stevanovic and Pavlovic 2018, 40). Demonstration of awareness, or defining the internal element of offenses, is central to discussions of AI as a legal entity, since AI is presumed to lack awareness (Charney 2015, 70). From philosophical perspective "mankind alone is the focus of criminal law; only individuals have the capacity of self-determination and the capacity for moral choice, and the essence of criminal liability relies upon a sum of physio-psychic factors unique to individuals" (De Maglie 2005, 548). Corporations faced that same challenge in corporate liability discussions, as they could be prosecuted only for offences not requiring intent until mens rea was established for LE in trials within the United States courts (Slapper and Tombs 1999, 28). The same issue is still important for establishment of corporate criminal liability. Criminal liability for legal entities in the 21st century is already established in most European countries but the argument that a legal entity lacks a mental state and an independent will or intent still dominates in academic literature (Mishra and Tatiya 2009, 70; Panainte 2014, 140; Khanna 1996, 1485–1486). The question remains if the arguments used for culpability of a corporation are transferable to AI.

Introduction of criminal liability for legal entities was originally based on two traditional theories of criminal law: deterrence and retribution (Robson 2010, 120). Retribution notions of criminal law (Sheyn 2010, 3) are more common and acceptable than deterrence, since retribution is sometimes argued to be "the only means of effectually controlling the subject-matter and correcting the abuses" (New York Central 1909, 496). The jurisprudence of the US courts, the cradle of criminal liability for legal entities, explains deterrence as effective control, correction of abuses, stimulation of maximum effort by owners and managers for assuring adherence by such agents to the requirements of the legislator (Sheyn 2010, 4). "Pragmatic considerations have led courts and legislatures to expand the panoply of corporate crime in order to deter conduct ranging from reprehensible, to undesirable, to merely annoying" (Robson 2010, 109). Criminal deterrence is thought to be the "ultimate deterrent" (Robson 2010, 109) for corporate crime, while encouraging (through special cultures or habits in the company) or neglecting to prevent individual employee wrongdoing when, for example, there is a lack of regulations or practical control for individual actions (O'Sullivan 2002 28–29).

Evaluating deterrence theory from a societal perspective is suitable for establishment of AI criminal liability with the main goal of regulating AI to prevent possible harm to society and provision of negative consequences to guilty persons (European Commission 2018, 3.3). However AI is not a moral agent and cannot understand or feel preventive aspects by itself as humans can. Therefore, deterrence could not be applied directly to AI until it is treated as an object or an instrument in the hand of humans as "the current capabilities of AI systems do not reach the legal standard of awareness and volition" (Charney 2015, 70) required by traditional concept of criminal law. However, there are no legal barriers to provide legal personality for AI (Singh 2017, 194) in case such legislative decision is grounded and necessary (Bryson, Diamantis, and Grant 2017, 273). For example, in the US state of Nevada, autonomous vehicles are already granted legal personhood with all the rights and duties provided to the driver (Sellwood 2017, 869). At this stage of AI development, deterrence theory could help to reach the natural persons that stand behind AI including users/owners, manufacturers, or software programmers, depending on their established fault. Most of these individuals would be charged with negligence as such under current jurisprudence.

Retribution theory could also bring charges to those behind AI, as it "brings to the criminal law an evaluative function, an announcement that there are some actions, so outrageous and heinous, that they merit punishment, not solely to deter future conduct, but to restore and reaffirm the community's core values" (Robson 2010, 144). This theory is also grounded in mens rea, which depends on moral and psychological issues of natural persons in the legal entity. Criminal liability, in such a case, "expresses the community's condemnation of the wrongdoer's conduct (which shows disrespect for certain generally accepted values) by emphasizing the standards for appropriate behavior" (Friedman 1999, 843). Some moral condemnation for the wrongful act as "application of the criminal law to an actor in society is a means to express a moral judgment about that actor's conduct" (Henning 2009, 1420).

In relation to corporate criminal law, retribution theory is less developed and controversial, as some scholars do not accept it because corporations as legal entities lack subjectivity and personality. Corporate liability was not always acceptable, and "from the beginning, legal scholars appeared to be uncomfortable with retribution as a goal of organizational criminal liability. Without consciousness and self- awareness, business organizations lacked the capacity to be morally blameworthy. Courts, while not totally dismissing the possibility of retributive sanctioning, continued to view organizational criminal law solely as a means of deterring corporate misconduct." (Robson 2010, 121). This same theory could be used to explain the guilty minds of the natural persons behind AI because the improper functioning of technology is related to the negligence of manufacturers, software programmers or improper usage of the technologies. But retribution theory does not help to explain the culpability of AI itself and does not address the future status of AI, when AI becomes fully autonomous and could cause serious damage.

A third theory or corporate liability, involving treating a juridical person as a separate subject, is inspired by organizational theory and states that "the corporation is no longer a collection of individuals, but a complexity of synergetic interactions that cannot be reduced to individual actions" (Diez 2011, 46). This theory provides some abilities to establish separate culpability of the legal entity and is a consequence of the role which corporations play in a modern society. As Henning (2009) argues, "corporations are as much a part of modern society as individuals, and so this expressive function of the criminal law is particularly worthwhile when applied to organizations that play such a significant role in the economy and indeed throughout society" (1420). The argument behind the theory is a pragmatic one, in that "...criminal laws can and must be applied to corporations to ensure that these organizations, which are growing in ever-increasing size and strength, are held accountable for their actions" (Dervan 2011, 19). Furthermore, corporations, "in modern life ... can be blamed for wrongdoing with a kind of moral assessment characteristic of judgments of criminality and on the basis of the organization's relationship to the wrong" (Buell 2006, 477). As a result, a company may have an intention which no individual who is associated with a company shares (Cavanagh 2011, 429). Still it looks like this theory also requires establishment of certain culpability of individuals, as certain corporate actions still originate from actions of individuals. This reliance on individual actions could challenge the establishment of direct liability to AI.

Following development of corporate criminal liability, scholars have argued that "the responsibility of an individual perpetrator must not be a prerequisite for corporate liability" (OECD 2015, 27). In some jurisdictions, such as the US, Finland, Switzerland, and Netherlands, there is no requirement for identification of an individual perpetrator in corporate criminal liability cases (OECD

2015, 27). Some jurisdictions do not require the mens rea for specific offences, while other jurisdictions use corporate culture as a proxy for mens rea. For example, in 1995, Australia adopted a modern and complex model of corporate culpability, expressly addressing the problem of a corporate culture as a fault element. According to the Criminal Code Act of Australia (1995), "corporate culture means an attitude, policy, rule, course of conduct, or practice existing within the body corporate generally or in the part of the body corporate in which the relevant activities takes place" (Section 12.3). This final theory could be useful for finding arguments to ground culpability of AI if AI is given legal personality.

The criminalization of corporate liability was a gradual process, starting with offences without establishment of mens rea, and then moving to finding an intentional corporate mind starting with managers (directing mind), continuing with lower categories of workers (controlling officers; Slapper and Tombs 1999, 29–32), and ending with corporation conviction without finding a concrete natural guilty person (OECD 2015, 27).

AI currently plays important role in society and, if AI becomes fully autonomous and is no longer fully controlled by manufacturer, IT, owner, or user, will require the introduction of a separate artificial legal person similar to other legal entities that shield natural persons behind a corporation. Future AI "[will possess] all the necessary elements typical of entities recognized as subjects of law: intelligence, autonomous decision making, the ability to learn form their own experience, memory, planning, complexity, formality and the ability to manipulate structures" (Čerka, Grigienė, and Sirbikytė 2017, 12). Therefore, there will be real possibilities for the establishment of AI culpability. Each corporate doctrine could be applied, to a certain extent, to AI and various stages of its development. While criminal liability of AI could be partially grounded in certain aspects of mens rea, a closer analysis of each corporate liability model and doctrine needs to be implemented in order to find the limits, advantages, and critics of each.

5 Legal Personhood Liability Theories and Their Suitability for AI

The main doctrines or models grounding criminal liability for legal persons could be extended to the development of AI criminal liability models. Legal scholars have previously discussed how to overcome the mens rea element for AI criminal liability. For example, Hallevy (2010) proposed three possible alternatives for criminal liability on Al entities: "The Perpetration-via-Another liability model; the Natural-Probable-Consequence liability model; and the Direct liability model" (Hallevy 2010, 174). The first two models are based on nominalist theory (Cavanagh 2011, 414) that explains criminal liability for programmers or manufacturers because of wrongdoing of AI in classical criminal law structures, but not holding AI itself liable. The direct liability model deals with the direct liability of AI, which is the basis of realistic theory (Cavanagh 2011, 414). In the analogy of AI to an employee model, the AI entity is considered as an innocent agent. (Solum 1992).

Scholars often use an analogy between existing legal liability models and new paradigms, as it helps to develop new theories or extend existing theories to new phenomena, such as AI. For example, LE models use the analogy of the relationship between AI and programmer/user and employee and employer. The drawback of this model that it gives ground for personal liability of humans behind the AI, not AI itself. Hallevy (2010, 174) proposes models based on already existing analogies from criminal law principles and applies them for criminal liability of user/programmer for the wrongdoings of AI. However, Hallevy (2010, 186) also recognizes that the autonomy of AI may be an obstacle for setting criminal liability for programmer/user since "the perpetration-via-another liability model is not suitable when an AI entity decides to commit an offense based on its own accumulated experience or knowledge" (2010, 181).

When dealing with very autonomous decisions, it is difficult to establish an appropriate analogy for direct criminal liability relating to AI actions. Legal models dealing with AI as analogy to employee/employer or controlled instrument/controlling person do not give satisfying answers to current challenges because of AI's autonomous nature. There is no cue to a situation when no human actor may be held directly liable because of inability to foresee the actions of AI. The classical model of nominalist theory, identification doctrine, can inform certain aspects of AI criminal liability.

5.1 *Identification Doctrine and AI*

"The Identification doctrine approach considers that a corporate offence occurs when an individual commits all the elements of the offence and he is sufficiently senior to be seen as the controlling mind of the corporation. Under the identification theory an employee is assumed to be acting as the company and not for the company" (Tolvanen 2009, 337). The identification doctrine could be used to demand criminal liability of AI when the person controlling AI (usually the operator) used it for illegal purposes. Without such doctrine, only the controlling actor and not AI would be liable, but the identification doctrine dictates liability consequences for all human actors related to AI, such as owners or even programmers, since AI itself could experience negative consequences such as activity restriction, reprogramming, or even destruction. For example, let us assume that an owner of AI rents AI to an operator, who then uses it for legal and illegal purposes that were unforeseeable by the owner due to the autonomous nature of the AI. Therefore, the owner cannot be liable as a person for the illegal acts, even when the owner benefits from the rental of the AI. Therefore, there is a material difference between the situation where only the operator is liable and between the situation where both the operator and AI are separately liable because in later case consequences are felt not only by operator, but by the owner too.

5.2 *Vicarious Liability Model and AI*

The criminal vicarious liability may be summed in word of judge Binghan LJ:

> "A company may be vicariously liable for the negligent acts and omissions of its servants and agents, but for a company to be criminally liable for manslaughter … it is required that the mens rea and actus reus of manslaughter should be established not against those who acted for or in the name of the company but against those who were to be identified as the embodiment of the company itself." (High Court of Justice, 1989, 16)

Similarly, Clifford (2016) stated that:

> "The basis or proposed basis of liability for corporate entities within those countries where liability exists (or is proposed) mostly rests on the premise that the acts of certain employees can be attributed to a corporate entity. The category of employees which can trigger corporate liability is limited in some jurisdictions to those with management responsibilities and the act must generally occur within the scope of their employment activities. The act must also generally be done in the interests of or for the benefit of the corporate entity." (5)

In a broad interpretation of vicarious liability, Tolvanen (2009) stated that the principle "is often invoked to establish corporate manslaughter" (338). Where an employee commits a crime within the sphere of their employment and with the intention of benefiting the corporation, their criminality can be imputed to the company. Basically it is respondent superior – the person acts for the company, but their acts are attributed to the company. There is quite a considerable resistance among legal scholars to this theory being applied to criminal liability, "since criminal responsibility is attributed to the company in this manner, whether or not the company was at fault in any way or not, vicarious liability was seen as unjust, and lacking in defensible penal rationale" (Tolvanen 2009, 339). Additionally, "one goal of holding a corporation vicariously liable for the torts committed by its agents relates to the corporation's ability to best compensate the injured party." (Ellis and Dow 2014, 183). This theory could be extended to AI in two ways: holding AI to be an employee of the operator of the AI (as presented by Hallevy, 2010) or hold the operator to be an employee or agent of the AI. In the former case, a person would be held liable under classical criminal law, even when they acted through other persons. The opposite approach, the consider the operator as an employee of AI, would result in a granting the idea of personhood to AI while the consequences would be delivered to the operator and other actors, since the loss of AI will most seriously be felt by the owners.

5.3 *Aggregation Doctrine and AI*

Aggregation theory, which is much less used in practice for grounding criminal liability of legal person, can be used for explaining criminal liability of AI. "The aggregation doctrine/collective intent approach aggregates all the acts and mental elements of various company employees and finds the offence if all the elements of a crime are made out, though not necessarily within a single controlling mind" (Tolvanen 2009, 337). In United States v. Bank of New England, the Bank of New England was convicted for willfully violating the Currency Transaction Reporting Act, although there was no proof that any individual employee had willful intent at the moment of the actus reus. Thus, individually, none of the employees had committed an offence or willfully failed to file a report (Ragozino 1995, 434). The rationale is based upon the difficulty in identifying an individual offender who operates within a complex and decentralized structure. Under these circumstances, a finding of the collective intent is "not only proper but necessary" (United States Court of Appeals, UNITED STATES of America v. BANK OF NEW ENGLAND, N.A., 1987, 56). De Maglie (2005) further argues that "any other rule would allow a company to compartmentalize information and thereby avoid criminal liability. Thus, if one assumes a crime with elements A, B, and C, where A is known to officer A, B is known to officer B, and C is known to officer C, then, for the purpose of criminal liability, all elements are known by the corporation" (557).

The aggregation doctrine could be applied to the liability of persons related to AI by holding the knowledge element as joint knowledge of programmer, manufacturer, owner, and operator as of one joint person. Therefore, under the aggregation doctrine, even if none of the separate individuals related to AI could foresee wrongdoing, criminal liability would apply to the AI if the crime could be done from the perspective of joint knowledge, and consequences of the crime would be felt by all. For example, if there was a programmed error that was used by operator and distributed by manufacturer, who then both benefited from illegal activity but were unable to understand program error itself, criminal liability could not be applied to all actors. However, when considering the same example under the aggregation doctrine, AI could be held liable and consequences would be felt by most of the actors. In addition, AI could be held liable if some defect could be found in the intents of the programmer, producer, owner and controller of AI. This liability would hold when none of the AI-related persons had any possibility to foresee the wrongful choices of AI but the illegal activity could have been avoided if the related persons had been acting with due care. This theory could be used for solving the problem of separately and individually criminally charging each actor related to AI, since neither of them separately would be able to foresee the wrongful act of AI. In accepting the aggregation doctrine for AI, the possibility to have collective mens rea of all human actors related to AI could be introduced.

5.4 *Reactive Corporate Fault Theory and AI*

The reactive corporate fault theory of corporate criminal liability, which is getting more popular in common law countries, can be described as "supporting [the] idea that where an individual has committed, e.g. actus reus of manslaughter, a court should have the power to order the employing corporation to institute measures to prevent further recurrence and should face criminal prosecution should they fail to do so" (Tolvanen 2009, 338). This theory could be extended for establishing liability of AI when human actors, even though not liable for an error itself, were unable to stop the error afterwards or correct it in a timely manner. By this theory, personal liability applies to the responsible programmers who had a duty to correct the AI, but there would also be consequences for all human actors, including the owner, through loss of the AI.

It is evident that the two first alternatives provided by Hallevy are actually classical and already existing criminal law principles applied to AI as an employee or controlled instrument. These two perspectives cannot help us in situations where AI decisions are so autonomous that no human actor was able to control or foresee them. Therefore, alternative perspectives are necessary. A similar problem, corporate liability, was solved by applying liability of corporate subjects when there existed situations where none of the human actors (like shareholders or even managers or employees) were able to control or foresee the criminal act. These persons where actually the main benefiters of such activities, therefore criminal liability of legal person was an effective remedy because, by punishing the legal entity, there was an established legal possibility to indirectly reach a "pocket" of shareholders and managers. Therefore, legal person liability doctrines may provide an analogy for AI liability.

When considering legal persons criminal liability models, there should be effective separation between nominal theoretical models, considered above, and realistic models. In realistic model approaches, the company itself is seen as a separate person. The two most popular models from this group are the management failure model and the corporate culture model, which are interrelated since "the offences of corporate manslaughter and corporate homicide respectively will also attribute criminal responsibility to companies on the basis of collective management failure which can be evidenced by failures of corporate culture" (Belcher 2006, 8).

5.5 *The Management Failure Model, the Corporate Culture Model, and AI*

With the management failure model approach, liability may be imposed when it is found that the individual who committed the unlawful act had a reasonable basis for believing that an authoritative member of the corporation would have "authorized or permitted the commission of the offence" (Tolvanen 2009, 339). The management failure model is related to the corporate culture model, where it is not necessary to find an individual responsible for the unlawful act in order to impose liability upon the corporation. Under the corporate culture model, "the organization must be in breach of that duty of care as a result of the way in which certain activities of the organization were managed or organized by its senior managers. This introduces an element of 'senior management failure'" (Mishra and Tatiya 2009, 69). According to this view, the corporation as a whole is responsible for the unlawful act, not simply the individual who may have independently carried out the act. For example, a corporate entity in France may be convicted on the basis of negligence resulting from carelessness and/or defective organization of the company, even if the fault cannot be attributed to a representative or an employee to whom the corporate entity has delegated functions.(Clifford 2016, 10). Under the corporate culture model, the corporate "mens rea approach is to accept the legal fiction of corporate personality and to extend it to the possibility of corporate mens rea, to be found in corporate practices and policies. This approach has been widely advocated in the U.S. as the corporate ethos standard" (Tolvanen 2009, 340).

Models of management failure and corporate culture could be extended to AI by introducing the demand to implement an AI safety control mechanism to prevent possible wrongful actions and setting the standard of the defective program itself. Today, there is no liability for having defective program except in situations where harm was caused by such program. Introducing a defectiveness standard for AI would create liability for defectiveness itself. We could imagine that in some situations real problems of wrongdoing lie not in knowledge or intentions related to AI human actors, but in the defectiveness of AI itself. Therefore, extension of these two models provides consequences of criminal liability by the owner when safety control systems were not able to stop AI from wrongful action, even when the action itself was not possible to foresee.

We have already shown that criminal liability of AI is possible based on the same arguments as provided by theories of criminal liability of corporal entities. The main purpose of all these theories is to balance justice interests toward human actors behind the curtain, be they corporate or AI. Although such AI criminal liability is possible, such liability is not always necessary. Most legal scholars argue that justice interests are achieved by the strict liability of the owner of AI (as the owner of a possible dangerous item) or by the administrative sanctions model as a simplified criminal liability model. Opponents to the idea that corporate criminal liability should be introduced argue that penal code is based on the notion of individual culpability, and therefore corporate entities may not be held criminally liable as they lack the capacity to act in the criminal law sense.

Not all countries, such as Germany, use criminal liability of legal person coupled with punishment (Beck 2010, 11) achieved by an administrative penalty to punish legal entities (employees, shareholders, etc.). A move towards criminal liability for corporate entities in Germany is considered inconceivable at present. Nevertheless, the imposition of regulatory fines and the siphoning off of economic benefits are tools used frequently as practical solutions to sanction corporate entities for wrongdoing. For example, in 2007, Siemens AG received a regulatory fine of EUR 201 million. We should mention that the civil strict liability model holds owners of AI liable for wrongful acts of AI without the possibility to foresee the actions of AI, but civil liability only allows compensation for all actual damage and does not allow a penalty for the owner for the incident itself per se. Therefore, the civil liability model lacks a motive to prevent persons to risk with possibly wrongful AI. For example, since not all wrongful acts will be proven, a person could be willing to experiment with wrongful software, thinking that the possible benefit from their actions would not be lost, since only theoretically civil liability would be proven for all actions. This would be a weak side of the civil strict liability model, but introducing the civil liability model does not need to introduce the complexity of AI legal personhood.

6 Conclusions

The historical development of criminal liability of legal entities and the arguments supporting the defense of this idea should be reapplied to the development of AI legal personality recognition. This would serve as a response to the challenge of balancing the legal interests of all the humans who are related to AI. AI legal personhood is one of the possible solutions for establishing criminal liability of AI when none of human actors, who relate to and benefit from wrongdoings of AI, cannot be directly charged. Introducing the legal personhood to AI from the criminal liability institute perspective is possible when none of the natural human persons could be held criminally liable. Arguments for legal personhood of AI are motivated by the actions of AI, and therefore the AI criminal liability model is not related to the philosophical issue of AI personhood. This model makes it possible to create some liability for persons who benefit from AI activity, even when they cannot directly be held criminally liable. Lastly, if AI legal personhood model is introduced into the system of criminal law, the comparison of legal persons criminal liability and AI criminal liability could and should be considered, since those theoretical models may be extended to criminal liability of AI, though other alternatives for AI criminal liability may also exist.

References

Atabekov, A., Yastrebov, Oleg. 2018. "Legal Status of Artificial Intelligence Across Countries: Legislation on the Move." *European Research Studies Journal* XXI(4): 773–782.

Beck, Susanne. 2010. "Mediating the Different Concepts of Corporate Criminal Liability in England and Germany." *German Law Journal* 11: 1093–1114.

Beck, Susanne. 2016. "Intelligent agents and criminal law - Negligence, diffusion of liability and electronic personhood." *Robotics and Autonomous Systems* 86: 138–143.

Bekey, George. 2005. Autonomous Robots: From Biological Inspiration to Implementation and Control. Cambridge: MIT Press.

Belcher, Alice. 2006. "Imagining How A Company Thinks: What Is Corporate Culture?" https://ojs.deakin.edu.au/index.php/dlr/article/viewFile/234/240.

Bryson, Joanna J., Mihailis E Diamantis, and Thomas Grant. 2017. "Of, For, and By the People: The Legal Lacuna of Synthetic Persons." *Artificial Intelligence and Law* 25(3): 273–291.

Buell, Samuel W. 2006. "The Blaming Function of Entity Criminal Liability." 81 *Indiana Law Journal* 81 (473): 473–537.

Cavanagh, Neil. 2011. "Corporate Criminal Liability: An Assessment of the Models of Fault" *The Journal of Criminal Law* 75: 415–440.

Čerka, Paulius, Jurgita Grigienė, and Gintarė Sirbikytė. 2017. "Is it Possible to Grant Legal Personality to Artificial Intelligence Software Sytems?" *Computer Law & Security Review*: 1–15.

Charney, Rachel. 2015. "Can Android Plead Automatism – A Review of *When Robots Kill: Artificial Intelligence under the Criminal Law* by Gabriel Hallevy." *University of Toronto Faculty Law Review* 73(1): 69–72.

Clifford Chance. 2016. "Corporate Liability in Europe." https://www.generalcounsel.nl/wp-content/uploads/2016/05/Corporate_Criminal_Liability_April_2016_6032092.pdf.

Coffee, John C. 1999. "Corporate Criminal Liability: An Introduction and Comparative Study." In *Criminal Responsibility of Legal and Collective Entities. International Colloquium, May 4–6, 1998, Berlin*, edited by Albin Eser, Günter Heine and Barbara Huber, 9–37. Freiburg: Max-Planck-Institut für ausländisches und internationals Strafrecht. https://pure.mpg.de/rest/items/item_2640412/component/file_3012698/content.

Criminal Code Act Australia. 1995. "Criminal Code Act." https://www.legislation.gov.au/Details/C2004C03212.

De Maglie, Cristina. 2005. "Models of Corporate Criminal Liability in Comparative Law." *Washington University Global Studies Law Review* 4(3): 547–566.

Dervan, Lucian E. 2011. "Reevaluating Corporate Criminal Liability: The DOJ's Internal Moral Culpability Standard for Corporate Criminal Liability." 41 *Stetson Law Review* 41(7): 7–20.

Dyson, Matthew, and Benjamin Vogel, eds. 2018. *The Limits of Criminal Law. Anglo-German Concepts and Principles.* United Kingdom: Intersentia.

Drakšas, Romualdas. 2008. *Baudžiamoji atsakomybė ir jos realizavimo formos.* [Criminal responsibility and its forms of implementation]. Monograph. Vilnius: Justitia.

Duff, P.W. 1927. "The Personality of an Idol." *Cambridge Law Journal* 3(1): 42–48.

European Commission. 2018. "Artificial Intelligence for Europe." Communication from the Commission to the European Parliament, the European Council, the Council, the European Economic and Social Committee and the Committee of the Regions. Brussels, 25.4.2018 COM(2018) 237 final. https://ec.europa.eu/digital-single-market/en/news/communication-artificial-intelligence-europe.

Ellis, Nan S., and Steven B. Dow. 2014. "Attaching Criminal Liability to Credit Rating Agencies: Use of the Corporate Ethos Theory of Criminal Liability." *University of Pennsylvania Journal of Business Law* 17(1): 167–183.

Freitas, Pedro Miguel, Francisco Andrade, and Paulo Novais. 2014. "Criminal Liability of Autonomous Agents: From the Unthinkable to the Plausible." In *AI Approaches to the Complexity of Legal Systems*, edited by Casanovas Pompeu, Pagallo Ugo, Palmirani Monika and Sartor Giovanni, 145–156. AICOL International Workshops. Revised selected papers. Springer.

Friedman, Lawrence. 1999. "In Defense of Corporate Criminal Liability." *Harvard Journal of Law and Public Policy* 23: 833–858.

Future of Life Institute. "National and International Strategies." https://futureoflife.org/national-international-ai-strategies/?cn-reloaded=1.

Gomez-Jara Diez, Carlos. 2011. "Corporate Criminal Liability in the Twenty-First Century: Are All Corporations Equally Capable of Wrongdoing." *Stetson Law Review* 41: 41–61.

Gordon, John-Stewart. 2018. "What Do We Owe to Intelligent Robots?" *AI & SOCIETY*, 1–15.

Hallevy, Gabriel. 2010. "The Criminal Liability of Artificial Intelligence Entities – from Science Fiction to Legal Social Control." *Akron Intellectual Property Journal* 4(2): 171–201.

Henning, Peter J. 2009. "Corporate Criminal Liability and the Potential for Rehabilitation." *American Criminal Law Review* 46: 1417–1436.

HIGH COURT OF JUSTICE. HM Coroner for East Kent, Exparte Spooner & Others (1989) 88 Cr App R 10 (CA) 16.

Holder, Chris, Vikram Khurana, Joanna Hook, Gregory Bacon, and Rachel Day. 2016. "Robotics and Law: Key Legal and Regulatory Implications of the Robotics Age (Part II of II)." *Computer Law & Security Review* 32: 557–576.

Horsman, Yasco, and Frans-Willem Korsten. 2016. "Introduction: Legal Bodies: Corpus/Persona/Communitas." *Law & Literature* 28(3): 277–285.

Keiler, Johannes, and David Roef, eds. 2019. *Comparative Concepts of Criminal Law*. United Kingdom: Intersentia.

Khanna, Vikramaditya S. 1996. "Corporate Criminal Liability: What Purpose Does it Serve?" *Harward Law Review* 109(7): 1477–1534.

Kim, Susanna M. 2000. "Characteristics of Soulless Persons: The Applicability of the Character Evidence Rule to Corporations." *University of Illinois Law Review* 763(3): 763–812.

King, Thomas, Nikita Aggarwal, Mariarosaria Taddeo, and Luciano Floridi. 2018. "Artificial Intelligence Crime: An Interdisciplinary Analysis of Foreseeable Threats and Solutions." *SSRN*, May 22, 2018: 1–36; https://papers.ssrn.com/sol3/papers.cfm?abstract_id=3183238 http://dx.doi.org/10.2139/ssrn.3183238.

Koops, Bert-Jaap, Angela Di Carlo, Luca Nocco, Vincenzo Casamassima, and Elettra Stradella. 2013. "Robotic Technologies and Fundamental Rights: Robotics challenging the European Constitutional Framework." *International Journal of Technoethics* 4(2): 15–35.

Krainska, Agnieszka. 2018. "Legal Personality and Artificial Intelligence." https://newtech.law/en/author/agnieszka-krainska/.

Leroux, Christophe et al., eds. 2012. *Suggestion for a Green Paper on Legal Issues in Robotics. The European Robotics Coordination Action*. https://www.unipv-lawtech.eu/files/euRobotics-legal-issues-in-robotics-DRAFT_6j6ryjyp.pdf.

Lima, Dafni. 2018. "Could AI Agents Be Held Criminally Liable: Artificial Intelligence and the Challenges for Criminal Law." *South Carolina Law Review* 69: 677–696.

Marchant, Gary E. et al. 2015. "International Governance of Autonomous Military Robots." In *Handbook of Unmanned Aerial Vehicle*, edited by Valavanis Kimon P. and Vachtsevanos George J., 2879–2910. Springer.

Mishra, Vinay, and Siddharth Tatiya. 2009. "Corporate Criminal Liability: Lessons Learnt and to Be Learnt." *Asian Business Law* 4(69): 70–92.

Naučius, Mindaugas. 2018. "Should Fully Autonomous Artificial Intelligence Systems be Granted Legal Capacity?" *Teisės apžvalga/ Law review* 1(17): 113–132.

OECD Anti-Corruption Network for Eastern Europe and Central Asia. 2015. "Liability of Legal Persons for Corruption in Eastern Europe and Central Asia." https://www.oecd.org/corruption/ACN-Liability-of-Legal-Persons-2015.pdf.

OPEN LETTER TO THE EUROPEAN COMMISSION ARTIFICIAL INTELLIGENCE AND ROBOTICS. 2018. Accessed March 15, 2019. https://g8fip1kplyr33r3krz5b97d1-wpengine.netdna-ssl.com/wp-content/uploads/2018/04/RoboticsOpenLetter.pdf.

Orland, Leonard, and Charles Cachera. 1995. "Corporate Crime and Punishment in France: Criminal Responsibility of Legal Entities (Personnes Morales) under the New French Criminal Code (Nouveau Code Penal)." *Connecticut Journal of International Law* 11(111): 111–168.

O'Sullivan, Julie Rose. 2002. "Professional Discipline for Law Firms – A Response to Professor Schneyer's Proposal." *Georgetown University Law Center* 16(1): 1–90.

Panainte, Rodica. 2014. "Considerations on Corporate Criminal Liability." *Journal of Public Administration, Finance and Law*: Special Issue (1): 140–145.

Paraschiv, Daniel-Stefan. 2013. "Corporate Criminal Liability in Comparative Law." *Contemporary Readings in Law and Social Justice* 5(1): 100–105.

Ragozino, Anthony. 1995. "Replacing the Collective Knowledge Doctrine with a Better Theory for Establishing Corporate Mens Rea: The Duty Stratification Approach." *Southwestern University Law Review* 24: 423–471.

Risse, Mathias. 2018. *Human Rights and Artificial Intelligence. An Urgently Needed Agenda.* Cambridge: Carr Center for Human Rights Policy.

Robson, Regina A. 2010. "Crime and Punishment: Rehabilitating Retribution as a Justification for Organizational Criminal Liability." *American Business Law Journal* 47(1): 109–144.

Sellwood, Michelle. 2017. "The Road to Autonomy." *San Diego Law Review* 54: 829–873.

Sheyn, Elizabeth R. 2010. "The Humanization of the Corporate Entity: Changing Views of Corporate Criminal Liability in the Wake of Citizens United." *University of Miami Law Review* 65(1): 1–44.

Singh, Shubham. 2017. "Attribution of legal personhood to artificially intelligent agents." *Bharati Law Review*: 194–201.

Slapper, Garry, and Steve Tombs. 1999. *Corporate Crime*. UK: Longman.

Solaiman, Sheikh M. 2017. "Legal Personality of Robots, Corporations, Idols and Chimpanzees: A Quest for Legitimacy." *Artificial Intelligence and Law* 25(2): 155–179.

Solum, Lawrence B. 1992. "Legal Personhood for Artificial Intelligences." *North Carolina Law Review* 70(4): 1231–1287.

Stevanovic, Aleksandar, and Zoran Pavlovic. 2018. "Concept, Criminal Legal Aspects of the Artificial Intelligence and its Role in Crime Control." *Journal of Eastern-European Criminal Law* 31: 31–45.

Supreme Court of the State of New York. 2015. "Matter of Nonhuman Rights Project, Inc v Stanley." NY Slip Op 31419(U). https://cases.justia.com/new-york/other-courts/2015-2015-ny-slip-op-31419-u.pdf?ts=1438809087.

The European Commission's HIGH-LEVEL EXPERT GROUP. 2018. "On Artificial Intelligence a Definition of AI: Main Capabilities and Scientific Disciplines." Brussels, December 18, 2018. https://ec.europa.eu/digital-single-market/en/news/definition-artificial-intelligence-main-capabilities-and-scientific-disciplines.

Tolvanen, Matti. 2009. "Trust, Business Ethics and Crime Prevention – Corporate Criminal Liability in Finland." *Jurisprudencija/Jurisprudence* 1(115): 335–358.

United States Court of Appeals. 1987. "United States of America, Plaintiff, Appellee, v. BANK OF NEW ENGLAND, N.A., Defendant, Appellant, 821 F.2d 844, 56 USLW 2042, 23 Fed. R. Evid Serv. 417 No. 86-1334."

U.S. Supreme Court. 1909. "New York Central R. Co. v. United States, 212 U.S. 481 (1909), 496." https://supreme.justia.com/cases/federal/us/212/481/.

Wessels, Johannes. 2003. *Baudžiamoji teisė. Bendroji dalis*. [*Criminal Law. General Part*]. Vilnius: Eugrimas.

PART 4

AI and Information Technologies

CHAPTER 12

Responsibility by Design?! – On the Standardisation of "Smart" Systems

Kai Jakobs

1 Brief Introduction, Some General Background and a Bit of Motivation

By way of an introduction this section will briefly look at the terms in the chapter's subtitle – "standardisation" and "smart systems."

Standards – in a very broad sense – have been with us for quite some time now. About 4,000 years ago the first alphabets emerged, enabling new forms of communication and information storage. Around the 7th century BC the Lydians invented the first coin-based currency; it established the basis for easier inter-regional trading. The advent of the railroad in the 19th century resulted in a need for technical standards to enable compatibility between individual parts of technical artefacts, defining e.g. the width of railway gauges, the diameter of screws, etc. This was once more reinforced when mass production generated a demand for interchangeable parts. Similarly, telecommunication (the transmission of information of any nature) is not a modern invention. According to the Greek author Aischylos the news about the Greeks' victory over Troy was communicated to Athens over a distance of more than 500 km via a chain of signal fires.

In the early nineteenth century the invention of the electric telegraph triggered the dedicated development of standards in the field of electrical communication technology. About 50 years later, in 1865, the International Telegraph Union – to become the International Telecommunication Union in 1932 (ITU; today a UN agency) – was founded. The other major international standards setting bodies, the International Organization for Standardization (ISO) and the International Electrotechnical Commission (IEC), were founded in 1947 and 1906, respectively.

Today, ISO defines a standard as document, "*established by consensus and approved by a recognized body, that provides, for common and repeated use, rules, guidelines or characteristics for activities or their results, aimed at the achievement of the optimum degree of order in a given context*" (ISO 2017, 138).

 | DOI:10.1163/9789004437876_014

The fact that standards are established "by consensus"[1] distinguishes them from legislation.

In the field of Information and Communication Technologies (ICT) international standards are now the major single mechanism to ensure interoperability between systems, i.e. their ability to meaningfully exchange information. Nevertheless, (ICT) standards are not just technical documents; they may have ramifications well beyond technology. For instance, they frequently describe a commonly agreed platform upon which innovations can be based. Moreover, standards in general are a valuable means of technology transfer. They may also be used as policy tools – in Europe, for example, standards are a major enabler of the Single Market. Their potential economic implications must also not be under-estimated. A new standard may be used to extend a market, or help open up a new one. Intellectual Property Rights (IPR), mostly patents, also play a major role here.[2] That is, standards also have a strong economic dimension. Accordingly, firms may well also consider them as strategic tools.

A quote from a document published by the European Commission sums it up nicely: "*Standards are not only technical questions. They determine the technology that will implement the Information Society, and consequently the way in which industry, users, consumers and administrations will benefit from it*" (CEC 1996, 1). That is, those who develop standards today at the same time shape much of the future environment we all will live in. This implies that standards must meet the needs of all stakeholders (including e.g. individuals, communities, NGOs, unions, businesses and governments) to provide benefits to society as a whole.

Smart systems are a fairly recent phenomenon that is becoming increasingly popular. Prominent examples include the Smart Grid, Smart Manufacturing, Intelligent Transport Systems, Smart Homes and Smart Cities. They utilise a smart communication infrastructure, based on the Internet of Things (IoT). The "smartness" of these systems is the result of the integration of ICT into "traditional" technologies, thus providing them with advanced control, communication and amenability functions previously available only in the virtual world.

A unique characteristic is smart systems' true ubiquity. Recent estimates[3] predict more than 75 billions of IoT-connected devices for 2025 (up from

1 Consensus is achieved through a general agreement and the absence of sustained opposition to substantial issues. Consensus does not imply unanimity.

2 Just think of the recent patent war between Apple and Samsung; see e.g. https://www.bbc.com/news/technology-44108835. See also e.g. Paik and Zhu 2016.

3 https://www.statista.com/statistics/471264/iot-number-of-connected-devices-worldwide/.

23 billions in 2018). Today, the Internet is frequently said to be "ubiquitous." However, come to think of it, this holds at best for the access network, i.e. you can access the Internet from almost everywhere. One limitation here is the "almost." There are still numerous white spots on a Internet connectivity map. Moreover, these days interaction with an application via the Internet still remains a deliberate activity in most cases. That is, you need to explicitly switch on a computer, type or say something and wait for a response. With smart systems, interaction will be (almost) continuous and "implicit"; billions of sensors, interconnected to form the IoT, will collect data and transmit it to some central entity, which computes it and possibly draws conclusions and makes recommendations or takes action. Either way, the outcome of the analyses of this massive amount of data will have major socio-economic and ecological ramifications and directly impact citizens as well as businesses. On top of that, a considerable part of this data will be personal or private, its collection and processing will have considerable legal ramifications, many of which link to security and privacy issues. While this is not a phenomenon exclusively related to smart systems, their ubiquity adds a completely new dimension.

Depending on your point of view, this ubiquity may border on inescapability (think George Orwell and Aldous Huxley). The IoT in combination with big data analytics and machine learning approaches has the inherent potential for both fostering the good of humankind and enabling the emergence of a surveillance society. Against this background it makes perfect sense to try and make sure that the development of smart systems is guided not just by technical feasibility aspects and by economic considerations (as ICT developments frequently are). Rather, decidedly non-technical (e.g. societal, legal, environmental and ethical) aspects also need to be taken into account. To do so in a credible way requires the active participation of stakeholders with associated interests and expertise during at least some stages of the system development process.

To put it very bluntly: I would argue that it is people's fundamental right to have a say in how their future will look like. At the political level, we've got democracy (at least in some parts of the world). Unfortunately, this does not really extend to technical development. Most Research and Innovation (R&I) activities are carried out by private companies and are thus mostly off-limits for external parties. And even in publicly funded technical research many stakeholders, specifically societal ones, are hardly ever directly involved, despite (at least theoretical) efforts to the contrary, see e.g. Owen et al. 2012.

Using smart systems as an example this chapter suggests to exploit the (international) standardisation process to include a broader base of stakeholders in an early stage of technology development. Ultimately, this should contribute to the emergence of standards, and thus eventually of new technologies, that are not just technically sound and economically interesting but also ethically, societally and environmentally desirable. However, I also have to admit that this chapter is not so much about proposing an immediately working and deployable solution, but about providing some food for thought especially for those stakeholders who are not normally involved in standards setting but (perhaps) should be.

More specifically, the remainder of this chapter is organised as follows. You will find a very brief discussion of smart systems and their relevant (in the context of this paper) characteristics, along with an outline of the associated current standardisation environment, in sect. 2. This is followed by a discussion of the concept of "Responsible Research and Innovation" (RRI), which emerged in the early 2000s in response to the Lund Declaration (2009), complemented by a discussion of two related approaches, in sect. 3. The idea of "Responsible Standardisation" is based on RRI principles and aims to overcome some of its practical shortcomings; it is discussed in sect. 4. Sect. 5 offers some concluding remarks and a proposal.

2 Smart Systems and Standardisation – Some More Background

2.1 *Smart Systems*

The integration of ICT instils smartness into more "traditional" applications. Convergence of formerly separate technologies has been an ongoing trend over several decades. For instance, the integration of (tele)communication and information technology led to ICT. Examples of smart application areas include, among others, Intelligent Transport, Smart Manufacturing, the Smart Grid and Smart Cities.

One of the defining characteristics of these smart applications is their multidisciplinarity. Table 12.1 shows (some of) the technical disciplines that contribute to individual smart applications.

"Smart Cities" arguably represent the broadest application area; I will use them as example in this section. Nevertheless, or perhaps precisely because of that, so far no generally agreed definition of what exactly establishes a smart city has emerged. Albino et al. (2015) found 23 definitions in the literature. The one by Nam and Pardo (2011) says that smart cities are about "*leveraging interoperability within and across policy domains of the city (e.g. transportation, public safety, energy, education, healthcare and development). Smart City strategies require innovative ways of interacting with stakeholders, managing resources and providing services.*" This definition highlights the importance of interoperability

TABLE 12.1 Disciplines contributing to different smart applications

Smart manufacturing	e-Health	Intelligent transport systems	Smart grid	Smart buildings
Production Engineering, Computer Science, Robotics Control Engineering, Telecommunication	Medicine, Computer Science, Mechanical Engineering, Telecommunication	Transport Telematics, Traffic Engineering, Power Engineering, Automotive Computer Science, Telecommunication	Power Engineering, Computer Science, Telecommunication	Architecture, Civil Engineering, Computer Science, Telecommunication

ADAPTED FROM JAKOBS (2016A)

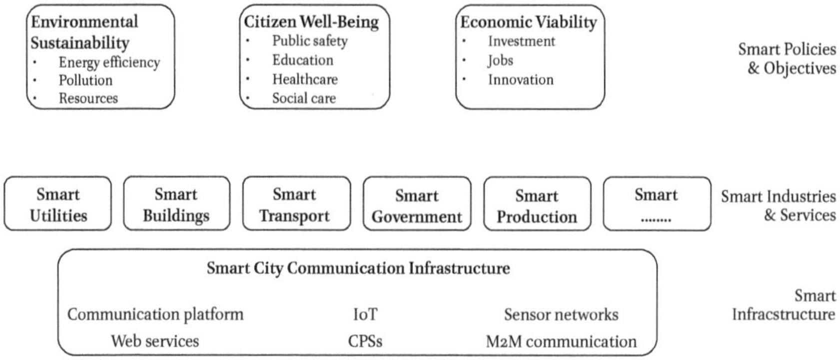

FIGURE 12.1 Smart city model
ACCORDING TO AND ADAPTED FROM NAVIGANT (2011)

at different levels, necessarily including the technical one. The importance of technology for smart cities is further underlined in Bakıcı et al. (2012), who note that "*Smart city as a high-tech intensive and advanced city that connects people, information and city elements using new technologies in order to create a sustainable, greener city, competitive and innovative commerce, and an increased life quality.*" Figure 12.1 shows a popular model of a smart city. On the one hand, this model shows that a number of other smart applications contribute to the objectives of smart cities. That is, in a way, a smart city represents a superset of "smart" applications. On the other hand, it highlights the relevance of a smart (ICT) infrastructure as the basis of a smart city.

Standards for "smart policies and objectives" (see Figure 12.1) provide guidance to (city) leadership for the development of an overall smart city strategy, the identification of priorities, the development of a practical implementation roadmap and for an effective approach to monitoring and evaluating progress. It is primarily here that fundamental civic rights may be at stake. Standards represent a form of industry self-regulation (see e.g. Kallestrup 2017). The associated lack of a democratically legitimated oversight raises issues of the legitimacy of the of the standard itself, the underlying process and of the issuing Standards Setting Organisation (SSO;[4] see e.g. Werle and Iversen 2006; Hofmann 2016). A wide range of stakeholders actively involved in the process would be one way to address this problem. Accordingly, especially when policies and objectives are concerned (and, to a lesser but still considerable degree, smart industries and services) special care should be taken to ensure broad

4 This term denotes both "formal" bodies like ISO and IEC (referred to as Standards Developing Organisations (SDOs)) as well as private standards consortia like the W3C or OASIS.

and active stakeholder participation as a way to increase the level of legitimacy. To this end, contributions relating to e.g. societal, environmental and ethical issue will be essential.

Virtually all smart applications have strong security requirements (see e.g. Li et al. 2016). Accordingly, a vast array of SSOs are active in the fields of cybersecurity. Security encompasses a set of services, including authentication, authorisation, integrity and confidentiality. Moreover, privacy needs to be guaranteed. Obviously, in this domain as well human rights are at stake.[5] Like security, privacy has several dimensions, including a technical one. This technical dimension is very much based on standards. Yet, as Cantero Gamito (2018, 397) observes "*standards are not politically neutral, ... international standards serve as a tool for policy and regulatory diffusion.*" So, here as well active participation specifically of societal stakeholders will be important to ensure that technology actually provides the desired levels of security and privacy, without any backdoors.

For several smart applications and, particularly, for the smart ICT infrastructure the number SSOs active in these fields has skyrocketed over the past eight years (see e.g. Jakobs 2016a). This may be seen as an indicator of the increasing perceived (market) relevance of smart systems. It also represents a problem for those stakeholders that do not have the financial means to participate in all (potentially) societally relevant activities, the outcomes of which may well have significant ramifications for them nonetheless. Please keep this issue in mind, it will be addressed in sect. 5.

2.2 *Just Briefly – The Standardisation Environment*

Already over 30 years ago Andrew S. Tanenbaum (1988, 254) noted that "*The nice thing about standards is that you have so many to choose from.*" This observation is still true today, probably more so than ever. The major reason for this is the proliferation of SSOs. This holds particularly for the ICT sector (specifically for telecommunication). This sector is characterised by a number of national/regional bodies and by more than 250 private standards setting consortia. The huge number of SSOs and the complex links that exist between them yield an almost impenetrable maze. See e.g. Jakobs (2017) for a rough idea of this complexity.

As ICT is one crucial element of smart systems, the complexity of its standardisation environment extends to this field as well. In fact, the complexity here is even higher, as the SSOs active in the respective "traditional" sector also

5 Article 12 of the United Nations' Universal Declaration of Human Rights identifies the right to privacy as a human right.

play a role. Figure 12.2 shows an excerpt of the situation in the smart manufacturing sector (without any claim for completeness).

Individual SSOs differ widely in terms of e.g. coverage (both topical and geographical), processes, by-laws, voting procedures and membership base. They are also subject to different jurisdictions, which implies that the legal status of their respective output may differ (it may e.g. be legally binding or strictly voluntary). The nature of the links that exist between them (i.e. the level of co-operation) is equally diverse. It may take the form of an information exchange about planned new work items to the (rare) joint development of standards or anything in between. It may also be non-existent. And given the multitude of entities involved there is bound to be competition in some case: "*Political competition between standards group is of course a reality especially when it comes to hype topics such as M2M or IoT, since every organization want to claim a dominant role in such areas*" (Jakobs and Wehrle 2017, 62).

The processes adopted by most major SSOs more or less follow a set of "*...principles and procedures that should be observed, when international standards ... are elaborated...*" (WTO, 1995, 24) that have been defined by the World Trade Organization (WTO). Most major consortia do so although formally these principles are only relevant for national and international SDOs (Standards

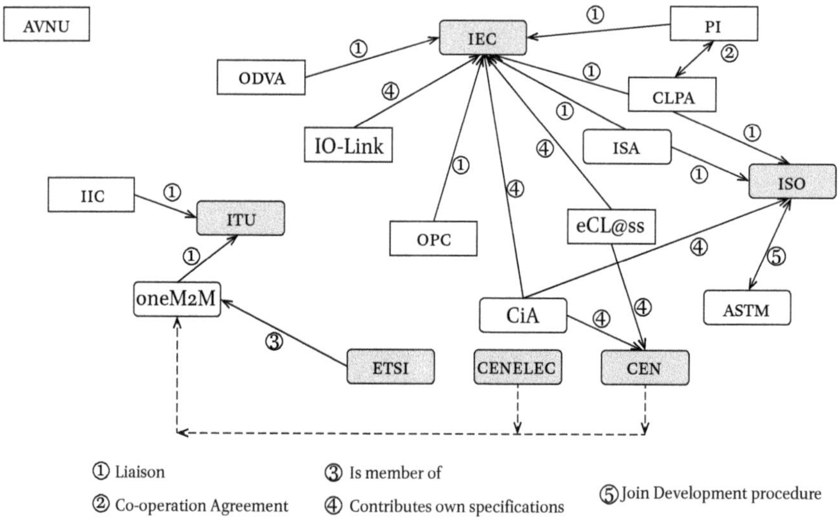

FIGURE 12.2 Links between SSOs in the smart manufacturing sector today (The SSOs depicted in grey include the European Standards Organizations (ESOs; CEN, CENELEC, ETSI) and the World Standards Cooperation (WSC; ISO, IEC, ITU) are fully meshed.)
ADAPTED FROM JAKOBS (2017)

Developing Organisations; i.e. those with some sort of official legitimation, such as ITU or ISO; the WTO principles are not binding for private standards consortia). These WTO principles include transparency, openness, impartiality and consensus, effectiveness and relevance, coherence and to address the concerns of developing countries. They are supposed to make sure that standards meet the needs of as many groups of stakeholders as possible.

2.3 *Who Is Represented in Standards Setting and How?*

The European Union as well highlights the need to include as many different stakeholders as possible: "*European standardisation organisations shall encourage and facilitate an appropriate representation and effective participation of all relevant stakeholders, including SMEs, consumer organisations and environmental and social stakeholders in their standardisation activities*" (EU 2012, Article 5).

Today, the most dominant stakeholders in the ICT sector are (large) companies with a strong economic interest in the technology to be standardised. For example: The European Telecommunications Standards Institute (ETSI) is the European[6] SDO for the telecommunication sector. Out of it's around 900 members about 400 are manufacturers (typically of network equipment). The other member categories include (among others) network operators (ca. 60), service providers (ca. 80), universities and research institutes (ca. 120), users (ca. 40, many of which are large companies) and around 50 government agencies. It is safe to say that other SDOs in this sector will have a similar membership base and thus an equally similar distribution of power.

European 'consumers, environmental and societal stakeholders,' specifically "*the European stakeholder organisations receiving Union financing*[7]," are represented by ANEC, ECOS and ETUI, respectively. The fourth such organisation is Small Business Standards (SBS), representing SMEs. Apparently, from the European Commission's point of view these are the stakeholders that also need be represented – consumers, environmentalists, workers and SMEs.

This is a somewhat limited selection, representing an equally limited view.[8] For one, all organisations – understandably – have a strong focus on the "technical" aspects of their respective domain,[9] where they aim at representing

6 With members from over 60 countries and five continents.

7 The European consumer voice in standardisation (ANEC), the European Environmental Citizens' Organisation for Standardisation (ECOS) and the European Trade Union Institute (ETUI).

8 This is not to say that the work of these organisations is not valuable – it is.

9 See https://www.cencenelec.eu/societal/Pages/default.aspx for a list of committees and working groups where the three organisations are represented.

their respective constituency. Yet, this implies that non-technical aspects (e.g. the ethical and societal dimensions) of standards are likely to be frequently ignored. On the brighter side, ANEC and ECOS are active in the areas of smart systems and of data security and privacy; in both of which fundamental civic rights are potentially at risk. Accordingly, societal stakeholders need to be adequately involved in the standardisation processes in these areas.

On the other hand, the general need for wide stakeholder participation may be contested. Egyedi (2003), for example, argues that "democratic" standardisation is not necessarily a value per se and that it depends on the type of standard at hand. For compatibility standards, for example, she states that "*…non-consensus consortium standards would seem preferable if seen from the 'democratic' viewpoint*" (Egyedi 2003, 33). This is echoed in Jakobs (2005), who argues that for comparably short-lived, purely technical standards like e.g. those for PC interfaces (think USB) trying to bring in everyone at all costs is counter-productive. On the other hand, for other types of standards – Egyedi gives health and safety as examples; smart systems would be another one – democratic procedures (including *adequate* stakeholder representation) should strictly be followed and be monitored systematically.

In addition, the issue of how stakeholders should be represented remains open. Jakobs (2015) argues that umbrella organisations should not automatically be the representatives of choice. This is because stakeholder diversity implies that in very many cases a common ground to be presented by such an umbrella organisation will not exist. This holds particularly for SMEs, for the other groups it is a subject for further study.

A very different aspect that also needs to be taken into account in this context is the real distribution of power within the Working Groups (WGs), where the actual technical decisions are being made. To a considerable extent this distribution is based on the knowledge and abilities of the individual members of a WG (see e.g. Jakobs 2014). After all, a standard emerges as the outcome of the deliberations of these members.

Most WGs, at least in the ICT sector, are primarily populated by engineers and computer scientists with a strong technical background. The natural assumption would thus be that a technically sound proposal, representing or going beyond the state-of-the-art stands the best chances to become the new standard – i.e. that members' technical proficiency is their most important trait. Yet, these members are, after all, humans. Thus, their motivations, attitudes, views and prejudices are very likely to also have an influence on the outcome of a standards setting process. Moreover, there is evidence that other such rather non-technical aspects may well be important. For one, their diplomatic, negotiation and rhetoric skills will play a crucial role. In fact, speaking out at meetings for or against a proposal has been named the most important single factor that influences technical decisions (Jakobs et al. 2010). That is, even superior proposals will hardly be considered if nobody is available to explain or defend them at meetings. Continuous active participation by the same capable representative(s) thus seems to be the most promising approach. This will also lead to an accumulation of social capital, which will, in turn, strengthen the position and influence of the representative (see also e.g. Isaak 2006). This may be one avenue to explore for those stakeholders that cannot afford a numerically strong participation in the process (like the societal ones). A quote from a survey reported in Jakobs et al. (2001) may serve to highlight this "human factor": *"Oddly enough, it's been my experience that _individuals_ dominate ISO. Sometimes the individual will have a powerful multinational corporation or government/national interest on their side, but the bully pulpit is controlled by individuals, and only those with a strong sense of purpose survive."*

Even if this is not necessarily always the case, the influence a strong-minded individual may have on the outcome of the standards setting process must not be underestimated. Umapathy et al. (2007, 296) put it quite nicely: "*The human dimension of standards setting is an important component of the consensus-based process...*".

3 Responsible Research and Innovation

3.1 *A Brief Introduction*

Innovation is not inherently good. Nor is it just about the innovator's economic gain. Accordingly, the perceived need for closer collaboration between the Social Sciences and Humanities (SSH) on the one hand and the STEM subjects (Science, Technology, Engineering, Maths) on the other is a major driver for Responsible Research and Innovation (RRI). This is also noted by Grunwald (2011, 17), who adds the ethical dimension and observes that "*...Responsible Innovation unavoidably requires a more intense inter- and trans-disciplinary cooperation between engineering, social sciences, and applied ethics.*" RRI today aims to "*...go beyond a purely technological understanding of innovation*" (Burget et al. 2017, 4), i.e. to guide research and innovation to address the various challenges society faces.

Ideas that relate to RRI may be traced back a couple of decades. In 1947, Detlev Bronk testified before a congressional committee that "*...that competent social scientists should work hand in hand with the natural scientists, so that problems may be solved as they arise, and so that many of them may not arise in the first instance*" (quoted in Steelman 1947).

In 1972, the Office of Technology Assessment (OTA) was founded in the US; similar entities were subsequently established in many European countries. The definition of Technology Assessment (TA) used during the early years of the OTA expresses the need for pro-active analysis and evaluation of technological choices: "*TA is the systematic identification, analysis and evaluation of the potential secondary consequences of technology in terms of its impacts on social, cultural, political and environmental systems and processes*" (OTA 1976, 312). A number of different TA approaches emerged, including e.g. Constructive TA, Integrative TA, Participatory TA and real-time TA. The latter is supposed to "*…inform and support natural science and engineering research, and it can provide an explicit mechanism for observing, critiquing, and influencing social values as they become embedded in innovations*" (Guston and Sarewitz 2002, 94).

Definitions of RRI emerged largely from the policy domain, most notably in the EU. One example would be by v. Schomberg (2011, 9), who defines RRI as "*a transparent, interactive process by which societal actors and innovators become mutually responsive to each other with a view to the (ethical) acceptability, sustainability and societal desirability of the innovation process and its marketable products (in order to allow a proper embedding of scientific and technological advances in our society).*" Along similar lines, Stilgoe et al. (2013) propose four dimensions of RRI:

– Anticipation

This dimension aims to adopt a view that takes into account opportunities, risks, environmental concerns, etc. I.e. researchers are supposed to think about "*what is known, what is likely, what is plausible and what is possible*" (Stilgoe et al. 2013, 1570), about contingencies and about what is desirable and what is not. It thus links to (Constructive) TA. "*Anticipation … aims at envisioning the future of research and innovation and understanding how current dynamics help design the future*" (Burget et al. 2017, 12). Especially the aspect of designing the future also links to the Social Shaping of Technology (SST) approach.[10] Generally, the identification of societal and ethical impacts, including any potential negative influences of products and services, should be considered during the development process. All, in all, anticipation should contribute to a better and more desirable future.

– Inclusion

Not least to overcome the risk of a "tunnel vision" inclusion refers to the approach of adding additional stakeholders that may challenge long-standing views and thus to identify socially more desirable outcomes of an innovation

10 For a good introduction to this see Williams and Edge 1996.

activity. This is perhaps *the* archetypical dimension of RRI. The basic underlying idea is that public involvement will lead to socially desirable outcomes (see e.g. Owen et al. 2012). Typical tools here include citizen panels, focus groups, consensus conferences and science shops (Chilvers 2010).

– Responsiveness

Responsiveness implies the ability to react to new knowledge and to newly emerging views, norms and circumstances. Specifically, RRI needs to address ethical issues and to be responsive to societal challenges. To this end, its focus must include an open organisational culture emphasising innovation,[11] creativity, interdisciplinarity as well as commitment to public engagement and to taking account of the public interest[12] (Macnaghten and Chilvers 2014; Stilgoe et al. 2013). That is, innovation policy needs to make sure that the purpose of research and innovation is not limited to triggering economic growth.

– Reflexivity

Not unlike responsiveness, reflexivity also links to public dialogue, science/public collaboration and thus inclusion. It aims at placing research into its wider (societal and ethical) context by reflecting on the values and beliefs during research and development. Schuurbiers (2011) reports about studies where laboratory engagement of social scientists and philosophers caused natural scientists to acknowledge the relevance of reflection about the broader socio-ethical context of their work. Reflexivity may also be linked to anticipation. Robinson (2009, 1230), for example, advocates the stimulation of "*a process of reflexive anticipation through controlled speculation based on exploring the underlying dynamics of emergence.*" This should help avoid the risk of making wrong (strict) predictions, especially in the early stages of a technological development.

RRI emphasises that not just technical and economic, but also e.g. societal, ethical and environmental aspects shall be taken into account during the innovation process. This, in turn, means that a broad range of stakeholders has to be involved in the process, notably those that are not normally involved in research and innovation (i.e. e.g. consumers, NGOs, but also e.g. social scientists, humanists, etc.). The notion that "stakeholders" – as opposed to "actors" – need to be involved is also reflected in a definition by the European Commission (2013, 3), which reads "*Responsible Research and Innovation (RRI) refers to the comprehensive approach of proceeding in research and innovation in ways that allow all stakeholders that are involved in the processes of research and innovation at an early stage (A) to obtain relevant knowledge on the consequences of the*

11 Think Open Innovation, e.g. Chesbrough 2006.

12 This also links to "Inclusion."

outcomes of their actions and on the range of options open to them and (B) to effectively evaluate both outcomes and options in terms of societal needs and moral values and (C) to use these considerations (under A and B) as functional requirements for design and development of new research, products and services."

Adopting a broader perspective, Stilgoe et al. (2013, 1570) state that "*Responsible innovation means taking care of the future through collective stewardship of science and innovation in the present.*" Along similar lines, Stahl (2013, 5) defines RRI as "*RRI is a higher-level responsibility or meta-responsibility that aims to shape, maintain, develop, coordinate and align existing and novel research and innovation-related processes, actors and responsibilities with a view to ensuring desirable and acceptable research outcomes.*" Especially this view implies that e.g. technology assessment needs to be applied in order to detect potentially undesired outcomes of the deployment process and to take appropriate counter-measures as early as possible. This holds particularly for innovations in new and hardly understood technologies such as nanotechnology and smart technologies, for which standards should be essential not only as technical "guarantors" of interoperability, but also as socio-ethical frameworks.

3.2 *Related Concepts*

Two recent paradigms in the field of innovation share some characteristics with responsible innovation – "open innovation" and "social innovation." They will be briefly considered in the following, with a focus on the question whether or not they may be of relevance, e.g. by complementing RRI – for a socially beneficial standardisation of smart systems.

The idea underlying open innovation (OI; see e.g. Chesbrough 2006) is that today an individual firm cannot innovate in isolation. Rather, it has to bring in ideas – and utilise resources – from outside as well[13] – "*firms can and should use external ideas as well as internal ideas, and internal and external paths to market, as the firms look to advance their technology*" (Chesbrough 2003, XXIV). Typically, though, "external" refers to other firms or research organisations "*from the individual inventor or high-tech start-up in Silicon valley to research facilities of academic institutions, to spin-offs from large, established firms*" (Chesbrough 2006, 2). Other actors are not normally considered, with the exception of users, whose knowledge and experience may well be exploited for product and service development (see e.g. Bogers et al. 2010). Another exception, albeit much less popular, are (online) communities (see e.g. Dahlander et al. 2008). However, "*More commonly, firms play a central role in creating and organizing innovation communities,*" at least for economically valuable innovations (West and

13 In reverse conclusion this implies that some of the firm's own ideas and resources will also need to flow outside.

Lakhani 2008, 226). This suggests that frequently such communities are also used for service or product development; most notably in the field of open source software development. Nevertheless, the idea of open innovation (OI) may also be applied in the public sector. Martins et al. (2015) report on efforts in this sector aiming to crowdsource ideas and how this has been used for innovations. These efforts, however, are indeed limited to the public sector and the co-operation between government and citizens. In the field of smart systems, Schaffers et al. (2011) discuss the integration of Living Labs – as user-driven OI ecosystems – with the European Future Internet (FI) research and innovation process. The focus, however, is rather on the technical side, identifying FI/multi-disciplinary researchers, users and enterprises as main stakeholders. Dahlander and Gann (2010) ask how open "open innovation" actually is. They identify four different types of openness, out of which the type "Inbound innovation sourcing" is the most relevant one within the context at hand. They note that typically firms collect external ideas and knowledge from suppliers, customers, competitors, consultants, universities, public research organisations, etc. This as well point in the same directions as does the above – it's (mostly) about technical knowledge.

Social innovation (SI) has been defined as "*innovative activities and services that are motivated by the goal of meeting a social need and that are predominantly developed and diffused through organisations whose primary purposes are social*" (Mulgan et al. 2007, 8). As the name already suggests, the focus here is on the social/societal, environmental and also the ethical aspects of innovations (as opposed to the more technical focus of OI). Accordingly, social innovations[14] are not necessarily driven by the desire for profit-making (Pol and Ville 2009), which sets them apart from (typically for-profit) business innovations.

SI is a distributed process, frequently enabled by today's global information and communication infrastructure like social networking tools (Murray et al. 2010). It also involves stakeholders that are not normally represented in OI (i.e. societal actors). For the case of smart cities, for example, these stakeholders may include local governments, research institutions, grassroots movements, technology vendors, property developers, etc.[15] They are of particular relevance when it comes to what Angelidou (2014) calls the "soft infrastructure" of the city (i.e. e.g. social and human capital; knowledge, inclusion, participation,

14 Examples of social innovations include e.g. microcredits, fair trade and e-learning (Sharra and Nyssens 2010). While they do have a business dimension, making money was not the idea behind these innovations.

15 I.e. mostly those that are not normally represented in either business innovation or, indeed, standardisation.

social innovation, social equity) as opposed to city's hard infrastructure systems (i.e. e.g. transport, water, waste, energy).

3.3 *Discussion*

With respect to the problem at hand – i.e. the responsible development of smart systems – all approaches discussed above have their respective specific strengths and weaknesses.

Starting with SI we may note that one of its remarkable aspects is the focus those stakeholders that are not normally taken into account in the context of business innovation and open innovation (which in ICT typically include (large) manufacturers, operators, service providers). While most of these (societal) stakeholders are highly unlikely to be in a position to contribute any technical knowledge and expertise to the development of smart systems or their underlying communication infrastructure, they certainly are well positioned to contribute non-technical requirements (see (Jakobs 2002)). In contrast to the majority of work done by most SSOs the standardisation of smart systems – particularly of the topmost layer of the smart city model; see Figure 12.1 above – will require extensive input beyond the purely technological. On the other hand, SI focuses rather more on smallish projects and very concrete social welfare than on anything else. And while this would also relate to e.g. smart cities' objectives of "Citizen Well-Being" and "Environmental Sustainability" (see Figure 12.1) it would cover only a small part of these rather general objectives. That is, SI, or perhaps its underlying ideas, do have a place in smart system standardisation. But it will only be able to cover a small part of it.

In contrast, OI is typically concerned with (purely) technical innovations, open source software development being a good example. Moreover, in many cases open innovation communities are triggered, if not governed, by companies. Dahlander and Magnusson (2005) look at four cases of open source software communities and companies; in three cases the community was established by a company (the ensuing type of relationship between both entities is a different aspect; see Dahlander and Magnusson 2005). This is in line with the observed central role companies play in many innovation communities, at least in those that are of (potential) economic interest (see e.g. West and Lakhani 2008). Accordingly, it may be assumed that economic considerations are at the heart of many such OI initiatives. Along with the typically technical focus and the resulting limited number of groups of stakeholders involved (virtually excluding societal stakeholders in practice) this renders the OI approach of limited relevance for the development of "responsible" smart systems.

On paper, RRI has all the necessary ingredients. It highlights broad stakeholder participation (with a focus on those that are not normally represented in research and development), includes the idea of forecasting or technology assessment and specifically highlights the need to reflect about potential impacts innovations under development may have and to adapt to any changes in the broader environment. The one severe downside is the fact that in the vast majority of cases R&I activities are off-limits for those who could represent "responsibility" (i.e. the social sciences and the humanities, but also citizens and NGOs). This holds primarily, though not exclusively, for the private sector. Here, in 2015 business expenditure on R&I amounted to more than €29 billion in the EU, compared to an estimated €6.3 billion of public funding.[16] It appears safe to assume that, despite the increasing popularity of OI, this research did not include many non-technical stakeholders. Looking at publicly funded R&D in the EU, the situation does not seem to be that much better. With a lack of hard facts and figures about the level of participation of e.g. the social sciences and humanities in technical programmes some rather more anecdotal evidence will have to suffice: The EU's ICT Work Programme (WP) 2018–2020, which covers areas like "Technologies for Digitising European Industry" and "European Data Infrastructure: HPC, Big Data and Cloud technologies" does not even mention social sciences or humanities,[17] despite ample opportunities for contributions from these disciplines here. Almost the same may be said about the WP "Secure societies – Protecting freedom and security of Europe and its citizens,"[18] which is about one of the identified Societal Challenges.

In summary, the theoretical and practical aspects associated with RRI and related approaches suggest that a different mechanism is needed to actually (or at least better) enable societal stakeholders to have their voices heard in the early stages of the development of smart applications. So let us now turn to a slightly different approach towards the responsible development of smart applications.

4 Responsible Standardisation

The discussion above shows that
- participation of an adequate range of stakeholders and an element of "foresight" are crucial ingredients of RRI;
- current R&I activities (both private and public) do not adequately consider this.

16 See https://ec.europa.eu/jrc/en/news/latest-statistics-ict-sector-and-its-rd-investment-available.

17 http://ec.europa.eu/research/participants/data/ref/h2020/wp/2018-2020/main/h2020-wp1820-leit-ict_en.pdf.

18 http://ec.europa.eu/research/participants/data/ref/h2020/wp/2018-2020/main/h2020-wp1820-security_en.pdf.

We may also note that "*The shaping process* [of a technology] *begins with the earliest stages of research and development*" (Williams and Edge, 1996, 874).[19] That is, to achieve the goal of developing technologies that are not just technically sound but also environmentally, economically, societally and ethically desirable, RRI principles need to be applied as early as possible in the development process.

Especially in the ICT sector (and thus also for smart systems) standardisation represents (one of) these earliest stage(s) of (technical) development, during which a future technology is shaped.[20] Moreover, and in contrast to corporate R&I efforts, almost all major SSOs are, in principle, open to participation of all kinds of stakeholders.[21] It thus represents a potential arena where those stakeholders that are not normally represented in the development of an emerging technology (see above) may have the chance to exert influence.[22] Such broader participation (see also sect. 2.3), in turn, would be an important step towards responsible standardisation.

The remainder of this section will, therefore, propose the deployment of an alternative mechanism that may help achieve the goal of giving ideally everyone a direct or indirect say in the development of the future we will have to live in. More precisely, this mechanism should give stakeholders the opportunity to make their voices heard in the development of smart systems that are technically sophisticated, societally and environmentally beneficial and ethically acceptable. This mechanism could be the responsible standardisation process.

4.1 *Responsibility and Standardisation*

Three orthogonal types of approaches to link RRI to standards and standardisation may be identified:

– Deploying standardisation to attract more, and more diverse, stakeholders to RRI.

A report issued by the European Commission on "Options for Strengthening Responsible Research and Innovation" observes that one option would be "*…linking research projects with standardisation activities, because national and European standardisation bodies provide the platform for the involvement of all interested stakeholders to participate in the development of standards, which*

19 This point is made by those who advocate the Social Shaping of Technology (SST) approach; this also links to the discussion about "Reflexivity" above.

20 See e.g. von Burg (2001) or Lemstra et al. (2010) for fascinating accounts of the standardisation – and the shaping – of Ethernet and WiFi, respectively.

21 Albeit typically for a fee.

22 See e.g. Jakobs (2005) for the potential practical problems here.

reflect not only the interests of industry and research, but also representatives of societal interests,..." (Jacob et al. 2013).

– Standards *for* RRI.

A number of standards have been developed that could be of relevance for RRI. At the more general level, these include e.g. the ISO 26000 series on Corporate Social Responsibility and the ISO 31000 series on Risk Management. Closer to the field of smart systems, (ITU-T 2015), for instance, offers an in-depth discussion of the roles of the different groups of stakeholders in the standardisation of smart cities. Moreover, a number of other ITU-T Technical Reports discuss aspects relating to sustainable cities.

– Applying RRI principles in standards setting.

Typically, only some groups of relevant stakeholders are on board in standards setting. In particular, this holds for the ICT sector. Rather, e.g. users, consumers and NGOs are dramatically under-represented, as are the social sciences and humanities (forming the "Third Estate" in standards setting; Jakobs 2005).

In the following the focus will be on the latter.

Although it has so far hardly been discussed in the literature the standardisation process is certainly of relevance from the "responsibility" angle. Wickson and Forsberg (2015) refer to standardisation as a "*...kind of interstitial space because it effectively occupies a location somewhere between what are currently recognized as key domains, e.g. those of science, policy, civil society and industry.*" This space, or rather its output, has a tremendous albeit usually overlooked impact in almost all sectors. The aptly titled paper "IT standardization: The billion dollar strategy" (Hurd and Isaak 2008) gently hints at the economic importance of standards and the underlying standardisation process. Accordingly, Stilgoe et al. (2013, 1573) note that *"Responsible innovation .... demands explicit scrutiny of the tensions and governance mechanisms within .... technological standards, which often act to close down innovation in particular ways, and other norms, pressures and expectations that reinforce particular path dependencies and lock-ins."* While this may indeed happen in certain cases, the at one time popular perception that standards largely hamper innovation has been disbanded. Swann and Lambert (2010), for example, show that standards can do both, hamper and enable innovation and that the latter effect is much more pronounced.

The legitimacy of the process (and its outcome) also links to responsibility in that both rely on broad stakeholder participation. Standardisation's legitimacy has long been a matter of concern. From a social sciences' point of view, Werle and Iversen (2006) noted that SDOs were moving away from their erstwhile focus on input legitimacy (which is desirable but hard to achieve) to output legitimacy. The former relates to the standardisation process as such (basically the WTO principles, especially broad stakeholder participation), the latter focuses on the standard (not all interests are necessarily directly represented, external monitoring by interest groups). They conclude that both "*…share the aim to stimulate participants in standards developing processes to consider potential non-market and non-technical impacts of standards.*"

"Openness" is another term that is closely linked to legitimacy and responsibility. It also links to the level of trust a standard or, rather, an SSO enjoys. Krechmer (2008) offers the perhaps most comprehensive and workable definition of open standards. He identifies ten characteristics they need to posses, along with an indication for whom (creator, implementer, user) they are of relevance. Jakobs (2016b) extends this list with a view towards more trustworthy ("legitimate") standardisation. Table 12.2 summarises the most relevant criteria to be met in the context of responsible standardisation.

Except for criterion #4 (which is of lower relevance here) those criteria above that were identified by Krechmer are met by most major SSOs. The same holds for #5. Criteria 6–8 are more tricky; each is met only by some SSOs; I am not aware of one that meets them all.

Criterion #1 has been discussed quite intensively in the literature. Not unlike RRI, it is met on paper. In practice, however, the development of most ICT standards[23] – certainly of those with high enough economic ramifications – is dominated by (large) manufacturers, service providers and network operators. Users, NGOs, citizens and many other stakeholders are almost always excluded. Accordingly, much of the literature assumes that standards reflect and materialise existing power relations (see e.g. Graz 2018); after all they are part of "*the technical, political, social, economic, and ethical infrastructure that constitutes human societies*" (Busch 2011, 13). This observation is interesting in more than one respect. For one, it highlights the crucial importance of standards way beyond the technical and the economic dimension. Second, it explicitly includes the ethical dimension. This, in turn, reinforces the notion that the standards setting process is indeed a place where also issues relating to fundamental civic rights (that relate to the standard in question, if applicable) should be addressed. In conjunction with the typical power distribution in standards WGs this once again emphasises the need for broader stakeholder participation (within limits; see sect. 2.3).

Along similar lines, Graz and Hauert (2019) discuss the heterogeneity and overlap of lay knowledge and expert knowledge, both of which are, or rather

23 This holds primarily for telecommunication standards.

TABLE 12.2 The most relevant criteria for open (trustworthy) standardisation

		Relevance for trust by	
#	Criterion	Implementer	User
1	Open Meeting All may participate		xxx
2	Consensus Agreements found without sustained opposition		xxx
3	Due Process Fair proceedings; appeal process		xxx
4	Open IPR Low or no royalties for IPR necessary to implement the standard	xxx	
5	Open Rules Publicly available bylaws, regulations etc.		xx
6	Transparent Membership WG Membership (entities and individuals) is published		xx
7	Open Roles Names and affiliations of those holding leadership roles are made public		xx
8	Open History All SSOs associated with a standard are known		xx

CRITERIA 1–4 ARE FROM KRECHMER (2008); 5–8 ARE FROM JAKOBS (2016B)

should be, available during standards development. They observe that it is very complicated to actually mobilise lay knowledge for the daily work of SSOs' committees or working groups. They also note that the scientific ("expert") knowledge inherent in standards offers the justification for regulators and governments to reference standards in law, thus further stressing the relevance of such knowledge. Plus, earlier research found that at least back then "*Committee members have also named technical sophistication on the side of the user rep-*

resentatives as a major prerequisite for meaningful participation" (Jakobs et al. 2001, 106), which quite conveniently ignores that there are other important aspects to standards than just purely technical functionality.

However, I would argue that in the case of smart systems, with their entirely unclear ramifications for society (again, think e.g. George Orwell) the distinction between lay knowledge and expert knowledge becomes void. Here, what may be seen as lay knowledge for e.g. technical deliberations may become expert knowledge when it comes to the consideration of societal and/or ethical impacts of the technology to be developed (and vice versa). In this case, only expert knowledge, albeit from different domains, needs to be available. And it should be of equal value.

5 By Way of Discussion and Conclusion – A Proposal

Multi-disciplinary co-operation (e.g. between SSH and STEM) is a critical prerequisite for the standardisation of smart systems. This holds particularly for smart applications.[24] Yet, it brings along a number of problems, most importantly that of communication. Piso et al. (2016, 648) quite rightly observe that "*The interdisciplinary challenge for a collaborative effort is to foster and maintain open communication about disciplinary perspectives, with each investigator learning what they need to know about the contributions of the others to properly situate their own project contributions.*" In standards setting, problems to be associated with multi-disciplinarity are further reinforced by the fact that the SSOs active in the more "traditional" disciplines work under very different boundary conditions as those from the ICT sector (e.g. different product life cycles, different levels of economic importance; see e.g. Jakobs 2016b).

This multi-disciplinarity has two dimensions. On the one hand, technical experts from different application areas (see also Table 12.1) and from different backgrounds need to co-operate to integrate ICT with the respective applications. On the other hand, the true ubiquity of smart systems mandates that societal and ethical issues also have to be taken into account during standards development. To this end, "exotics" (in the standardisation world) including citizens, NGOs, unions as well as politicians, sociologists and philosophers will

24 The "smart infrastructure," to which most of the ongoing standards setting activities belong, offers rather more limited scope for societal and ethical consideration.

need to have a say in one way or another. In an environment with such diverse actors, many of whom have never co-operated before and where a certain level of initial mutual scepticism may be assumed trust and collaboration become all the more important. Rosenkopf et al. (2001, 754) note that *"frequent and sustained face-to-face meetings among firm representatives engender trust and collaboration among alliance partners."* The importance of face-to-face meetings is further corroborated, albeit from a slightly different angle in (Kramer and Cook 2004), who note that *"Trust theorists have long argued the benefits of face-to-face interaction and direct experience with others in the trust-building process"* (12).

Unfortunately, today's standardisation environment is not that well suited to support multi-disciplinarity. This may at least partly be attributed to history. Most WGs are populated by the "usual suspects," i.e. computer scientists and (telecommunication) engineers.[25] That is, they have a – frequently long standing – track record of technical standards setting where the contributing disciplines have been limited to exactly these two. Specifically, non-technical stakeholders have, in the vast majority of cases, not been represented at all.

As an aside: This strong technical focus has another, perhaps not so obvious detrimental effect: The scientific insights supposed to underpin standards and the notion that technical expertise on the side of all WG members are decisive for being taken serious reinforce the techno-centricity of standards and need to be overcome.

The non-participation of those stakeholders that form the "Third Estate" in (ICT) standards setting has different reasons. Generally speaking, they have to do with a lack of knowledge about the importance of standards (including those aspects beyond the purely technical ones) and a resulting lack of interest plus, perhaps most importantly, a lack of funding (see also e.g. Jakobs 2005). For smart systems, the latter increases with the number of SSOs that are of potential relevance (see also Figure 12.2) and will eventually form an insurmountable barrier for those without a strong economic interest (and deep pockets).

To overcome these problems and in conjunction with the general observation that smart systems' eventual true ubiquity and the associated major societal and ethical ramifications warrant extra efforts to make sure that the best possible systems will be designed, I would like to propose the creation of a

25 This holds for all WGs I'm aware of.

dedicated, central new SSO to deal exclusively with the technical, societal, environmental and ethical aspects of these systems.[26]

This new entity could adopt a process of "Responsible Standardisation" to ensure that standardisation takes into account not just technical and economic, but also e.g. ethical, environmental and societal aspects. To this end, privacy-by-design ideas could be applied to integrate these aspects into the process.[27] This would enable "responsibility by design," a standardisation process that incorporates Anticipation, Inclusion, Responsiveness and Reflexivity (see sect. 3.1[28]).

Moreover, technology assessment could be applied, as could be principles of the social shaping of technology, in order to detect potentially undesired outcomes of the technology under standardisation and to take (or at least suggest) appropriate counter-measures as early as possible.

Such a dedicated "Smart Systems SSO," sponsored by as many SSOs as possible, could lead, co-ordinate, align and at least partly perform relevant standardisation activities. Such an entity would probably not solve all co-ordination problems, but if it were sponsored by at least the major SSOs in the field their combined expertise and "market power" should be a unique selling point and attract the big corporate players.

This entity's adopted process would have to cater to a broad variety of stakeholders. Given their necessary variety of expertise and knowledge it might be helpful to adopt a two-tier process, where the technical issues and the non-technical ones are considered separately by different groups and where primarily the latter informs the former.[29] Last but not least, it will probably be necessary to provide (government) money to ensure participation of societal stakeholders.

Figure 12.3 shows a rough sketch of a model of such a process. The "traditional" life cycle of an ICT standard is depicted on the right hand side; primar-

26 This is despite the fact that in a recent study the experts quite unanimously opposed the idea of any new SSOs on the grounds that the number of SSO is large enough as it is. This argument indeed carries some weight; a new entity may well increase the existing complexity without adding much value (Jakobs 2018).

27 Basically, privacy by designs mandates that organisations need to consider privacy at the initial design stages and throughout the complete development process of new products, processes or services that involve processing personal data; see also e.g. Langheinrich 2001.

28 See also e.g. EC (2015) for a discussion of related aspects in the IoT context.

29 An information flow in the reverse direction will also be necessary if certain requirements jut cannot be met on technical grounds.

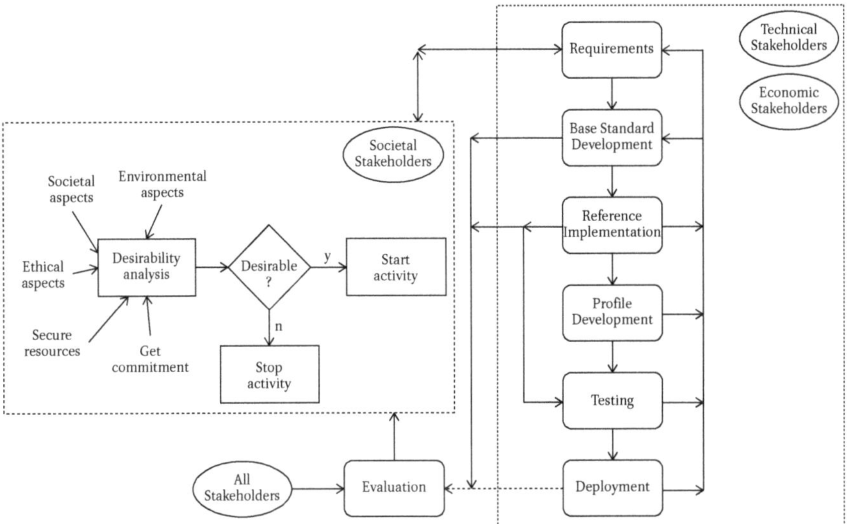

FIGURE 12.3 A simple model of the proposed two-tier standardisation process for smart systems

ily technical and economic stakeholders are involved here. This is preceded and complemented by an analysis of the societal, environmental and ethical desirability of a technical proposal, which is done by societal stakeholders (depicted on the left hand side). Based on the outcome of this analysis a go / don't go decision is made. In case of a "go," a list of requirements to be met and of boundary conditions not to be violated are provided.

One benefit of such a process is that the "normal" (i.e. technical) part will remain largely unchanged Another one is that the communication between the technical and the societal world will largely be limited to one well-defined interface (the contribution and explanation of societal requirements). Benefit number three is that such a split will significantly reduce the necessary level of involvement of the societal stakeholders (and thus the associated costs), which will encourage a broader variety of them to become active in the process. This, in turn, will contribute to a stronger societal representation and thus to a louder voice to demand the consideration of fundamental civic rights and to denounce any potential violation at a very early stage of technical development.

An evaluation of the actually deployed system would be highly desirable and should also be performed. However, even if necessary changes to the standard are identified here it will probably be very difficult to change existing working systems accordingly.

Smart systems standardisation is still at a comparably early stage, so it should not be too late to devise an environment that adequately caters for the standardisation of a technology which has the potential to dramatically change out lives – for better or worse.

References

Albino, Vito, Umberto Berardi, and Rosa Maria Dangelico. 2015. "Smart Cities: Definitions, Dimensions, Performance and Initiatives." *Journal of Urban Technology* 22(1): 3–21.

Angelidou, Margarita. 2014. "Smart City Policies: A Spatial Approach." *Cities* 41: 3–11.

Bakıcı, Tuba, Esteve Almirall, and Jonathan Wareham. 2012. "A Smart City Initiative: The Case of Barcelona." *Journal of the Knowledge Economy* 2: 1–14.

Bogers, Marcel, Allan Afuah, and Bettina Bastian. 2010. "Users as Innovators: A Review, Critique and Future Research Directions." *Journal of Management* 36(4): 857–875.

Burget, Marjam, Emanuele Bardone, and Margus Pedaste. 2017. "Definitions and Conceptual Dimensions of Responsible Research and Innovation: A Literature Review." *Science and Engineering Ethics* 23(1): 1–19.

Busch, Lawrence. 2011. *Standards: Recipes for Reality*. Cambridge: The MIT Press.

Cantero Gamito, Marta. 2018. "Europeanization Through Standardization: ICT and Telecommunications." *Yearbook of European Law* 37: 395–423.

CEC. 1996. "Communication from the Commission to the Council and the Parliament on 'Standardization and the Global Information Society: The European Approach.'" COM (96) 359 final. https://op.europa.eu/en/publication-detail/-/publication/ca944132-f4a3-4f36-ad31-95e5f7bfda77/language-en.

Chesbrough, Henry. 2003. *Open Innovation: The New Imperative for Creating and Profiting from Technology*. Boston: Harvard Business School Press.

Chesbrough, Henry W. 2006. *Open Innovation: The New Imperative for Creating and Profiting from Technology*. Boston: Harvard Business Press.

Chilvers, Jason. 2010. "Sustainable Participation? Mapping out and Reflecting on the Field of Public Dialogue in Science and Technology." University of East Anglia. https://ueaeprints.uea.ac.uk/37545/1/Chilvers_Sustainable_Participation_report.pdf.

Dahlander, Linus, Lars Frederiksen, and Francesco Rullani. 2008. "Online Communities and Open Innovation." *Industry and Innovation* 15(2): 115–123.

Dahlander, Linus, and David M. Gann. 2010. "How Open is Innovation?" *Research Policy* 39(6): 699–709.

Dahlander, Linus, and Mats Magnusson. 2005. "Relationships Between Open Source Software Companies and Communities: Observations from Nordic Firms." *Research Policy* 34(4): 481–493.

EC, eds. 2013. "Options for Strengthening Responsible Research and Innovation." https://wbc-rti.info/object/document/11932/attach/options-for-strengthening_en.pdf.

EC, eds. 2015. "Internet of Things – IoT Governance, Privacy and Security Issues." http://www.internet-of-things-research.eu/pdf/IERC_Position_Paper_IoT_Governance_Privacy_Security_Final.pdf.

Egyedi, Tineke M. 2003. "Consortium Problem Redefined: Negotiating Democracy in the Actor Network on Standardization." *International Journal of IT Standards and Standardization Research* 1(2): 22–38.

EU. 2012. "Regulation (EU) No 1025/2012 of the European Parliament and of the Council on European Standardisation." https://eur-lex.europa.eu/LexUriServ/LexUriServ.do?uri=OJ:L:2012:316:0012:0033:EN:PDF.

Graz, Jean-Christophe. 2018. "Global Corporations and the Governance of Standards." In *Handbook of the International Political Economy of the Corporation*, edited by Andreas Nölke and Christian May, 448–461. Cheltenham, UK: Edward Elgar Publishing.

Graz, Jean-Christophe, and Christophe Hauert. 2019. "Translating Technical Diplomacy: The Participation of Civil Society Organisations in International Standardisation." *Global Society* 33(2): 163–183.

Grunwald, Armin. 2011. "Responsible Innovation: Bringing Together Technology Assessment, Applied Ethics, and STS Research." *Enterprise and Work Innovation Studies* 31: 10–19.

Guston, David H., and Daniel Sarewitz. 2002. "Real-time Technology Assessment." *Technology in Society* 24(1–2): 93–109.

Hofmann, Herwig C. 2016. "A European Regulatory Union-The Role of Agencies and Standards." In *Research Handbook on the Law of the EU's Internal Market*, edited by Panos Koutrakos and Jukka Snell, 460–478. Elgar Publishing.

Hurd, John, and Jim Isaak. 2008. "IT Standardization: The Billion Dollar Strategy." In *Standardization Research in Information Technology: New Perspectives*, edited by Kai Jakobs, 20–26. Hershey: IGI Global.

Isaak, Jim. 2006. "The Role of Individuals and Social Capital in POSIX Standardization." *International Journal of IT Standards and Standardization Research* 4(1): 1–23.

ISO, eds. 2017. "ISO/IEC Directives, Part 1, Consolidated ISO Supplement." https://www.iec.ch/members_experts/refdocs/iec/isoiecdir-1-consolidatedIECsup{ed13.0}en.pdf.

ITU-T 2015. "Focus Group on Smart Sustainable Cities – Technical Report – Setting the Stage for Stakeholders' Engagement in Smart Sustainable Cities." https://www.itu.

int/en/ITU-T/focusgroups/ssc/Documents/website/web-fg-ssc-0113-r9-SSC-stakeholders.docx.

Jacob, Kai, et al. 2013. Options for Strengthening Responsible Research and Innovation: Report of the Expert Group on the State of the Art in Europe on Responsible Research.

Jakobs, Kai. 2002. "A Proposal for an Alternative Standards Setting Process." *IEEE Communications Magazine* 40(7).

Jakobs, Kai. 2005. "The Role of the 'Third Estate' in ICT Standardisation." In *The Standards Edge: Future Generation*, edited by Sherrie Bolin. The Bolin Group.

Jakobs, Kai. 2014. "The Role of The Individual in ICT Standardisation." In *Modern Trends Surrounding Information Technology Standards and Standardization within Organizations*, edited by Kai Jakobs. Hershey: IGI-Global.

Jakobs, Kai. 2015. "Exploiting Standardisation to Shape the Cloud – Potential Avenues to Success for the 'Third Estate.'" *Proceeding of the 14th STS Conference Graz on Critical Issues in Science, Technology and Society Studies.*

Jakobs, Kai. 2016a. "Standardisation of E-merging IoT-Applications: Past, Present and a Glimpse into the Future." *Proceeding of the 4th IEEE Conference on Future Internet of Things and Cloud.*

Jakobs, Kai. 2016b. "Trust in Open Standardisation?! Management in a Digital World: Decisions, Production, Communication." *Proceeding of the 28th Sinergie Annual Conference*, 21–33.

Jakobs, Kai. 2017. "ICT Standardization." In *Encyclopedia of Information Science and Technology*, edited by Khosrow-Pour, 4679–4691. Hershey: IGI Global.

Jakobs, Kai. 2018. "'Smart' Standardisation." Presented at the 82nd IEC General Meeting. (Available upon request from the author).

Jakobs, Kai, Rob Procter, Robin Williams. 2001. "The Making of Standards: Looking Inside the Work Groups." *IEEE Communications Magazine* 39(4): 102–107.

Jakobs, Kai. et al. 2010. "Creating a Wireless LAN Standard: IEEE 802.11." In *The Innovation Journey of WiFi*, edited by Wolter Lemstra, John Groenewegen and Vic Hayes. Cambridge: Cambridge University Press.

Jakobs, Kai, and Wehrle, K. 2017. *Cyber-Physical Systems – Future Standardisation for Europe. Final Project Report.* Available upon request.

Kallestrup, Morten. 2017. "Stakeholder Participation in European Standardization: A Mapping and an Assessment of Three Categories of Regulation." *Legal Issues of Economic Integration* 44(4), 381–393.

Kramer, Roderick M., and Karen S. Cook, eds. 2004. *Trust and Distrust in Organizations: Dilemmas and Approaches.* Russell Sage Foundation.

Krechmer, Ken. 2008. "Open Standards: A Call for Action. Innovations in NGN: Future Network and Services." *Proceedings of the 1st ITU-T Kaleidoscope Academic Conference*, 15–22.

Langheinrich, Marc. 2001. "Privacy by Design – Principles of Privacy-aware Ubiquitous Systems." *International conference on Ubiquitous Computing*, 273–291. Berlin, Heidelberg: Springer.

Lemstra, Wolter, John Groenewegen, and Vic Hayes, eds. 2010. *The Innovation Journey of WiFi*. Cambridge: Cambridge University Press.

Li, Shancang, Theo Tryfonas, and Honglei Li. 2016. "The Internet of Things: A Security Point of View." *Internet Research* 26(2): 337–359.

Lund Declaration. 2009. "Europe Must Focus on the Grand Challenges of our Time." Swedish EU Presidency. Lund, Sweden. https://era.gv.at/object/document/130/attach/1lund_declaration_final_version_9_july.pdf.

Macnaghten, Phil, and Jason Chilvers. 2014. "The Future of Science Governance: Publics, Policies, Practices." *Environment and Planning C Government and Policy* 32(3): 530–548.

Martins, Teresa Christina Monteiro, Paulo Henrique de Souza Bermejo, and Wagner de Souza. 2015. "Open Innovation for Citizen Coproduction." In *International Conference on Electronic Government and the Information Systems Perspective*, edited by Andrea Kö and Enrico Francesconi, 177–188. Cham: Springer.

Mulgan, Geoff, et al. 2007. "Social innovation: What it is, Why it Matters and How it can be Accelerated." Oxford Said Business School. http://eureka.sbs.ox.ac.uk/761/1/Social_Innovation.pdf.

Murray, Robin, Julie Caulier-Grice, and Geoff Mulgan. 2010. *The Open Book of Social Innovation* (2). London: National Endowment for Science, Technology and the Art. http://kwasnicki.prawo.uni.wroc.pl/pliki/Social_Innovator_020310.pdf.

Nam, Taewoo, Theresa A. Pardo. 2011. "Conceptualizing Smart City with Dimensions of Technology, People, and Institutions." In *Proceedings of the 12th Annual International Digital Government Research Conference: Digital Government Innovation in Challenging Times*, 282–291. ACM.

Navigant, eds. 2011. "*Smart City*" *Report*. Navigant Research.

OTA, eds. 1976. "Technology Assessment Activities in the Industrial, Academic, and Governmental Communities." https://www.princeton.edu/~ota/disk3/1976/7622/762208.PDF.

Owen, Richard, Phil Macnaghten, Jack Stilgoe. 2012. "Responsible Research and Innovation: From Science in Society to Science for Society, with Society." *Science and Public Policy* 39(6): 751–760.

Paik, Yongwook, and Feng Zhu. 2016. "The Impact of Patent Wars on Firm Strategy: Evidence from the Global Smartphone Industry." *Organization Science* 27(6): 1397–1416.

Piso, Zachary, et al. 2016. "The Production and Reinforcement of Ignorance in Collaborative Interdisciplinary Research." *Social Epistemology* 30(5–6): 643–664.

Pol, Eduardo, and Simon Ville. 2009. "Social Innovation: Buzz Word or Enduring Term?" *The Journal of Socio-economics* 38(6): 878–885.

Robinson, Douglas K. 2009. "Co-evolutionary Scenarios: An Application to Prospecting Futures of the Responsible Development of Nanotechnology." *Technological Forecasting and Social Change* 76(9): 1222–1239.

Rosenkopf, Lori, Anca Metiu, and Varghese P. George. 2001. "From the Bottom Up? Technical Committee Activity and Alliance Formation." *Administrative Science Quarterly* 46(4): 748–772.

Schaffers, Hans, et al. 2011. "Smart Cities and the Future Internet: Towards Cooperation Frameworks for Open Innovation." In *The Future Internet*, edited by John Domingue et al., 431–446. Berlin, Heidelberg: Springer.

Schuurbiers, Daan. 2011. "What Happens in the Lab: Applying Midstream Modulation to Enhance Critical Reflection in the Laboratory." *Science and Engineering Ethics* 17: 769–788.

Sharra, Romeo, and Marthe Nyssens. 2010. "Social Innovation: An Interdisciplinary and Critical Review of the Concept." Université Catholique de Louvain Belgium, 1–15.

Stahl, Bernd C. 2013. "Responsible Research and Innovation: The Role of Privacy in an Emerging Framework." *Science and Public Policy* 40(6): 708–716.

Steelman, John R. 1947. "Science and Public Policy: A Report to the President." https://babel.hathitrust.org/cgi/pt?id=mdp.39015017123830;view=1up;seq=7.

Stilgoe, Jack, and Richard Owen, and Phil Macnaghten. 2013. "Developing a Framework for Responsible Innovation." *Research Policy* 42: 1568–1580.

Swann, G.M. Peter, and Lambert, Ray. 2010. "Why do Standards Enable and Constrain Innovation?" In *Services Standardisation. Proceedings of the 15th EURAS Conference. EURAS Contributions to Standardisation Research, Vol. 2*, edited by Jean-Christoph Gratz and Kai Jakobs. Mainz Academic Publishers.

Tanenbaum, Andrew S. 1988. *Computer Networks: 2nd Edition*. Prentice-Hall, Inc. Upper Saddle River.

Umapathy, Karthikeyan, Sharoda Paul, Sandeep Purao, John W. Bagby, and Prasenjit Mitra. 2007. "Avatars of Participants in Anticipatory Standardization Processes." In *The Standards Edge – Unifier or Divider?* Edited by Sherrie Bolin, 295–302. Chelsea, MI: Bolin Group.

von Burg, Urs. 2001. *The Triumph of Ethernet: Technological Communities and the Battle for the LAN Standard*. Stanford: Stanford University Press.

von Schomberg, René. ed. 2011. *Towards Responsible Research and Innovation in the Information and Communication Technologies and Security Technologies Fields*. Luxembourg: Publications Office of the European Union.

Werle, Raymund, and Eric J. Iversen. 2006. "Promoting Legitimacy in Technical Standardization." *Science, Technology and Innovation Studies* 2: 19–39.

West, Joel, and Karim R. Lakhani. 2008. "Getting clear about communities in open innovation." *Industry and Innovation* 15(2): 223–231.

Wickson, Fern, and E.M. Forsberg. 2015. "Standardising Responsibility? The Significance of Interstitial Spaces." *Science and Engineering Ethics* 21(5): 1159–1180.

Williams, Robin, and David Edge. 1996. "The Social Shaping of Technology." *Research Policy* 25(6): 865–899.

WTO, ed. 1995. "Agreement on Technical Barriers to Trade." https://www.wto.org/english/docs_e/legal_e/17-tbt.pdf.

CHAPTER 13

The Shift from Traditional Computing Systems to Artificial Intelligence and the Implications for Bias

Vladislav V. Fomin

> Algorithmic decisions are not automatically equitable just by virtue of being the products of complex processes, and the procedural consistency of algorithms is not equivalent to objectivity.... Algorithms may be mathematically optimal but ethically problematic. (Osoba and Welser 2017, 2)

1 Introduction

In her seminal 1985 paper, Shoshana Zuboff (1985, 5) writes: "Technology defines the horizon of our material world as it shapes the limits of what is possible and what is barely imaginable." The technology cannot be considered neutral, Zuboff continues: "it is brimming with valence and specificity in the opportunities that it creates and forecloses" (Zuboff 1985, 5).

While the technology is extending the horizons of possibility, it is the human activity of choice, which ultimately shapes the present day uses and the future designs of technology (Star 1999). People make choices on domains and contexts for technology use, as well as on specific technology designs.

Artificial intelligence (AI) has already proven it can extend the horizons of possibility in general, and it has been enjoying a massive uptake in business[1] (Chui and Malhotra 2018). Machine learning (ML) is fast becoming the future of enterprise data science (Burt 2018). The advent of the new paradigm of computing has also sparkled discussion on whether it should be *the technology of choice* in a number of domains where the *faultlessness* and/or predictability of technology-mediated outcomes is crucial (Executive Office of the President 2014). As reliance on artificial agents continues to grow, we must ask, what are the consequences and risks of such dependence (Osoba and Welser 2017)? "Ethical debates are well under way about what's 'right' and 'wrong' when it comes to high-stakes AI applications" (Burkhardt, Hohn, and Wigley 2019, 2).

1 According to McKinsey Global Institute research, AI could deliver an additional global economic output of $13 trillion per year by 2030 (Cheatham, Javanmardian, and Samandari 2019).

 | DOI:10.1163/9789004437876_015

In this chapter, we examine the concept of bias, which is central to both how information technologies (IT)[2] are designed and how we put trust in results produced with mediation of IT. The intention is to provide the reader with an overview of the changing paradigm of computation – the shift from the traditional (algorithmic) IT to IT powered by ML and referred to as AI – and the general implications of this shift for scientists and for society at large.

The overview of the principles governing the design and control of IT in this chapter is presented against the background of the human tradition of reflective practice (Schön 983) for governing domains and tasks requiring complex cognitive skills – law, medicine, finance, to name a few. Machines can help us to extend the limits of what is possible in everyday routine, but can we imbue the new generation of IT with human (ethical) judgement to make it follow the rules and norms that guide human systems?

2 The Origins of Computing and Bias

Like any other (type of) system, any IT system can be abstractly presented as having three main elements: the input provided to the system, the process(ing) part of the system, and the output generated by the system (see Figure 13.1). IT systems typically execute some sort of computational processes, where data are provided as an input and data are produced as an output (the results of the computation). The output from one system can be an input to another system.

The *process* of the computation is based on an algorithm. In simple terms, an algorithm can be defined as a sequence of steps and instructions that can be applied to (input) data (Executive Office of the President 2014, 46). Among the properties an algorithm must possess, the following are relevant:

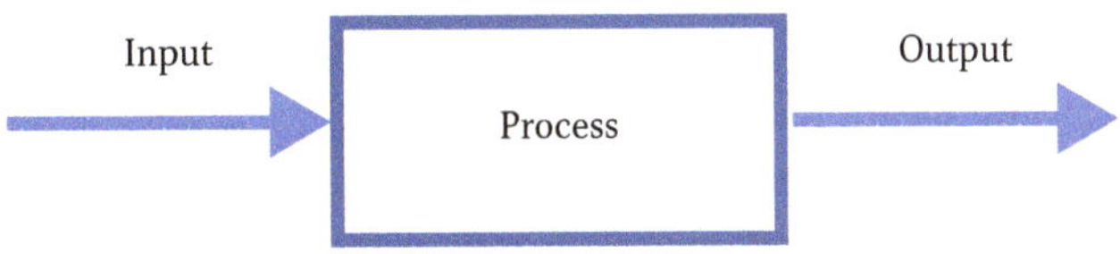

FIGURE 13.1 A GENERIC REPRESENTATION OF AN IT SYSTEM

2 In this paper, IT is used as a generic reference to both information systems (as software) and information technologies (as a combination of software and hardware), except for where specific references are made to AI and/or ML systems as particular types of IT.

- *Definiteness*: each step must be precisely defined; the actions to be carried out must be rigorously and unambiguously specified for each case.
- *Output*: an algorithm has one or more outputs, which have a specified relation to the inputs.

Programming languages are used to transform the algorithm into computer [machine] code. When implemented, IT systems can perform operations on data as programmed.[3]

The definitiveness of algorithms and the possibility of programming systems with certain algorithmized logic has led humans to strive to automate certain tasks (Edwards 1998; Srnicek and Williams 2015; Zuboff 1985). Since the very first uses of computers, people have typically developed and adopted IT systems because they believed such systems would allow them to accomplish more, more quickly, and at lower cost (Zuboff 1985, 7). IT systems were built to substitute for humans not only in tasks where fast and vast calculations were required, but also where jobs were perceived to be too complicated or dangerous for humans. Importantly, computation allowed the introduction of control where humans did not have the capability to control.

Having laid out the principles of computational systems, we can explain the relationship between a computational system and bias. Key to automation of decisions is knowing the rules to arrive at a decision given the circumstances (Simon 1962). Following the computational model as presented in Figure 13.1, this requirement for automation implies that there can be no ambiguity or distortion in the decision logic: an IT system is expected to *produce the same result* (output) *every time the same input* (data) *is fed in.*

This intolerance of distortion in traditional computational systems establishes a tight connection with the concept of bias, as implied in the meaning of the latter:[4]

1. "inclination or prejudice for or against one person or group, especially in a way considered to be unfair," and
2. "a systematic distortion of a statistical result due to a factor not allowed for in its derivation."

A few important observations must be made with regard to the definition of the word bias and its application to the computational model, as summarized in Table 13.1. First, the prejudice or inclination may be the result of distortion: i.e., a faulty *process* (see Figure 13.1) will produce faulty result (output). Second,

3 While by no means every IT system is used to assist in human decision making, Simon (1977) introduced the concept of *programmed decisions* to refer to decisions, the logic of arriving at which can be represented by an algorithm and subsequently programmed into IT.

4 According to the Oxford Dictionary: https://en.oxforddictionaries.com/definition/bias.

TABLE 13.1 Possible sources of distortion in a computational system

Input →	**→ Process →**	**→ Output**
Proper	Proper	Proper
Proper	Distorted	Distorted
Distorted	Proper	Distorted

incomplete or faulty *input data* will stall the *process* or result in a faulty *output*. Given that IT systems are usually complex, encompassing nesting and chaining of *input-process-output* systems, where the output from one (sub-)system is fed as an input to another, the third important observation must be made – a faulty input or process in one (sub-)system is likely to escalate the distortion in connected systems.

Understanding the cycle by which IT is developed and the way it is deployed is crucial in our discussion on bias. The development starts from the codification of *human knowledge*, and the deployment is always accompanied by the *creation of information* (Zuboff 1985, 8). This information about the underlying processes (automated by technology or monitored by technology) allows fault detection (distortion or bias) in the execution or in the results, leading to the creation of knowledge (Simon 1977), i.e., learning on human behalf, and it allows subsequent introduction of modifications (improvements) to the IT to *eliminate the undesired distortion and biases*.

Thus, we can summarize, that in their striving to automate certain tasks (or to control them), people developed and deployed IT systems that had to be distortion- and bias-free. Controlling for distortion and bias in computational systems required human oversight over the processes and the results, as illustrated through a few historical examples below.

What is believed to be the first commercial application of computers – LEO[5] – was built and implemented to automate tasks in which errors and ambiguities could not be tolerated: tracking stock, calculating the salaries of employees, and performing financial accounting functions. Financial miscalculations have notoriously unpleasant consequences for companies. Deliberate financial misreporting practices, as in the well-known examples of Tesco and Enron,[6] result in large-scale scandals and shakeouts of markets.

5 https://warwick.ac.uk/services/library/mrc/explorefurther/digital/leo/story/.
6 https://www.theguardian.com/business/2017/mar/28/tesco-agrees-fine-serious-fraud-office-accounting-scandal. https://www.britannica.com/event/Enron-scandal.

The origins of the modern Internet also reveal an error-intolerant foundation. The origins of the Internet are in the development of a computer-based network for the anti-missile defense system ARPANET[7] during the Cold War. Flows in calculations or communication patterns within the system would undermine the defensibility of the system, and hence the whole purpose of building one. While the developers of defensive and offensive systems would naturally strive to keep the systems free of any computational faults, recent history provides examples of deliberate introduction of distortion to the system by third parties, such as the alleged disruption of GPS navigation signals over northern Scandinavia by Russia during NATO military exercises (Fouche and Adomaitis 2018).

The marriage of the two epitomes of modern life – a car and a miniature computer device – is to be found in the first-ever computer-based control system in cars – the Lambda probe[8] pollution control system – which was introduced in Volvo 240/260 models in the U.S. market in 1976. The probe was used to monitor for departures from optimal chemical conditions in the carburetor-exhaust system to minimize CO_2 emissions. While originally intended to govern narrowly defined technical/chemical processes in a car, in 2015 the posterity of the Lambda probe erupted into a Dieselgate[9] scandal in U.S. society – apparently car manufacturers were introducing intentional distortions into the way CO_2 emissions were reported to the controlling agencies, thus trying to bypass strict national norms for pollution prevention.

By accepting the dictionary definition of the word bias, and by reviewing the examples provided above, we have to agree that distortion or bias in a computational system can be intentional or unintentional, with respect to both cause and effect.

Distortion in computational systems can be introduced intentionally, for example when seeking to create inclinations or prejudices towards a certain group or person, or to mislead users. Thus we can extend Table 13.1 to show the categorization of intentional and unintentional computational faults with respect to the *input* to the system (data) or the *processing* of data in the system (the algorithm), as well as the positive or negative connotations of such distortions (see Table 13.2).

While the intentional introduction of distortion or bias in most contexts may be interpreted as having a negative connotation, the concept of affirmative

7 https://www.darpa.mil/about-us/timeline/arpanet.
8 https://blog.bosch.com/history/en/2016/07/20/939/.
9 https://phys.org/news/2018-06-vw-dieselgate-scandal.html.

TABLE 13.2 Examples of intentional and unintentional distortion and bias in computational systems

		Intentional	Unintentional
Data (input)	Bias	*Connotation*: Negative *Example*: Political propaganda (at its target)	*Connotation*: Negative *Example*: Overlooking certain historical or systemic dependencies in the input data
	Distortion	*Connotation*: Negative *Example*: Financial misreporting *Connotation*: Positive *Example*: Political propaganda (at its origin)	
Process (algorithm)	Distortion	*Connotation*: Negative *Example*: Dieselgate scandal	*Connotation*: Negative *Example*: *ERROR*
	Bias	*Connotation*: Positive[a] *Example*: Affirmative action	*Connotation*: Negative *Example*: *ERROR*

a This is not the connotation in the UK, where affirmative action is illegal.

action is one common example of an intentional bias intended to foster positive changes.

Affirmative action is the policy of promoting the education and employment of members of groups that are known to have suffered from discrimination. As an example, affirmative action has been (and continues to be) deployed by several U.S. universities, resulting in (substantially) higher admission rates for certain groups of the population. In many cases, the U.S. Supreme Court has supported such intentional biases,[10] allowing educational institutions to consider race as a factor when admitting students, thus creating more favorable conditions for one specific racial group over others.

The case of affirmative action emphasizes the importance of *human* decision making, or the precedence of human over algorithmic (machine-made)

10 https://library.law.hawaii.edu/2017/01/30/gratz-v-bollinger-case-summary/.

decisions. It is through the reflective action of experts that realization of the need for biased admissions came about – an otherwise faultless computational model had to be superseded by additional criteria to cater for a more balanced ethnic composition of university students to minimize the imperfect, unwanted racial bias in a social system.

With respect to unintentional distortion, traditionally, computational systems have been deployed to detect it faster, at lower cost than humans, but especially in environments in which humans cannot or do not want to perform the monitoring/detection task by themselves.

In the context of IT, computational/algorithmic distortion and the resulting bias (if any) have always been treated as an error that must be eliminated through the reflective action of humans (identification of causes) and the subsequent new codification of human knowledge into machine code (or the data).

Approaching the end of the first quarter of the 21st century, more than half a century down the road from the first introduction of computers in business, the military, and the everyday lives of citizens, computers have become inseparable and invisible elements of modernity (Edwards 1998). The paramount reasons for the development and implementation of IT have not, however, changed since the dawn of the computer era: *to control and to inform humans to enable control.*[11] Neither have the thresholds for distortion and bias tolerance changed. What has changed with the advent of the new paradigm of intelligent (learning) IT, however, is the *ability of humans to detect computational faults* and to control for bias-free results of computation.

3 The New Paradigm of Computing and Bias

The advent of the new computational paradigm – the one based on ML – brings forth a new conceptualization of bias in the context of IT. Before we start elaborating on causes and risks of bias, we must present the new paradigm of intelligent computing. The name of this paradigm is AI.

3.1 *What Is AI*

AI is a collective term for different ML techniques. AI usually refers to the development of computer systems able to perform *tasks normally requiring human intelligence*, such as decision-making, visual perception, speech recognition,

11 Zuboff (1985, 7) distinguishes one more objective for the use of IT: to increase the continuity of processes. But the continuity dimension functions as a subpart of the other two: control and information provision.

etc. More broadly, AI is typically defined as the ability of a machine[12] to perform cognitive functions, such as perceiving, reasoning, learning, interacting with the environment, problem solving, and even exercising creativity (McKinsey Analytics 2018). The definition of AI, as proposed by the European Commission, refers to "systems that display intelligent behaviour by analysing their environment and taking actions – with some degree of autonomy – to achieve specific goals." (EC AI HLEG 2019a).

AI-based systems can be purely software-based, such as the voice assistants or image analysis and face recognition software found in our smartphones or in cloud services (Press 2017). AI can also be embedded in hardware devices, such as autonomous cars or drones (EC AI HLEG 2019a).

If traditional (algorithmic) computational systems were built on the foundation of strict control over the system (low-to-zero tolerance to distortions in algorithmic logic and the computational output, as discussed above), then *ML algorithms, which power AI, are radically different,* in that they do not receive explicit programming instructions. Instead, ML systems receive datasets containing what in the traditional IT systems would be denoted as *input data* and *output data*: the data upon which the calculation must be performed and the correct results of that calculation, accordingly. The ML system then tries to figure out how the *output* can be derived from the *input*. That "trying" is the "learning" of the system.

The key feature of AI – the ability to learn – constitutes the second radical distinction between AI and traditional computational systems, which *never learn* (Lyytinen, King, and Nickerson 2020), as they are programmed with fixed codified human knowledge. This key feature of AI – learning – is what makes AI more bias-prone than traditional IT systems.

3.2 *The Nature of Bias in ML Systems*

In the context of ML, the term bias was introduced by Mitchell (1980) to mean "any basis for choosing one generalization [hypothesis] over another, other than strict consistency with the observed training instances" (Dietterich and Kong 1995, 1).

Today the concept of machine bias is often associated with data, which are used to train ML systems: "programming that assumes the prejudice of its creators or data" (Shadowen 2017, 5, referring to Garcia 2016). Thus we can distinguish between two types of causes for machine bias, as per the definitions (see Table 13.3): bias as a result of algorithmic processing, as per Mitchell (1980), and bias as an effect of "biased" data, as per Garcia (2016).

12 The American Heritage Dictionary defines intelligence as "the ability to acquire and apply knowledge and skills."

TABLE 13.3 Two facets of the concept of machine bias

Input →	→ Process →	→ Output	Source of the concept
Proper	Biased	Biased	Mitchell (1980)
Biased	Proper	Biased	Garcia (2016)

Comparing the nature of distortion and bias in traditional and ML-based computational systems, as depicted in Table 13.1 and Table 13.3, we can conclude that *input data* can be a source of bias for both generations (types) of computational systems. With regard to the *process,* in the traditional IT system we used the term *distortion,* referring to erroneous codification of human knowledge,[13] whereas in ML-based systems, algorithmic bias can be the product of machine intelligence (the outcomes of the machine learning process).

While the two sources of bias in AI systems can be and often are interlinked, this interdependence is more complicated than the simple cause-effect relationship shown in Table 13.3. The High-Level Expert Group on Artificial Intelligence provides a good overview of causes and effects of bias in AI systems:

> Bias is an inclination of prejudice towards or against a person, object, or position. Bias can arise in many ways in AI systems. For example, in data-driven AI systems, such as those produced through machine learning, bias in data collection and training can result in an AI system demonstrating bias. In logic-based AI, such as rule-based systems, bias can arise due to how a knowledge engineer might view the rules that apply in a particular setting. Bias can also arise due to online learning and adaptation through interaction. It can also arise through personalisation whereby users are presented with recommendations or information feeds that are tailored to the user's tastes. It does not necessarily relate to human bias or human-driven data collection. It can arise, for example, through the limited contexts in which a system is used, in which case there is no opportunity to generalise it to other contexts. Bias can be good or bad, intentional or unintentional. In certain cases, bias can result in discriminatory and/or unfair outcomes, indicated in this document as unfair bias. (EC AI HLEG 2019b, 36)

13 Irrespective of whether a traditional or an ML-based computational system is concerned, bias can be introduced by a human in the *process*: "bias can arise due to how a knowledge engineer might view the rules that apply in a particular setting" (EC AI HLEG 2019b, 36).

The following subsections explore data- and process-originating bias in AI systems in more detail.

3.2.1 Process-Originating Biases in AI Systems

The learning element is what makes AI different from traditional computational models. The learning determines both how the subsequent cycles of computation will unfold, and it implies some sort of optimization of the process with respect to the produced results. If in the traditional computational systems the results could have always been checked against the fixed set of instructions – the algorithm[14] – in the machine-learning AI system this kind of control cannot be (easily/at all) established. In other words, for the users of the system, it becomes impossible to control for algorithmic bias. One reason for this control-inability is the pace with which the machines learn – they learn faster than we can adapt (Lyytinen et al. 2020).

ML-based AI detects patterns in large data sets, and it learns how to make predictions and recommendations by processing data and experiences. *Technical properties influencing learning remain ultimately unknown* (Lyytinen et al. 2020), and the data-processing-for-learning cycle of AI may eventually render the AI system's algorithmic logic considerably different from the guidelines their programmers initially gave them (Etzioni and Etzioni 2016, 30) – another reason for the control-inability of AI systems by humans.

Citing Jenna Burrell from the School of Information at UC-Berkeley, Etzioni and Etzioni (2016, 30) introduce the concept of opacity with regard to the AI algorithms: "machine learning and/or the number of different programmers involved renders an algorithm opaque even to the programmers." The system becomes a black box,[15] and the traditional human reflective action to determine algorithmic biases becomes inapplicable:

> You can't just look inside a deep neural network to see how it works. A network's reasoning is embedded in the behavior of thousands of simulated neurons, arranged into dozens or even hundreds of intricately interconnected layers. The neurons in the first layer each receive an input, like the intensity of a pixel in an image, and then perform a calculation before

14 Except for the cases when the algorithm is protected by patent, which limits the openness of the computational system to verification (Rachel 2018).

15 Two more sources of opacity are being distinguished, and they similarly hold for traditional computational systems: Intentional opacity, for example with proprietary algorithms that a government or corporation wants to keep secret, and technical illiteracy, where the complexity and function of algorithms is beyond the public's comprehension (Etzioni and Etzioni 2016, 30).

> outputting a new signal. These outputs are fed, in a complex web, to the neurons in the next layer, and so on, until an overall output is produced. Plus, there is a process known as back-propagation that tweaks the calculations of individual neurons in a way that lets the network learn to produce a desired output. (Knight 2017)

3.2.2 Data-Originating Biases in AI Systems

As ML-powered AI systems rely on large amounts of data to perform well, bias in training data will likely result in AI systems not being able to generalize well and possibly making unfair decisions that can favor some groups over others (EC AI HLEG 2019a). Bias in data can occur due to historical, political, or technical factors that were at play when data were collected, resulting in data being not balanced or inclusive enough.

While scholars understand the risks of impartial data well, as seen for example through the prism of the concept of primary and secondary sources in collecting research data, the general public may not be aware of the difference. In the case of ML training, when a company does not have a sufficiently large dataset to train an AI system, the use of third-party data, especially repurposed data, is likely to result in biased data for training (Burkhardt et al. 2019, 4).

Whether original or third-party data are used, to date there is no possibility of automatically controlling for bias in training data, as there are no standards or solutions to categorize, label, or assess data according to how they were collected and whether they are accessible, searchable, findable, etc. (Kinder-Kurlanda 2019). This situation, on the one hand, emphasizes the importance of human reflective judgement in the process of building training data sets, and, on the other hand, gives scholars a reason to label raw data an *oxymoron,* as data at the same time become the fuel and the destroying agent for ML-engines of AI systems.

3.2.3 Explainability of Produced Results as a Factor Related to Bias

This concept of explainability is very important in the context of the overall discussion on the mission of computational systems in automating jobs, and in the light of the ongoing public debate and legal efforts on unprecedented (user) data-collection practices.

The aforementioned notion of a black box or the algorithmic opacity of AI implies not only the non-transparency of algorithmic logic, but also the inability of system users to trace the reasons for certain decisions of the system (EC AI HLEG 2019a, 5).

Full automation of the *input-process-output* cycle, which has been driving the deployment of traditional computational systems, is currently impossible

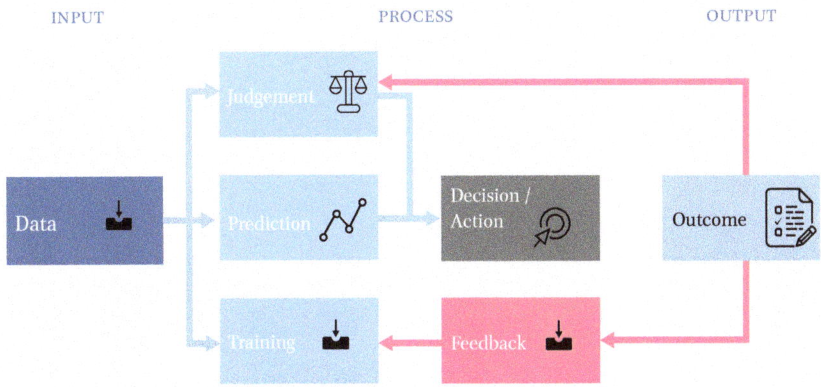

FIGURE 13.2 AI and human (non-automated) decisions
KÖSZEGI (2019), REFERRING TO AGRAWAL, GANS, AND GOLDFARB (2018).

to implement in AI systems without a risk of biased/distorted results. As depicted in Figure 13.2, the *output* of the system is produced through the reflective action/decisions of humans.

Human decisions can be fully automated when the underlying decision algorithmic logic is known, understood, and unambiguous (Simon 1977). While we already have (some) regulatory basis for the explainability of AI systems, such as GDPR, for example,[16] which states repeatedly that data subjects have a right to "meaningful information about the logic involved," the explainability of computational logic is likely to be increasingly hampered by patent protection of AI systems. As seen in Figure 13.3, there has been a dramatic growth of proprietary ML algorithms in recent years – and the trend is likely to be sustained.

Not having meaningful information about the logic involved, and facing situations where it is difficult, if not impossible, to explain the consequences of automated decision-making (Burt 2018), we are unlikely to see full automation of tasks becoming universal under the new computational paradigm, an outcome that represents a setback for the original goals of computation (Edwards 1998; Simon 1981; Zuboff 1988).

3.3 *The Original Bias of AI and the Possible Ways Forward*

Babel, Buehler, Pivonka, Richardson, and Waldron (2019) suggest a way forward to break out of the vicious circle of bias stemming from imperfect data and the associated biased learning of AI: "to address algorithmic bias, model-validation processes should be updated to ensure appropriate algorithms are selected in

16 Articles 13–15 of GDPR regulation. https://www.gdpr.org/regulation/index.html.

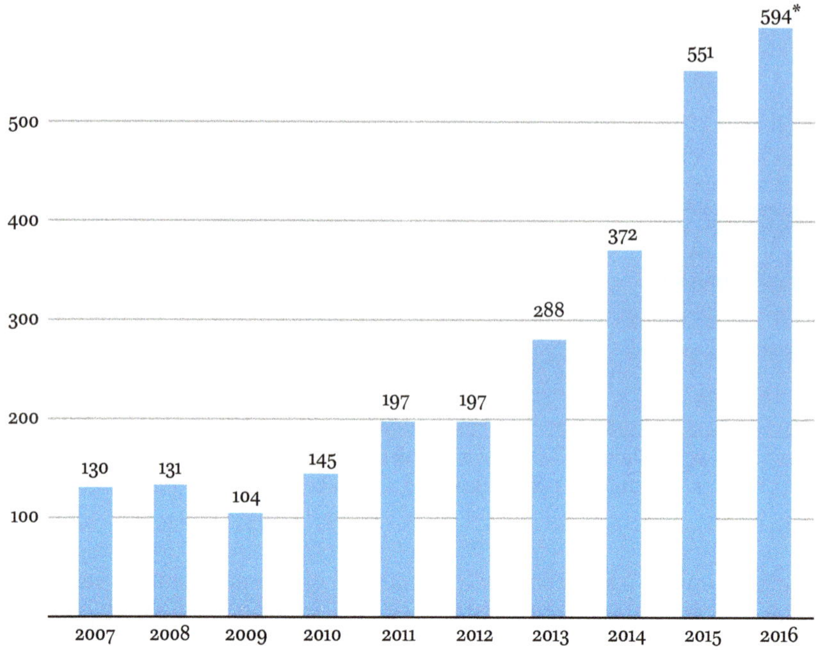

FIGURE 13.3 Patent applications for ML
SIMONITE (2018)

any given context.... Another approach is to develop 'challenger' models, using alternative algorithms to benchmark performance"[17] (Babel et al. 2019).

The offered solution to mitigate the risks of bias under AI appears to be no different from the way bias risks are mitigated under the traditional computational paradigm, where identification of distortions and biases and corrective action is based on reflective human action.

We find the difference between the traditional computing paradigm and the emerging AI paradigm in the level of algorithmic opacity – ML renders AI algorithms opaque to the observer, which in turn may lead to loss of trust in the results. Under the traditional computational paradigm, the results can be unambiguously verified against the computational model's algorithm,[18] whereas under AI, a competing/alternative computational model must be deployed and the results of the two computational efforts compared for consistency.

17 Both suggested approaches refer to the feedback arrows coming from "outcome" to "judgement" and "training" in Figure 13.2.

18 Except in cases when the algorithm is protected by patent, which limits the openness of the computational system to verification (Rachel 2018).

The suggested method for verification of AI-produced results by developing and deploying challenger models, and by using alternative algorithms to benchmark performance (Babel et al. 2019), however, has a number of limitations, as discussed below. The first is related to incomprehensibly short periods of time, the other to overly long periods of time, both rendering the verification task a challenge.

When an AI system is involved, the pace of its decision making is determined by machine time, which is incomprehensible to humans. An example to hand, as reported by Johnson et al. (2013), is that in an AI-governed high-frequency trading system 18,520 anomalies of less than 1.5 seconds duration were found in a five-year period. Anomalies occurring at this pace and in this volume cannot easily (or at all) be detected by humans. For humans, it is impossible to provide timely assessment of the consequences of those anomalies. Some sort of meta-system, a controlling entity operating at the same timescale as machines, must be introduced. *Alas, the system will be subject to the same bias-related challenges as the systems it is intended to control.*

Often, the success of bias-eliminating or corrective-bias-introducing efforts can be assessed only in the (very) long run. This long time required for humans to realize the presence or absence of bias has already been introduced above as the historical reason for having biased training data today.

4 Challenges for Mitigating Bias under AI

Traditionally, in domains and professions related to law, finance, and medicine, where complex cognitive skills are required, human experts have benefited from reflective practice (Lyytinen et al. 2020, referring to Schön 1983), i.e., undesired biases in source data (the input), rule-based governance/computational models, and their supplied results (the output), were identified using human expertise and judgement.

Human professionals have years of accumulated experience on which to reflect, as exemplified by cases such as affirmative action, where intentional bias has been introduced for what is believed to bring positive results in the long run.

Noticing change, considering the metalevel implications of change on the whole system, finding patterns, and adjusting the system accordingly has been a uniquely human endeavor (Lyytinen et al. 2020). While AI has certainly proven to have the capability of predictive assessment, to date we cannot tell whether machines can invent mechanisms for governance akin to human systems (Lyytinen et al. 2020).

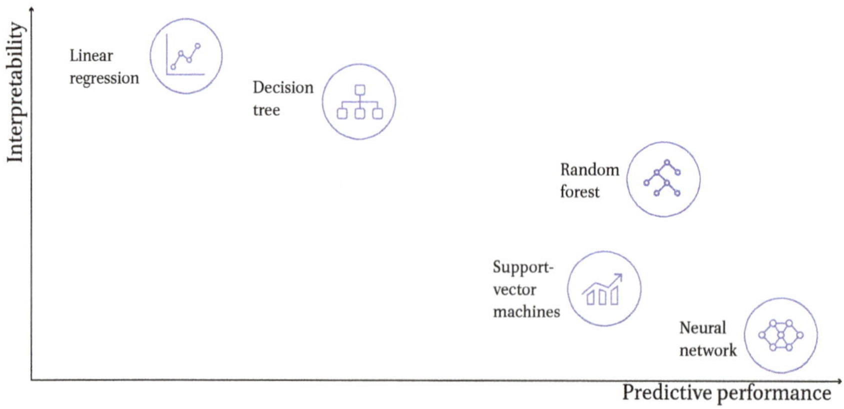

FIGURE 13.4 Interpretability vs. predictive performance trade-off in AI systems
BURKHARDT ET AL. 2019, 7

A second important question related to bias in AI is that of the explicability of the results. The design of reinforcement learning of AI is such that computing agents interact with and receive feedback from their environment – that is, they learn from experience – and the AI systems can and do create responses corresponding to such received experience, i.e., AI can identify patterns in data to match or even supersede humans. The challenge, however, remains how to relate the ML-produced outcomes to the rule-based systems (Lyytinen et al. 2020) with which humans have traditionally worked. How can AIs explain their decisions to the users, to develop trust that their operations and their results are bias-free? The advantages of AI-supplied predictive performance must be balanced against the requirement for interpretability of results (see Figure 13.4).

Data acquisition and use must also be managed by keeping a balance of system performance with transparency on how AI systems make their predictions (Burkhardt et al. 2019, 2).

5 Conclusions

Following the path of the first wave of computerization, the adoption of AI is rapidly taking hold across global business[19] (Chui and Malhotra 2018) and

19 According to the McKinsey Global Business Survey (Chui and Malhotra 2018), 47% of respondents say their companies have embedded at least one AI capability in their business processes, compared with 20% of respondents in a 2017 study who said their companies were using AI in a core part of their business or at scale.

military operations. As we have learned in the past, governments will follow with caution, as the consequences and risk of dependence on AI must first be assessed (Osoba and Welser 2017).

Large and complex *metahuman systems*, i.e., systems empowering humans to exert control over massive and complex tasks that could not be controlled without the use of AI, are already being introduced and used in critical societal tasks, including finance, energy, health, military, and transportation (Etzioni and Etzioni 2016; Lyytinen et al. 2018), but governance responsibility still rests with individual humans in authority.

Metahuman systems are unlikely to escape human control and responsibility unless and until (a) the larger regulatory and technological context evolves to deal with this issue (Lyytinen et al. 2018), and (b) trust in the governance principles and the results of metahuman systems is built. Both conditions are tightly related to the concept of bias, and how it can be detected and acted upon if detected.

For trust to be established, AI processes that are opaque to human observers must be made compatible with rule-based systems (Lyytinen et al. 2018) with which humans have traditionally worked to enable reflective analysis and explicability – this remains a challenge for both scholars and practitioners. Explainable AI (sometimes referred to as XAI) may be a way towards a future where the power of AI can be deployed without setbacks related to trust and avoidance of bias, thus making full automation of tasks possible, as once envisioned at the dawn of the computer age.

References

Agrawal, Ajay K., Joshua S. Gans, and Avi Goldfarb. 2018. *Prediction, Judgment and Complexity: A Theory of Decision Making and Artificial Intelligence*. No. NBER Working Paper 24243. https://www.nber.org/papers/w24243.pdf.

Babel, Bernhard, Kevin Buehler, Adam Pivonka, Bryan Richardson, and Derek Waldron. 2019. "Derisking Machine Learning and Artificial Intelligence." *McKinsey Quarterly. Business Technology Office*. https://www.mckinsey.com/business-functions/risk/our-insights/derisking-machine-learning-and-artificial-intelligence.

Burkhardt, Roger, Nicolas Hohn, and Chris Wigley. 2019. "Leading your Organization to Responsible AI." *McKinsey Analytics*. https://www.mckinsey.com/~/media/McKinsey/Business%20Functions/McKinsey%20Analytics/Our%20Insights/Leading%20your%20organization%20to%20responsible%20AI/Leading-your-organization-to-responsible-AI.ashx.

Burt, Andrew. 2018. "How will the GDPR Impact Machine Learning?" https://www.oreilly.com/ideas/how-will-the-gdpr-impact-machine-learning.

Cheatham, Benjamin, Kia Javanmardian, and Hamid Samandari. 2019. "Confronting the Risks of Artificial Intelligence." *McKinsey Quarterly*. https://www.mckinsey.com/business-functions/mckinsey-analytics/our-insights/confronting-the-risks-of-artificial-intelligence.

Chui, Michael, and Sankalp Malhotra. 2018. "Notes from the AI Frontier: AI Adoption Advances, but Foundational Barriers Remain." *McKinsey Analytics*. https://www.mckinsey.com/featured-insights/artificial-intelligence/ai-adoption-advances-but-foundational-barriers-remain.

Dietterich, Thomas G., and Eun Bae Kong. 1995. *Machine Learning Bias, Statistical Bias, and Statistical Variance of Decision Tree Algorithms*. Technical Report, Department of Computer Science, Oregon State University.

EC AI HLEG. 2019a. *A Definition of Artificial Intelligence: Main Capabilities and Scientific Disciplines*. https://ec.europa.eu/newsroom/dae/document.cfm?doc_id=56341.

EC AI HLEG. 2019b. *Ethics Guidelines for Trustworthy AI*. https://ec.europa.eu/newsroom/dae/document.cfm?doc_id=60419.

Edwards, Paul N. 1998. "Y2K: Millennial Reflections on Computers as Infrastructure." *History and Technology* 15: 7–29.

Etzioni, Amitai, Oren and Etzioni. 2016. "Designing AI Systems that Obey our Laws and Values." *Commun. ACM* 59(9): 29–31.

Executive Office of the President. 2014. *Big Data: Seizing Opportunities, Preserving Values*. https://obamawhitehouse.archives.gov/sites/default/files/docs/big_data_privacy_report_may_1_2014.pdf.

Fouche, Gwladys, and Nerijus Adomaitis. 2018. "Joining Finland, Norway says Russia may have Jammed GPS Signal in Arctic." Reuters, November 13, 2018. https://www.reuters.com/article/us-nordic-russia-defence/joining-finland-norway-says-russia-may-have-jammed-gps-signal-in-arctic-idUSKCN1NI267.

Garcia, Megan. 2016. "Racist in the Machine: The Disturbing Implications of Algorithmic Bias." *World Policy Journal*, 33(4), 111–117.

Johnson, Neil, Guannan Zhao, Eric Hunsader, Hong Qi, Nicholas Johnson, Jing Meng, and Brian Tivnan. 2013. "Abrupt Rise of New Machine Ecology Beyond Human Response Time." *Scientific Reports 3:* 2627.

Kinder-Kurlanda, Katharina. 2019. *Keynote Speech at the 18th Annual STS Conference Graz*. https://sts-conference.isds.tugraz.at/event/2/attachments/1/22/BoA2019_-_STS_Conference_Graz.pdf.

Knight, Will. 2017. "The Dark Secret at the Heart of AI." *MIT Technology Review*. https://www.technologyreview.com/s/604087/the-dark-secret-at-the-heart-of-ai/.

Köszegi, Sabine T. 2019. *Keynote Speech at the 18th Annual STS Conference Graz*. https://sts-conference.isds.tugraz.at/event/2/page/44-keynotespeaker.

Lyytinen, K., Nickerson, J. V., & King, J. L. 2020. Metahuman systems = humans + machines that learn. *Journal of Information Technology*, 0268396220915917. https://doi.org/10.1177/0268396220915917.
McKinsey Analytics. 2018. *An Executive's Guide to AI*. https://www.mckinsey.com/business-functions/mckinsey-analytics/our-insights/an-executives-guide-to-ai.
Mitchell, Tom M. 1980. *The Need for Biases in Learning Generalizations*. Department of Computer Science, Laboratory for Computer Science Research.
Osoba, Osonde A., and William Welser IV. 2017. *An Intelligence in our Image: The Risks of Bias and Errors in Artificial intelligence*. Rand Corporation.
Press, Gil. 2017. "Top 10 Hot Artificial Intelligence (AI) Technologies." *Forbes*, January 12, 2017. https://www.forbes.com/sites/gilpress/2017/01/23/top-10-hot-artificial-intelligence-ai-technologies/#a1c98f519287.
Schön, Donald A. 1983. *The Reflective Practitioner: How Professionals Think in Action*. London: Basic Books.
Shadowen, Ashley N. 2017. *Ethics and Bias in Machine Learning: A Technical Study of What Makes Us "Good."* Academic Works, CUNY. https://academicworks.cuny.edu/jj_etds/44.
Simon, Herbert A. 1962. "The Architecture of Complexity." *Proceedings of the American Philosophical Society*, 106: 467–482.
Simon, Herbert A. 1977. *The New Science of Management Decision*. Englewood Cliffs: Prentice-Hall.
Simon, Herbert A. 1981. *The Sciences of the Artificial*. Cambridge, MA: MIT Press.
Simonite, Tom. 2018. "Despite Pledging Openness, Companies Rush to Patent AI Tech." *Wired Magazine*, July 31, 2018. https://www.wired.com/story/despite-pledging-openness-companies-rush-to-patent-ai-tech/.
Srnicek, Nick, and Alex Williams. 2015. *Inventing the Future: Postcapitalism and a World Without Work*. Verso Books.
Star, Susan Leigh. 1999. "The Ethnography of Infrastructure." *American Behavioral Scientist* 43: 377–392.
Thomas, Rachel. 2018. "What HBR Gets Wrong About Algorithms and Bias." *Fast*, August 7, 2018. https://www.fast.ai/2018/08/07/hbr-bias-algorithms/.
Zuboff, Shoshana. 1985. "Automate/informate: The two Faces of Intelligent Technology." *Organizational Dynamics:* 14(2): 5–18.
Zuboff, Shoshana. 1988. *In the Age of the Smart Machine: The Future of Work and Power*. New York: Basic Books.

CHAPTER 14

Machine Bias and Fundamental Rights

Darius Amilevičius

1 Introduction

Artificial intelligence (AI) is rapidly transitioning from the realm of science fiction to the reality of our daily lives, and is taking the world by storm. Unlike more specialized innovations, AI is becoming a true general-purpose technology and is evolving into a utility that is likely to ultimately scale across every industry and sector of our economy, as well as nearly every aspect of science, society and culture. It fills human space silently and invisibly. The hype during the last few years relating to AI has helped society to understand that it is entering a new era of AI. However, as Clarke's third law states, any sufficiently advanced technology is indistinguishable from magic (Clarke 1962).

AI has evoked many issues in the past and will continue to do so. The way to go safe is to take precautions. A great deal of introspection, science, and capital are being invested in ethical research, economics, and the philosophy of AI. Researchers and industrialists have begun to recognize the appropriate role of AI in society. However, government policies have not yet attained a substantial stature. I agree with the opinion of Denise Feldner who stated that "now is the time for us to think about the crucial questions of being human in the Hybrid Age as we are, at the moment, profoundly ill prepared for future technologies."[1]

Stephen Hawking, Elon Musk, and some other entrepreneurs and scientists have expressed their concerns about the risks that AI systems might cause.[2] Denise Feldner is more specific and indicates that the biggest problem for now and for the near future with AI is not the science, but the applications[3] – machines are learning things and are able to make predictions, but they cannot understand them. The rapid industrialization of AI systems has an immense potential not only for the overall technological and well-being advancement of society but also for unethical practices. I agree with Atkinson's view

1 https://atlantic-community.org/2019s-problem-with-artificial-intelligence-isnt-the-science-its-the-application/.

2 https://futureoflife.org/ai-open-letter/.

3 https://atlantic-community.org/2019s-problem-with-artificial-intelligence-isnt-the-science-its-the-application/.

 | DOI:10.1163/9789004437876_016

that risk potential does not provide a reason for slowing AI development. No technology is an unalloyed good, but virtually all technologies that emerge in the marketplace produce benefits that are vastly in excess of their costs. AI is and will continue to be no different. As with any technology, the policy makers, developers, and managers have strong incentives to regulate, reduce, and manage risks (Atkinson 2016).

In this paper, I am not trying to demystify AI, or spread hype around it, or perform a review of the huge quantity of research papers on AI.[4] My main focus is to reflect rationally on where to implement the actual regulation of AI risk and harm in the context of the benefits of rapid AI development and industrialization. In the second section, I will decompose machine intelligence, evaluating how close the present unthinking and unintelligent machine is to its final scope, namely, general AI. In the third and in the fourth sections, I assess the level of accountability and responsibility of such a machine in relation to harm to humans produced by so-called "machine bias." In the fifth section, I review some major documents, considering fundamental human rights as the North Star and as the basis for commonly acceptable legal and ethical regulation norms for AI development, industrialization, and use.

2 Decomposition of Machine Intelligence

While some areas of AI remain only theoretical, others, such as machine learning, are already having an impact in real-world contexts. As Atkinson stated, although artificial intelligence has become commonplace, the public still has a poor understanding of the technology (Atkinson 2016). Up to now, there has been considerable confusion among technical and non-technical persons in differentiating the terms AI vs. Machine Learning (ML) vs. Deep Learning (DL). All three terms are closely related.

The term "AI" is exceptionally wide in scope. Even within the field of AI, it is not well defined and consists of everything from software that relies on many programmed rules (which is good-old-fashioned AI), all the way to systems that learn through trial and error and to futuristic technologies that may not be digital. Russell and Norvig indicated that the field of AI attempts not just to understand how humans think, but also to build intelligent entities (Russell and Norvig 2010). John McCarthy, widely recognized as one of the godfathers of

4 The number of research papers is growing rapidly. See, for example, Scopus AI Index 2018. http://cdn.aiindex.org/2018/AI%20Index%202018%20Annual%20Report.pdf.

AI, defined it as "the science and engineering of making intelligent machines."[5] Rich and Knight defined AI as "the study of how to make computers do things at which, at the moment, people are better" (Rich, Knight and Nair 2009, 3). Holistically, AI can be defined as algorithms and models targeted at thinking, perception, and action. As machines become increasingly capable, tasks considered to require "intelligence" are often removed from the definition of AI. In the past, playing chess and solving calculus used to be within the realm of AI. By today's standards, they most definitely are not. This constant shift in the definition and scope of AI is known as the AI effect and it is best described as Tesler's Theorem: "AI is whatever hasn't been done yet" (Hofstadter 1999, 601).[6]

AI is a moving target and its definition changes as its related technological advancements turn out to be further developed.[7] AI is any technology that showcases anything that resembles human intelligence. However, the biggest problem is that the word "intelligence" itself is not well defined at all. We do not have a unified standard of intelligence that can be applied to humans, animals, and machines. For this reason, it is not completely clear what kind of analogy we are trying to make through the term "AI."

We have no general mathematical theory of intelligence.[8] Without a general theory of "natural" intelligence trying to achieve the final scope – general AI – it is a bit like trying to do particle physics experiments with no standard model. Abstraction of biological neurons and insights based on fragments of biological networks topology do not constitute a general theory of intelligence. Yet I agree with Boden's opinion that, in actual situations, Turing's teasing discussion of the imitation game was aimed at avoiding endless definitions of intelligence, because philosophical industry is doing just that (Boden 2006).

Machine learning and its sub-branch deep learning are branches of AI that use mathematical models from data to make decisions. ML and DL are two of the ways through which we expect to achieve AI. Both rely on working with

---

5 http://www-formal.stanford.edu/jmc/whatisai/node1.html.

6 Larry Tesler (the author of the theorem) latter reformulated it: "Intelligence is whatever machines haven't done yet" (http://www.nomodes.com/Larry_Tesler_Consulting/Adages_and_Coinages.html).

7 For example, the European High-Level Expert Group has launched a nine-page long document containing only the definition of what AI is (High-Level Expert 2018).

8 Max Tegmark proposed the hypothesis that our physical reality is a mathematical structure (e.g., Tegmark 2014). Steven Pinker argued that intelligence does not come from a special kind of spirit or matter or energy, but from a different commodity, namely information. Information is a correlation between two things that is produced by a lawful process. Correlation is a mathematical and logical concept (Pinker 1998). However, Cathy O'Neil demonstrates the opposite that many algorithms are not inherently fair just because they have a mathematical basis (O'Neil 2016).

large data sets, by examining and comparing the data to find common patterns and explore nuances. In any case, DL is "magic," regardless of whether the public at times has issues observing its internal workings. In fact, while some tend to compare DL and neural networks to the way the human brain works, there are important differences between the two. DL learns from large amounts of data to make predictions. Neural networks are where DL "learns and trains" from a large set of data to determine the probable outcome of a situation. Without getting overly complicated, neural networks are where a computer would learn for thousands of hours to identify a person or animal in an image or even learn how to translate a language. Nevertheless, much of this process requires a human touch, meaning that a programmer has to do most of the heavy lifting. When DL "comes to life" and moves beyond simple programming, and can reflect and interact with people even on the most basic level, this is where AI comes into play.

AI is the step beyond DL. For now AI uses what it has gained from ML and DL to simulate intelligence. For now, instead of AI learning, a better term is ML, because the word "learning" in the context of ML is very well defined. For humans, learning is the accumulation of knowledge and skills through experience. For machines, the learning definition is very analogous because ML increases in performance through data by learning from past matches (supervised learning) or by playing against machine itself or other people (reinforcement learning). This is very analogous with human learning. More data, examples, and experience lead to improvement in performance. From this point of view, ML is not doing anything arcane or magical or even obscure like "thinking" or "feeling." It is simply doing something that is very common and mundane. However, with an equal number of experiences, machines are still very far away from the human level of learning.

Presently, AI does not have a free will, common sense, critical thinking capacity, and certainly is not conscious – the assumptions people tend to make when faced with advanced or overhyped technologies. The most advanced AI systems out there are merely products that follow processes defined by smart people. AI techniques (natural language processing and ML in the first place) enable private and public decision-makers to analyse big data in order to build profiles, which are used to make decisions in an automated way. Yet, for now, machines cannot make decisions on their own.[9]

9 Even in our hypertechnological world we cannot rely on computers to find moral answers. They can only assist humans in moral reasoning and decision-making, and in making better and more informed moral decisions. For this purpose, moral advisor AI systems already exist (see e.g., Giubilini and Savulescu 2018).

As technology advanced, people began reflecting on the questions of whether machines can think and whether machines will be able to make morally grounded decisions. The question of a machine's capability to think is highly relevant and there is growing interest in academia, but also at the level of non-academia.[10] However, at the current level of AI technological development, this question is comparable to whether submarines can swim or whether aircraft can fly. A submarine is doing something that is somewhat analogous to swimming and an aircraft is doing something that is somewhat analogous to flying. Both are imitating actions that are borrowed from nature. At present and for some decades to come, machine intelligence is its own form of intelligence with its own specialties compared to human intelligence or other animal intelligence. Possibly, within a few decades, today's innovative AI advancements will be considered as dull as calculators are to us right now. Though ML and DL are incredibly complex and clever techniques, they still do not possess any real human-like intelligence. The biggest problem with present ML and DL machines is that they are learning things, but they cannot understand them. None of actual neural networks understand what they are doing. They have been assigned a goal and have learned a way to accomplish it.

At the Brains, Minds, and Machines Symposium in 2011, famous linguist Chomsky derided researchers in machine learning who use purely statistical methods to produce behavior that mimics something in the world, but who do not try to understand the meaning of that behavior. Chomsky compared such researchers to scientists who might study the dance performed by a bee returning to the hive, and who could produce a statistically based simulation of such a dance without attempting to understand why the bee behaved that way. In the panel discussion, Chomsky said that "That's a notion of scientific success that's very novel. I don't know of anything like it in the history of science." Participants of the panel discussion N. Chomsky and B. Partee both made seminal contributions to the understanding of language by considering it as a computational rather than a purely cultural phenomenon. Both also felt that

10 If by thinking we basically understand "reason" (that is, conceiving and generating actions in a logical, sequential, and useful way), AI systems definitely can think. Human thinking is not only to logically combine existing knowledge and arrive at a solution to a problem or make a decision, but to reason and imagine outside of the established galaxy of knowledge in order to create a new kind of knowledge. A necessary precondition for human-like thinking is libertarian freedom – the ability to indeterministically choose between alternatives – though actual AI systems lack this ability. The actual thinking of machines is a capacity to solve problems using existing knowledge. If the notion of actual "unthinking machines" or "imitation (or simulation) of thinking" is unacceptable, it is possible to introduce the notion of "limited thinking" of actual AI systems.

understanding human language was the key to creating genuinely thinking machines. Partee has also indicated that really knowing semantics is a prerequisite for anything to be called intelligence.[11]

Gary Marcus analysed the relation between natural intelligence and AI in the context of the dilemma of nativism vs. empirism, concluding that it is time for AI to take nativism more seriously and suggesting that many aspects of human intelligence are native based. He believes that evolution (whether through natural selection or simulated artificial techniques) is a means towards building machinery with embedded prior knowledge, and not an alternative to prior knowledge (Marcus 2018). Alan Turing suggested that "Instead of trying to produce a programme to simulate the adult mind, why not rather try to produce one which simulates the child's? If this were then subjected to an appropriate course of education one would obtain the adult brain" (Turing 1950, 456).

Atkinson stated that there is, in fact, a fundamental difference between information processing and thinking (Atkinson 2016). Domingos is in agreement, noting that "Unfortunately, what we have so far is only a very crude cartoon of how nature learns, good enough for a lot of applications but still a pale shadow of the real thing" (Domingos 2018, 140). Marcus also concurred, stating that "Deep learning is only part of the larger challenge of building intelligent machines, [...] because such techniques [...] (are) still a long way from integrating abstract knowledge, such as information about what objects are, what they are for, and how they are typically used" (Marcus 2012). Nilson suggested that, even though we already have AI systems that can do some things better than humans can do them, there are still many, many human cognitive skills that we do not yet know how to automate (Nilson 2009).

Rodney Brooks evaluated actual artificial intelligence as an evolutionary process: the object-recognition capabilities of a 2-year-old child, the language capabilities of a 4-year-old child, the manual dexterity of a 6-year-old child, the social understanding of an 8-year-old child (Brooks, 2008). After one decade, Brooks' evaluation must be corrected only in two points: recent advances in image recognition systems (especially in 2D image semantic segmentation) with some benchmarks coming close to the human level and recent word embedding technologies (such as FastText[12]) with language capabilities that come close to a six-year-old child.

11 An article in *Technology Review* in 2011 summarizes Chomsky's and Partee's contributions. https://www.technologyreview.com/s/423917/unthinking-machines/.

12 For morphologicaly rich languages such as Lithuanian or Turkish, the word2vec algorithm is incapable of performing well since its atom level unit is the word and not the sub-word as with FastText.

The other requirement for basic robotic and ML is the autonomy process (decision-making and risk assessment included). Nevertheless, for now AI technologies are only in the operational morality stage so their autonomy is inversely proportional to their ethical sensitivity and responsibility (Wallach and Allen 2009).

Summarizing all of the above, modern machine intelligence is a combination of intelligence imitation (i.e., AI) and self-learning imitation (i.e., ML). The current generation of AI is a long away from the intelligent machines of science fiction and they completely lack common sense and critical thinking which are indispensable for evaluating data/information on their own. The current wave of advances in artificial intelligence does not actually bring us intelligence but instead a critical component of intelligence: prediction. As we know, prediction is a central input into decision-making and is based on data. In the absence of prediction, judgment might push decisions towards riskier actions. At present, a human has formal authority over any decision. However, if a decision is determined exclusively by a machine's prediction, then that authority may be abrogated; that is, the machine may have real authority. Theoretically, machines can learn judgment, but they need examples of judgment so that they can observe it in more environments and learn to mimic it. However, there are serious limitations to the ability of actual machines to learn judgment (Agrawal, Gans and Goldfarb 2018).

3 Machine Bias

A number of media articles, research papers, and books have raised awareness about unfair and unethical AI and AI ethics. One reason why we frequently use human-like terms to describe AI is that humans appear hardwired to anthropomorphize AI systems, especially robots. To refer to this phenomenon, there is wasteful amount of terms in use: bias, discrimination, inequality, inequitable, irresponsible, unethical, unfair, machine bias, machine learning bias, algorithmic bias, algorithm bias, etc. However, how are all of these terms connected to ML and AI? Cathy O'Neil asserts that when decisions are being made not by humans but by mathematical models, in theory this should lead to greater fairness, even if in reality the opposite is true (O'Neil 2016).

For the sake of simplicity, in this paper I use the term "machine bias" which encompasses all the terms listed above and takes various forms. As Paula Boddington affirmed, "getting rid of bias may be a great goal, but to understand what it means, and how to do it, is another matter and we can at least start with the bias that law requires us to eliminate" (Boddington 2017, 16).

The first question relates to what "bias" exactly is.[13] It should be noted that the English term "bias" is unique. In Lithuanian, I have not found a unique equivalent with such a vast array of meanings.

The term "bias" has a mathematically specific definition in machine learning, usually referring to errors in estimation or over/under representation of populations when sampling. In machine learning, there are two major sources of error: bias and variance. Roughly, bias is the error rate of an algorithm on a training set in the case of a very large training set. Variance instead refers to how much worse we do on the test set compared to the training set in some setting. When our error metric is mean squared error, we can write formulas specifying these two quantities and prove that "Total Error = Bias + Variance." There are a many ways in which bias and variance can be minimized.[14]

Another way to explain machine bias in scientific terms is by describing it as a "clustering" of data that is not inherently justified, where bias is one part of what engineers refer to as a "bias-variance" tradeoff. High bias can cause improper clustering. High variance can cause excessive data scattering. Engineers might refer to a system or result as "high bias, high variance" or "low bias, high variance," or some other combination.

The two main reasons for high bias are insufficient model capacity and underfitting because the training phase was not complete. For example, if we have a very complex problem to solve (e.g., image semantic segmentation, which my colleagues and I used in one recent project to build the cultural heritage monitoring system) and we use a model of low capacity, this model would have high bias as a result of the model not being able of grasp the complexity of the problem. The main reason for high variance is overfitting[15] on the training set.

There are ways of reducing both bias and variance in a ML model. For example the easiest way to achieve this is to obtain more data (in some cases even synthetic data helps, but this technique has some negative nuances that will be discussed later). However, even though having more data cannot hurt, unfortunately it does not always help as much as we might hope. Sometimes it could be a waste of time to work on procuring more data. Only a good

13 In philosophy and ethics, the notion of bias is best defined by the term "prejudice," while in technical literature it is denoted as "distortion," etc. Readers may be confused if they do not distinguish these important notions properly.

14 There are a number of scientific papers on this subject (see e.g., Emmert-Streib and Dehmer 2019).

15 In statistics, overfitting is the production of an analysis that corresponds too closely or exactly to a particular set of data, and may therefore fail to fit additional data or predict future observations reliably.

understanding of bias and variance will help to decide whether adding data as well as other tactics will improve performance.

What we tend to do in practice is to first increase the capacity of the model in order to reduce the variance on the training set as much as possible. In other words, we want to make the model overfit. This is done because we want to make sure that the model has the capacity to sufficiently understand the data. Then we try to reduce the bias. This is done through regularization techniques (early stopping, norm penalties, dropout, etc.)

The second question relates to whether we are using AI and ML in a responsible way which is human-centric, does not harm humans, and will benefit all humans, or whether AI is going to make our world more imbalanced, unequal, and more dangerous. The goals of AI based well-being systems are to understand how our digital experience affects our health and our quality of life, and to design well-being systems that put humans at the center. The important challenges of this research are how to quantify subjective things such as happiness, personal impressions, fundamentals, and personal values, as well as how to transform them into scientific representations with corresponding computational methods. Important keywords in understanding AI in human health and wellness are cognitive bias and social bias.[16]

Cognitive biases are systematic patterns of deviation from norms or rationality in judgment, whereby inferences about other people and situations may be drawn in an illogical fashion (Haselton et al. 2015). Individuals create their own subjective social reality from their perception of the input. An individual's construction of social reality, instead of the objective input, may dictate their behaviour and lead to perceptual distortion, inaccurate judgment, illogical interpretation, or irrationality (Eickhoff 2018). Cognitive biases are relatively common in the rapidly expanding field of ML. As models are given more and more power to make decisions, we are beginning to see that these biases have farther-reaching consequences.[17] Biased AI systems are simply mirrors of a biased society, but they demonstrate only the imitation of cognitive biases because they are based on a world view and data that humans have created.

---

16 Entire sections of the 2018 AAAI Spring Symposium have been dedicated to the topics "AI and Society: Ethics, Safety and Trustworthiness in Intelligent Agents" and "Beyond Machine Intelligence: Understanding Cognitive Bias and Humanity for Well-Being AI" and offer many research papers. https://www.aaai.org/ocs/index.php/AAAI/SSS18/schedConf/presentations.

17 See, for example, the COMPAS system case (https://www.propublica.org/article/machine-bias-risk-assessments-in-criminal-sentencing), or Amazon's Recognition software case (https://www.forbes.com/sites/monicamelton/2019/02/01/amazon-still-pushing-biased-facial-recognition-software-to-law-enforcement-military-mit-researcher-contends/).

Social bias entails "prejudicial attitudes toward particular groups, races, sexes, or religions, including the conscious or unconscious expression of these attitudes in writing, speaking, etc."[18] (Houston 2001, 299). In AI, social bias can "arise" in many ways. For example, in data-driven and machine-learning-based AI systems, bias in data collection and training can result in an AI system demonstrating bias. In logic-based AI (e.g., rule-based systems) bias can arise due to how a knowledge engineer might view the rules that apply in a particular setting. Bias can also arise due to online learning and adaptation through interaction. It does not necessarily relate to human bias or human-driven data collection. Bias can arise, for example, through the limited contexts in which a system in used. Bias can be intentional or unintentional. In certain cases, bias can result in discriminatory and/or unfair outcomes, that is, the actions based on those biases.

The idea that actual AI could function unaffected by cognitive and social bias involved in data reflects a misunderstanding of how the technology works. All machine intelligence is built upon training data that was, at least at some point, created by humans. This means that unfair data is the quantitative expression of human cognitive and societal bias, and that machine bias is a product of unfair data since, for machines, humans exist only as a collection of data. It follows that unmitigated or unchecked machine bias can perpetuate societal prejudice, but also that machine bias has the potential to be a tool to uncover biases and prejudices of which we are unaware.

As Atkinson argued, AI systems are not independent from their developers and, more importantly, from the organizations that use them. Illegal things are going to happen with and without computers (Atkinson 2016). If an algorithmic system produces unintended and potentially discriminatory outcomes, it is not because the technology itself is malicious; it is because it simply follows human instructions, or more often relies on data sets from the world that may reflect bias. In other words, as we have entered the infancy of the AI age, machine bias reflects the bias of its creators.

The developers do not always possess enough data for ML or do not always have sufficient processing power. For those reasons, actually become popular such techniques as learning transfer and artificial data augmentation, which can be qualified as Hanlon's razor[19] especially in text data cases. In most data set cases, there is a lack of good description of previously obtained knowledge to be transferred. For this reason, those cases become potential sources to

18 The term cannot confuse with various "Discrimination" terms, which refer to the actions based on those attitudes.

19 Never attribute to malice that which is adequately explained by stupidity.

augment bias in an exponential way.[20] Artificial data augmentation is good for image recognition purposes. Flipping, cropping, etc. of a cat image does not change the essential content of data. However, all solutions for augmenting text data (massive word replacement with synonyms in data sets, etc.) may lead to obtaining senseless data or, in a much worse case, change the semantics of the original text and lead to absurdity.

Some tools and solutions for automatically checking the content of data sets for the presence of biased data have actually emerged. At present, they are not sufficient to catch and neutralize all possible biased data cases. If in the future such tools are sufficiently developed, they can become a part of a machine critical thinking mechanism at least to some extent. However, for now, all responsibility for training data quality lies only with the data scientist.

Since decision-making processes are automated in today's world that uses machine learning and AI, there are many areas affected by those decisions, so we must be careful in examining, debiasing, or mitigating the bias of predictive models. Recognizing systematic biases in data collection efforts is an important step towards countering their effect on those systems that are trained on the basis of these data and is a key enabling factor for algorithmic fairness (Hajian et al. 2016). There are various techniques for mitigating bias or for debiasing data sets used for machine learning (e.g., see a debiasing techniques review in Kliegr et al. 2018). However, most of current debiasing or bias mitigation techniques have been tested only on English language data collections. My team and I have repeated an experiment described in a canonical NIPS research paper (Bolukbasi et al. 2016) on a huge amount of Lithuanian language text collections. We have used a corpus of Lithuanian web news texts (500 million words) and a corpus of Lithuanian social media texts (10 million words). In most cases, we did not find the gender bias cases that are presented in English. This is due to the fact that Lithuanian is a very morphologically rich and flexible language. Yet problems with gender bias emerged when we began to use multilingual systems. We have performed experiments with the Google machine translation system to translate sentences containing terms relating to some professions from Lithuanian into English and back to Lithuanian. The sentence "*Pirmoji* (f.g.[21]) *politikė* (f.g.) *gyveno seniai*" was translated "*The first*

20 For example, Facebook has published word vectors for 157 languages, trained on Common Crawl and Wikipedia using fastText. They are very useful data sources for developers that have no sufficient computing resources or text data to construct word vectors by themselves, or not enough time or qualification to do it by themselves. However, sufficient descriptions of these word vectors are lacking. https://fasttext.cc/docs/en/crawl-vectors.html.

21 Feminine gender.

politician lived long ago." The translation back into Lithuanian gave the result "*Pirmasis* (m.g.[22]) *politikas* (m.g.) *gyveno seniai.*" Masculine gender orientation was obtained in cases of 10 prestigious professions, that are traditionally attributed to males. Nevertheless, the sentence "*Pirmasis* (m.g.) *slaugytojas* (m.g.) *gyveno seniai*" was translated into "*The first nurse lived long ago,*" which in the back translation to Lithuanian resulted in "*Pirmoji* (f.g.) *slaugytoja* (f.g.) *gyveno seniai.*" Feminine gender orientation was obtained in cases of 10 less prestigious professions that are traditionally attributed to females.

However, there are cases involving data collections when debiasing or bias mitigation techniques must be avoided. In one current project, our team is developing a DL classifier for automated detection of hate/offensive speech in Lithuanian social media texts. To train our classifier, we have collected a corpus of Lithuanian social media texts containing 3000 examples of hate/offensive speech explicitly directed towards all principal minority groups. Our collection is not sufficiently large to obtain very good results, but Lithuanian is an under resourced language and we generally avoid artificial augmentation techniques with text data.[23] The clear division between hate and offensive speech is major problem *per* se in all languages. Another problem is to identify explicit hate speech. Our team has recruited two lawyers (one for the Lithuanian legal system and the other for the European Community's legal system) to prepare a legally grounded methodology for this purpose. The conclusion of the legal report was that it is not possible to automatically decide that some texts contain hate speech. It is not sufficient to rely only on linguistically oriented features. It is also important to consider a complex set of external criteria: the event in question, social reaction to the event, etc. If all criteria are not present, automatic decisions regarding hate speech detection may lead to the potential violation of the right to free expression and freedom of speech. A good example of when all element of e complex set are present is the case of the Pope's visit to Lithuania, when a dark-skinned little girl from a mixed family was dressed in the Lithuanian national garment and was part of a chosen small group of Lithuanian children to meet the Pope in the airport. On Facebook and other social media, there was a huge amount on nationalist oriented bias in reactions to this occasion. Unfortunately, to build such a complex system and to analyze a complex set of external criteria is an impossible mission. Moreover, it must be taken into account that actual AI systems completely lack the capacity for critical thinking. While the system build by *OpenAI* has shown

22 Masculine gender.

23 We have also used other techniques, but our technical solution will be presented in another research paper.

impressive results in producing fake news,[24] the next goal in building automated AI systems is the creation of the tools and solutions to enable machines to think critically. There is another good reason for collecting and analyzing biased data collections: our team was astonished by all the sophisticated ways used by humans to express explicit and implicit bias.[25]

The third question relates to the notion of AI systems as "black boxes." For the first time in human history, we have created a technology that we cannot explain and understand in its full extent. Algorithmic transparency is an emerging research area aimed at explaining decisions made by algorithmic systems. The transparency issue, as yet unsolved, is a potential liability when using deep learning for problem domains like financial trade or medical diagnosis, in which human users might like to understand how a given system made a given decision. Marcus rightly argued that deep learning systems have millions or even billions of parameters, identifiable to their developers not in terms of the sort of human interpretable labels that canonical programmers use ("last_character_typed"), but only in terms of their geography within a complex network (e.g., the activity value of the ith node in layer j in network module k). Although some strides have been made in visualizing the contributions of individual nodes, most observers would acknowledge that neural networks as a whole in some cases remain something of a black box. I agree with Marcus' conclusion that how much the black box problem matters in the long run remains unclear; if systems are robust and self-contained enough it might not matter (Marcus 2018).

However, I also agree with other scientists' argument that if the algorithms are intractable, i.e., we do not know how they are doing something, then they are not doing it in best way and they are not the appropriate model for a problem solving. Even if these algorithms produce a solution that appears to be promising, it is often not the best solution for the problem. Moreover, if the algorithms are intractable, it is not always is possible at the initial stage of some algorithm creation to predict what kind of unforeseen results we will have at the end.[26]

Yet how can we know the limits that can be pushed and to what extent we can use actual AI systems in a "black box" case? One of the solutions can be found in Clarke's second law: the only way to discover the limits of the possible is to venture a little way past them into the impossible (Clarke 1962). Moreover,

---

24 See, for example, https://www.theguardian.com/technology/2019/feb/14/elon-musk-backed-ai-writes-convincing-news-fiction.

25 We are preparing a separate research paper on this topic.

26 Recently, new solutions for transparency have been emerging (see e.g., Datta et al. 2016).

publicly known cases show that AI can be used by corporations to transform society into a test-bed for experiments without informed consent. However, it this legal?

4 Search Neutrality and (Machine) Bias

As new, more complex AI algorithms move into an increasingly wide array of products, like online ad services, search services and business software, or talking digital assistants like Apple's Siri and Amazon's Alexa, tech companies will be pressurised to guard against the unexpected biases that are being discovered in these products. Researchers have long warned of bias in AI-based systems that learn from large amounts of data, including the facial recognition systems that are used by the police and other government departments, as well as popular Internet services from tech giants like Google and Facebook.[27]

Sergey Brin, the co-founder of Google, famously suggested that "the perfect search engine would be the mind of God." For more than half a century, digital search engines have relied upon the humble keyword. Yet, over the past few years, search engines of all kinds have increasingly turned to deep learning-powered categorization and recommendation algorithms to augment and slowly to replace the traditional keyword search. In 2008, TinEye became the first image search engine to use image identification technology, and this was the beginning of the visual search. In 2010, the Google Goggles app allowed users to search the physical world with their phone cameras. Search engines also have increasingly turned to guided and crowdsourced search, in which queries are autocompleted based on common similar searches and where semantic hierarchies of AI-assigned topical tags are used to guide users along predefined paths through the information wilderness. Guided search attempts to mirror the assistance of information professionals like reference librarians, but their limited, predefined pathways and their reliance on the "wisdom of the crowd" means they frequently reinforce stereotypes, biases and falsehoods. This matters, because as search engines increasingly redefine our world through their idiosyncratic topical taxonomies, they are implicitly guiding us

27 Google's ads business is sprawling. According to eMarketer, Google will reach a milestone of surpassing 20% of all U.S. ad spending both online and offline this year. Google captures 74.6% of U.S. search ad spending. And the company is expected to lead the U.S. digital ad market with a 37.2% share, totaling $48.05 billion this year. Facebook is second with a 22.1% share (https://www.cnbc.com/2019/09/11/google-antitrust-investigation-to-focus-on-advertising-business.html). The EU censured Google because of its dominant position online.

towards predefined answers. The topical taxonomies used by search engines are not carefully constructed global expert taxonomies. Instead, today's search engine taxonomies represent the world as seen by today's primitive deep learning algorithms. Topical categories are selected not according to what best organizes available information, but rather according to the topics AI algorithms can best distinguish.

Search engines obviously have to sort their results by some kind of criteria. Google uses a machine-learning AI system called RankBrain to help to sort through its search results. RankBrain is part of Google's overall search algorithm, a computer program that can sort through the billions of pages it knows about and find the most relevant ones for particular queries. Shivaun Raff believes that search neutrality is a natural extension of net neutrality, which forbids Internet service providers from artificially speeding up, slowing down or blocking certain websites and services. "Search neutrality is saying, apply your best guess at relevance," she says, even for smaller search engines. "Don't artificially promote your own services in the search results and demote or exclude competing services."

Search neutrality proponents (Odlyzko 2009), though, want to see more transparency about how such algorithms work. And, even if the algorithms are necessarily subjective descriptors of relevance and value, they can at least be applied in a straightforward way (i.e., by not penalizing sites like Foundem and making it unclear how to escape the penalty box). Others who use the term are concerned about the power of search engines like Google to promote Google's other services (essentially an antitrust, not a search concern; Google has had plenty of these lately).

Search neutrality would require even-handed treatment in search results. It would prevent search engines from playing favorites among websites. Search engines are attention lenses; they bring the online world into focus. They can redirect, reveal, magnify and distort. They have immense power to help and to hide. Out of the many ways that search engines can cause harm, the thorniest problems of all stem from their ranking decisions. I agree with J. Grimmelmann, that the government does not get involved every time a business is harmed by a bad ranking – or Consumer Reports would be out of business. Instead, search-engine critics base their case for regulation on the immense power of search engines, which can "break the business of a Web site that is pushed down the rankings." They have the power to shape what millions of users, carrying out billions of searches a day, see. On that scale, search engines are the new mass media – or perhaps the new meta media – capable of shaping public discourse itself. And while power itself may not be an evil, abuse of power is (Grimmelmann 2011).

But the whole idea of search neutrality remains murky. Search algorithms, by definition, are not neutral – it is not even clear what "neutral" would mean in such a situation. The algorithms are designed to assign value and then to discriminate based on that value to produce a list of relevant results. Being discriminatory is the entire point of its task.

Websites have been fighting Google over search rankings almost since the company was founded. In 2002, a site called Search King sued Google over a suddenly demoted PageRank score. The courts dismissed that suit, and a handful of similar lawsuits did not fare any better. But in the late '00s, as Google's power grew, one complaint started to gain traction.[28] It came from Shivaun and Adam Raff, the creators of a price-comparison service called Foundem.

Traditionally, the balance between efficiency and fairness that was struck by policy makers depended heavily on cost considerations. When a service was expensive to provide, fairness was deemphasized. The idea of *neutrality* (net and search) – not just competition or relevant results – has flourished in the conservative backlash against Silicon Valley. A 2018 documentary called *The Creepy Line*,[29] for example, supported a radical version of search neutrality. It used complaints like Yelp's and Foundem's to argue that if Google had manipulated its results for commercial purposes, it probably did the same thing for political reasons, including a supposed attempt to rig the 2016 U.S. presidential election.

Behavioural research psychologist Robert Epstein in his Testimony "Why Google Poses a Serious Threat to Democracy, and How to End That Threat" wrote: "Data I've collected since 2016 show that Google displays content to the American public that is biased in favor [of] one political party – a party I happen to like, but that's irrelevant. No private company should have either the right or the power to manipulate large populations without their knowledge" (Epstein 2019). And he offers five "disturbing findings," among which is: "In 2016, biased search results generated by Google's search algorithm likely impacted undecided voters in a way that gave at least 2.6 million votes to Hillary Clinton (whom I supported). I know this, because I preserved more than 13,000 election-related searches conducted by a diverse group of Americans on Google, Bing and Yahoo in the weeks leading up to the election, and Google search results – which dominate search in the U.S. and worldwide – were significantly biased in favor of Secretary Clinton in all 10 positions on the first page of search results in both blue states and red states."

28 https://www.nytimes.com/2018/02/20/magazine/the-case-against-google.html.

29 The Creepy Line is a 2018 American documentary exploring the influence Google and Facebook have on public opinion, and the power the companies have that is not regulated or controlled by national government legislation (https://www.thecreepyline.com).

House Majority Leader Kevin McCarthy, in the announcement "Transparency & Accountability: Examining Google and its Data Collection, Use and Filtering Practices," wrote: "Google has created some of the most powerful and impressive technology applications. Unfortunately, recent reports suggest Google might not be wielding its vast power impartially. Its business practices may have been affected by political bias."[30]

Senior Google engineer Gregory Coppola has been suspended from his job at Google for saying in an interview that he believes Google News and Search results have a political bias. In his blog post, he reveals some details of the mechanics of opinion manipulation, using Google News as an example and choosing "Donald Trump" as a subject: "The most-used site, CNN, is selected in 20% of all articles! In other words, even with the millions of sites on the Internet, 1 out of every 5 stories about 'Donald Trump' from Google News is from CNN." Coppola transmits the message to all the non-programmers that he really does not believe in the idea that big tech is politically neutral, and he thinks this must be incorporated that into whatever strategy humanity use to have a democracy going forward.[31]

As mentioned above, visual search made entrance into search engines, and it is closely related to the efficiency and limitations of image-processing and facial-recognition AI-based algorithms. Facial recognition is becoming more and more common these days, but it is also one of the more controversial forms of biometrics, because of its privacy implications. There is, however, one other problem with face ID that rarely pops up in reports: gender and racial bias. The awkward workaround illustrates the difficulties Google and other tech companies face in advancing image-recognition technology, which the companies hope to use in self-driving cars, personal assistants and other products.

Researchers found that image-processing algorithms both learned and amplified gender stereotypes. For example, algorithms showed a tendency to associate women with shopping and men with shooting.[32] In the researchers' experiments, the three programs' (Microsoft, IBM, Face++) error rates in determining the gender of light-skinned men were never worse than 0.8 percent. For darker-skinned women, however, the error rates ballooned – to more than 20 percent in one case and more than 34 percent in the other two. These findings raise questions about how today's neural networks, which learn to

30 https://republicans-judiciary.house.gov/press-release/house-judiciary-committee-to-hold-hearing-with-google-ceo-2/.
31 https://medium.com/@gfcoppola/analyzing-google-news-introduction-85087aa220e9.
32 https://www.wired.com/story/machines-taught-by-photos-learn-a-sexist-view-of-women/.

perform computational tasks by looking for patterns in huge data sets, are trained and evaluated. For instance, according to the paper, researchers at a major U.S. technology company claimed an accuracy rate of more than 97 percent for a face-recognition system they had designed. But the data set used to assess its performance was more than 77 percent male and more than 83 percent white.

In 2015, Jacky Alciné revealed that the image classifier used by Google Photos was labelling black people as "gorillas."[33] Google apologized profusely and set to work on the bug. Two years later, Google has simply erased gorillas (and, it seems, chimps and monkeys) from the lexicon of labels its image classifier can apply or be searched with, rendering these animals unsearchable and in some sense invisible to the AI that powers Google's image searching capabilities. The ability to classify images as containing gorillas remains in some Google products, like Cloud Vision API. A Google spokesperson confirmed that "gorilla" was censored from searches and image tags after the 2015 incident, and that "chimp," "chimpanzee," and "monkey" are also blocked today. "Image labeling technology is still early and unfortunately it's nowhere near perfect," the spokesperson wrote in an e-mail, highlighting a feature of Google Photos that allows users to report mistakes. Google's caution around images of gorillas illustrates a shortcoming of existing machine-learning technology. With enough data and computing power, software can be trained to categorize images or to transcribe speech to a high level of accuracy. But it cannot easily go beyond the experience of that training. And even the very best algorithms lack the ability to use common sense, or abstract concepts, to refine their interpretation of the world, as humans do.[34]

That increases the risk of false identification of women and minorities. This is probably down to the data on which the algorithms are trained: if the dataset contains many more white men than black women, the algorithm will be better at identifying white men (Cossins 2018).

Since search queries are strongly related to natural language processing, Google unveiled a breakthrough AI technology called BERT[35] that has changed the way scientists build systems that learn how people write and talk. Google itself has used BERT to improve its search engine. In fact, BERT will impact 1 in

33 This tweet has since been deleted but references can be found in reputable newspapers online. A follow-up article was published by The Verge in 2018: https://www.theverge.com/2018/1/12/16882408/google-racist-gorillas-photo-recognition-algorithm-ai and Wired: https://www.wired.com/story/when-it-comes-to-gorillas-google-photos-remains-blind/.

34 https://boingboing.net/2018/01/11/gorilla-chimp-monkey-unpersone.html.

35 https://www.blog.google/products/search/search-language-understanding-bert/.

10 of all search queries. This is the biggest change in searching since Google released RankBrain.[36]

It could be picking up on biases in the way a child mimics the bad behavior of his or her parents. But scientists are still learning how technology like BERT, called universal language modelling, works. And they are often surprised by the mistakes their new AI is making. Researchers are only beginning to understand the effects of bias in systems like BERT. In a blog post, Dr. Munro describes how he examined cloud-computing services from Google and Amazon Web Services that help other businesses to add language skills into new applications. Both services failed to recognize the word "hers" as a pronoun, though they correctly identified "his." After Dr. Munro pointed out the problem, Amazon corrected it. Google said it was working to fix the issue (Metz 2019).

Having said this, we can draw some conclusions for the present and for the future.

For now, many machine-learning systems are essentially black boxes: their creators know they work, but they cannot explain exactly why they make particular decisions. All AI-based algorithms are in an early stage of development and they depend strongly on their data. They process vast tracts of data. And, as a result, the troubling hidden roots of some data are exposed. For this reason, algorithms can be unknowingly biased. The machine has no opinion, the program did not teach itself anything. But the situation taught the companies something important about what can safely be automated. The actual state of things requires adequate regulations and policies. Open questions remain about the societal impact on development of AI – all crowdsourced practices lie heavily on the culture of society.

On the other hand, if a product causes a consumer to suffer injury or loss, liability claims are brought under legal regimes. But currently there is no an effective liability framework specifically applicable to harm or loss resulting from the use of emerging technologies such as AI. I agree with the Expert Committee on human rights dimensions of automated data processing and different forms of artificial intelligence (MSI-AUT), that nowadays there is the radical asymmetry in power between those who develop and deploy algorithmic systems and the individual users who are subject to them. Existing human rights institutions may struggle to provide effective and meaningful protection (MSI-AUT 2019).

36 https://searchengineland.com/welcome-bert-google-artificial-intelligence-for-understanding-search-queries-323976.

5 Machine Bias and Fundamental Rights[37]

Industrialization of AI systems has an immense potential not only for the overall technological advancement of society but also for unethical practices. AI-based systems depend on the collection of data and the use of vast quantities of data to make predictions which, in numerous cases, have served to perpetuate existing social patterns of bias and discrimination. Yet therein lies the problem: to the extent that an AI accurately replicates past patterns of human decision-making, it will necessarily perpetuate existing social biases and prejudices as well. What is worse, unlike human decision-makers who have the agency and the free will to change their moral perspective over time, for the foreseeable future AI systems will not have any such capabilities of their own.

Determining the human rights impacts of AI systems is no easy feat, for these technologies are being introduced and incorporated into existing social institutions which are not rights-neutral. The positive and negative impacts of AI systems on fundamental human rights are not distributed equally throughout society. Some individuals and groups are affected more strongly than others. AI systems can easily perpetuate existing patterns of bias and discrimination, since the most common way to deploy these systems is to "train" them to replicate the outcomes achieved by human decision-makers. The larger point, however, is that unless AI systems are consciously designed and consistently evaluated for their differential impacts on different populations, they have the very real potential to hinder rather than help progress towards greater equity.

37 In general, "rights" refer to the moral or legal entitlement over something. The rights which are fundamental to the life of the citizens of a country are known as fundamental rights. Human rights imply the rights that belong to all human beings irrespective of their nationality, race, caste, creed, gender, etc. Fundamental rights and human rights are vital for the existence and development of individuals. In the European Union (EU), fundamental rights afford basic legal protection for political, social, and procedural rights to individuals and legal entities. They cover a wide range of issues from human integrity, property and privacy rights, rights to conduct business, free movement, equal treatment, children's rights, rights of persons with disabilities, rights of citizens in their dealings with the European Union's institutions, procedural safeguards and much more. Respect for the Charter of Fundamental Rights of the European Union in the European Commission's (EC) acts and initiatives is a binding legal requirement. For pragmatic reasons, the impact assessment of EC initiatives with only an internal EU dimension focuses on the analysis of fundamental rights (guaranteed by the Charter of Fundamental Rights), while for EU initiatives with an external dimension, the analysis focuses on human rights which may be different from the fundamental rights guaranteed by a EU partner country, and which arise from international treaties and customs (see "European Commission Staff Working Paper 'Operational Guidelines on taking account of Fundamental Rights in Commission Impact Assessments'" 2011 (https://ec.europa.eu/info/sites/info/files/opperational-guidance-fundamental-rights-in-impact-assessments_en.pdf).

Introducing AI decision-making and risk assessment systems into real life situations, together with their consequences and impacts on fundamental human rights, is a multifaceted endeavor and affects various fields: criminal justice, finance, healthcare, education, recruitment and hiring, etc.

Any attempt to formulate a set of ethics guidelines faces two significant challenges: on the one hand, the lack of ultimate consensus about ethical standards; on the other, the fact that guidelines are always only voluntary. However, when technology is rapidly advancing, the law might not be able to keep up, and professional bodies and others considering ethical aspects of that technology might well lobby for appropriate changes to the law (Boddington 2017). The reflections taking place to produce all kinds of guidelines or reports always offer a useful starting point to ponder rationally on where the world should go with regard to AI development. Yet political initiatives and policy instrument will never be sufficient to ensure that the design and deployment of AI systems is human-centric. To achieve that objective, AI development must be matched by the enforcement and development of adequate legal safeguards. While there are many different conceptions of fundamental human rights, from the philosophical to the moral, industrialization of AI systems must take a legal approach. Fundamental human rights must be viewed in terms of the binding legal commitments that the international community has articulated in the widely accepted formal documents (Raso et al. 2018). The fundamental human rights discourse is not only an alternative to moral values, but also contains and has its foundation in moral values (Haule 2006). Fundamental and international human rights provide a robust and global formulation of those values, since there is evidence that some AI systems are already violating fundamental rights and freedoms.

The European Commission High Level Expert Group (HLEG) indicated that "even after compliance with legally enforceable fundamental rights has been achieved, ethical reflection can help us understand how the development, deployment, and use of AI may implicate fundamental rights and their underlying values, and can help provide more fine-grained guidance when seeking to identify what we should do rather than what we (currently) can do with technology." (HLEG 2019, 10)

The disturbing possibilities of violating fundamental rights by AI systems have given rise to a worldwide movement seeking to embed ethical and fundamental human rights considerations into the development and deployment of AI systems. Industries and governments have recently started to recognize the importance of ethics in AI systems. *DeepMind* launched an internal team to scrutinize the societal impacts of the technologies it develops, and tackle key ethical challenges such as privacy, transparency, governance, and morality, among others. Companies such as Microsoft and Google have started to follow this trend internally as well. They are establishing ethics teams headed by top scientists and philosophers to ensure a safe development. Therefore, the global desirability for openness in AI systems is emerging. The technical community also is moving towards openness for AI systems source codes and data collections. Openness always has a great impact on software: it creates better programmers, detects bugs, and improves the quality of code. It also makes public science, data, safety techniques, and capabilities.[38]

The United Nations document "Guiding Principles on Business and Human Rights" (United Nations 2011) constitutes a starting point for the application of human rights in the tech sector. One of its foundational principles states, that "States must protect against human rights abuse within their territory and/or jurisdiction by third parties, including business enterprises. This requires taking appropriate steps to prevent, investigate, punish and redress such abuse through effective policies, legislation, regulations and adjudication."

In its Communications of 25 April 2018 and 7 December 2018, the European Commission set out its vision for AI, which supports "ethical, secure and cutting-edge AI made in Europe." Three pillars underpin the Commission's vision: (1) increasing public and private investments in AI to boost its uptake, (2) preparing for socio-economic changes, and (3) ensuring an appropriate ethical and legal framework to strengthen European values.

In the "Ethics Guidelines for Trustworthy AI" of HLEG,[39] the concept of European trustworthy AI was introduced and defined by the guidelines as a voluntary framework to achieve legal, ethical, and robust AI. The guidelines lay out a list of requirements and an assessment list to operationalise this concept. The guidelines tackle the first of these challenges by founding ethical principles on human rights. The four ethical principles, which they present as imperatives for those developing and deploying AI systems are the following:

38 Open Source, metadata, provenance of data (showing the origin and the quality of data sets), and transparency initiatives around the building and training of ML models move the R&D community towards solving the big problems in ML algorithm building. By building open source models and training data, transparency is increasing, thus providing more opportunity for hidden biases to be uncovered. This does not solve the problem of cognitive bias in machine learning as a whole, but it opens the door to collaboration and innovation. In contrast, the openness of AI is also highly debatable from an ethical point of view because such openness may cause serious issues if used by bad people. For example, these people can make use of the power of AI to commit evil actions more efficiently.

39 The first draft of this document was published on 18 December 2018. After having received more than 500 critical comments, HLEG published a revised version on 8 April 2019 and presented it to the Commission. I review the revised version.

(1) respect for human autonomy, (2) prevention of harm, (3) fairness, and (4) explicability. These principles can draw force from their explicit grounding in human rights and can be a way for those designing and deploying AI systems to reflect upon the potential ethical and human rights implications of their work. In this way, ethics and human rights can reinforce one another.

As I have indicated above, the challenge for any set of ethical guidelines lies in their voluntary nature. However, HLEG's approach is wise because the guidelines' ethical principles are based on fundamental rights as enshrined in the EU Charter of Fundamental Rights, so the move from voluntary adherence to solid regulation should be more straightforward in the future. Where ethics lacks the means of enforcement, international human rights law possesses well-developed standards and institutions, as well as a universal framework for safeguards. Ethical principles grounded in human rights can take advantage of this well-established structure to ensure that AI systems are developed, deployed, and used in a manner that respects human fundamental rights. It must be noted that, due to inadequate stakeholder representation,[40] the ethical-technical imbalance gave outsized influence to industry concerns. This must be rectified in the predefined steps to ensure the expert group's ability to formulate more human-centric recommendations. Beyond this issue, European trustworthy AI has three fundamental components: it should be lawful, ethical, and robust. While the guidelines discuss ethics at length, they miss important key issues concerning robustness and they entirely omit the legal component.

Regarding robustness, what the guidelines fail to address is the disturbing link between the recent scandals surrounding big tech and the advertising business model which dominates the industry – with or without the AI component. Although tech companies are eager to frame damaging or exploitative uses of their algorithms as the results of "hacks" or "bugs," the reality is that the very aspects and incentives of these algorithms which make them so profitable are also what invite and facilitate these abuses.

HLEG's focusing too much on ethics loses sight of the ultimate goal of formalising the principles of Trustworthy AI into enforceable laws. Whereas the violation of an ethical principle can perhaps be written off as collateral damage, human rights offer a set of principles which command respect and adherence in all circumstances.

The HLEG guidelines state that the assessment list is not a series of boxes that need to be ticked, but rather an opportunity for open and sustained

40 https://ec.europa.eu/digital-single-market/en/high-level-expert-group-artificial-intelligence.

ethical reflection. Such an open reflection has its place, but only after we have ticked all the boxes for legal and human rights compliance. What is important now is to know what boxes need to be ticked in all situations.

These requirements of the guidelines are applicable to different stakeholders partaking in AI systems' life cycle: developers, deployers and end-users, as well as the broader society, which have different roles to play in ensuring that the requirements are met. However, in the guidelines the responsibility for unethical AI comportment is not indicated and not distributed across stakeholders, and the question of data in ML is not treated at all.

In May 2018, Amnesty International, Access Now, and a handful of partner organizations launched the Toronto Declaration,[41] which is set of principles for protecting the right to equality and non-discrimination in machine learning systems, aiming to put human rights front and centre in the development and application of machine learning technologies. In other words, the declaration is a landmark document that seeks to apply existing international human rights standards to the development and use of machine learning systems or actual AI systems. The declaration sets out the duties of states to prevent discrimination in the context of designing or implementing machine learning systems in public contexts or through public-private partnerships. The first section of the declaration includes principles to identify risks in the use of machine learning systems, and to ensure transparency and accountability. In the second section the declaration outlines the responsibilities of private actors in the context of the development and deployment of ML systems. It is based on the human rights due diligence framework (originally defined in the above-described United Nations document). These responsibilities include identifying potential discriminatory outcomes through mapping and assessing risks, taking effective action to prevent and mitigate discrimination (including by submitting systems to third party independent audits where there is a significant risk of human rights abuses), and being transparent, including through publishing technical specifications, samples of the training data used, and the sources of data. The third section of the declaration asserts the right to an effective remedy and that those responsible for abuses must be held to account. The declaration calls on governments to ensure standards of due process for the use of machine learning in the public sector, to act cautiously in the use of machine learning systems in the justice system, to outline clear lines of

---

41 The Toronto Declaration was drafted after discussions and interviews with dozens of experts in AI, human rights, and business, among others. Extensive consultations on the draft were held and it was adopted on 16 May at the start of RightsCon 2018 (*one of world's leading summits on human rights in the digital age*).

accountability for the development and implementation of ML applications, and to clarify which bodies or individuals are legally responsible for decisions made through the use of such systems.

The declaration sets out principles for policy and practice, but to make them effective we need practical implementation guidelines that help engineers and product managers in applying the principles in their work. As in case of the HLEG guidelines, the Toronto Declaration is important for asserting the centrality of human rights in the debate on AI ethics. It sends an important signal that there is a strong expectation for tech companies to also endorse the declaration and commit to ensuring that existing human rights are not diluted by new technologies. More outreach to the technology community in order to establish a dialogue between human rights practitioners and engineers will help to embed human rights in the use of AI development. The Toronto Declaration also outlines the responsibilities of both states and private sector actors with respect to the use of machine learning systems, including mitigation of discriminatory effects, transparency, and provision of effective remedies to those harmed.

In April 2018, Data & Society launched a report entitled *"Governing Artificial Intelligence: Upholding Human Rights & Dignity"*[42] (Latonero 2018). In general, the report draws the connections between AI systems and human rights, reframes recent AI-related controversies through a human rights lens, and reviews current stakeholder efforts at the intersection of AI systems and human rights. It is intended for stakeholders – including technology companies, governments, intergovernmental organizations, civil society groups, and academia – that are looking to incorporate human rights into social and organizational contexts related to the development and governance of AI systems. At very beginning, the report states that, in order for AI to benefit the common good, at the very least its design and deployment should avoid harm to fundamental human values and international human rights should provide a robust and global formulation of those values. A human-rights-based frame could provide AI systems developers with the aspirational, normative, and legal guidance to uphold human dignity and the inherent worth of every individual, regardless of country or jurisdiction.

The report focuses on five examples of human rights areas: nondiscrimination, equality, political participation, privacy, and freedom of expression. Human rights cannot address all the present and unforeseen concerns pertaining

---

42 The report was animated by the ideas generated at the workshop on Artificial Intelligence & Human Rights, held at Data & Society in April 2018 and published by Data & Society in October 2018.

to AI systems. Near-term work in this area should focus on how a human rights approach could be practically implemented through policy, practice, and organizational change.

The report provides six recommendations: two of them are addressed to technology companies, one to governments, one to academics, one to United Nations investigators and special reporters, and one is collective for various stakeholders (human rights lawyers, policy makers, social scientists, computer scientists, and engineers).

The report provides analyzed examples of unethical AI comportment (Facebook, etc.). It states that those working directly on AI need regulation, or "hard" laws, along with technical standards, social norms, and market incentives to effectively incorporate respect for human rights into their business models, policies, and practices. At the same time, those working on human rights need to be actively engaged in AI governance and monitoring. Because some areas of AI remain only theoretical, others, such as machine learning, are already having an impact in real-world contexts. Machine learning systems process large amounts of historical training data and "learn" from these examples to detect patterns that can be useful in decision-making. All machine learning algorithms contain some level of statistical bias, which produces incorrect decisions some of the time. However, if the historical data are incomplete or are not representative of a specified population, these biases can scale quickly and inexplicably across AI systems. Such systems can further entrench discriminatory outcomes in people's lives.

The report asks the following question: To what extent should a society allow ML systems to influence human decision-making or even make decisions on their own? It suggests that these concerns should be at the heart of AI debates and one of the most hotly debated topics in AI should be discriminatory algorithms and systems. This includes empirical research on facial recognition systems that cannot "see" people, particularly women with darker skin due to a lack of adequate training data or to faulty models, and therefore reproduce culturally engrained biases against people of color.

The report states that "while AI "ethics" is the topic *du jour* for the tech industry, human rights is beginning to emerge as an additional perspective" (p. 18). In the privacy section, the considerations of the report follow the HLEG guidelines very closely and give special attention to disability rights.

In March 2018, the World Economic Forum (WEF) launched the report entitled "How to Prevent Discriminatory Outcomes in Machine Learning" (WEF 2018). It contains three main sections. The first section deals with issues around data. The main question addressed is: "What are the sources of risk around training data for machine learning applications?." The report indicates two main sources: data availability and biased or error-ridden data. The second section of the report focuses on concerns around algorithm design. In third section, the report considers the responsibilities of business, which aligns very closely with the HLEG guidelines. The report challenges companies to prioritize compliance with human rights standards and to perform rights-based due diligence. Among the recommendations is a call for companies to actively include a diversity of input and norms in systems design. Companies are also encouraged to provide mechanisms of access and redress that make developers responsible for discriminatory outputs.

In 2018, the report entitled "The Malicious Use of Artificial Intelligence: Forecasting, Prevention, and Mitigation" was launched (Brundage et al. 2018). The report does not deal with AI in relation to fundamental rights, but it emphasises the dual-use nature of AI (civilian and military) and raises the main question: How can we forecast, prevent, and (when necessary) mitigate the harmful effects of malicious uses of AI? The report defines four priority research areas: learning from and with the cybersecurity community, exploring different openness models, promoting a culture of responsibility, and developing technological and policy solutions. The same problems are outlined also in the report entitled "AI in the UK: ready, willing and able?" indicating that an expert group "encountered several serious issues associated with the use of artificial intelligence that require careful thought, and deliberate policy, from the Government. These include the issue of determining legal liability, in cases where a decision taken by an algorithm has an adverse impact on someone's life, the potential criminal misuse of artificial intelligence and data, and the use of AI in autonomous weapons systems."

Today's AI debates are searching for principles to govern emerging and future technological systems for the common good. If the "good" involves upholding a human-centric society and human dignity, then the international human rights system is fit for this purpose. If AI researchers, developers, and designers work to protect and respect fundamental human rights, they could open the path for broad social benefit. To disregard human rights would be to close off that path. Policy makers, researchers, developers, and designers need to promote communicating about AI technologies in a balanced way as technologies that can be used for the good of society, but can also be abused. Presenting AI in either dystopian or utopian visions is not conducive to a sound debate, and it will distract stakeholders from addressing the real risks and from taking advantage of opportunities to use AI beneficially. Special attention must be paid to ethical AI questions because, as one of Niven's laws states, ethics can change with technology. Prior to teaching AI how to be ethical, people need to think about what makes themselves ethical (Pavaloiu and Kose 2017). As stated above, while the ultimate consensus about universal ethical standards is lacking, in the industrialization of AI systems a legal approach must be taken. Fundamental human rights can be viewed in terms of binding legal commitments and safeguards.

6 Conclusion

AI algorithms make art but do not understand it. Addressing the most significant questions in the universe will reveal the real potential of AI. But it is not AI *per se* that deserves respect or hate in the first place. Instead, it is the will and actions of humanity that determine the course of our future. At first glance, it appears that bias may have stemmed from AI itself. However, when we speak of AI as intelligence demonstrated by machines, we often forget that the algorithms on which AI runs were designed and implemented by humans. It is the responsibility of humans to ensure fairness. Minimizing AI bias relies on three factors: the quality of input data, the prediction algorithm, and the testing of the correctness of the output prediction. Actual bias has been introduced into AI by humans. Moreover, the quality of the training data set is one of the major contributors to the accuracy of the prediction output. Fortunately, many organizations already recognize that AI bias can occur and are taking active remedial measures to avoid it. Since there is no ultimate consensus about ethical standards, some other ground must be found to protect AI against big data, humans against AI, humans against humans (developers against developers, developers against data scientists, data scientist against data producers,[43] etc.). In this situation, fundamental human rights could be the North Star and the basis for commonly acceptable legal and ethical regulation norms that we need urgently for responsible AI development, industrialization, and use. If the solutions and regulations are not found and introduced in the short term, research on how biased actual AI is can teach us to be better humans.

References

Agrawal, Ajay, Joshua S. Gans and Avi Goldfarb. 2018. "Exploring the Impact of Artificial Intelligence: Prediction Versus Judgment." *Rotman School of Management Working Paper* 3177467.

43 For more on the quantity of data generated per minute, see Domo's annual research "Data Never Sleeps" https://www.domo.com/learn/data-never-sleeps-6.

Armstrong, Stuart. 2014. *Smarter than Us: The rise of Machine Intelligence*. Machine Intelligence Research Institute.

Athiwaratkun, Ben, Andrew Gordon Wilson and Anima Andkumar. 2018. "Probabilistic FastText for Multi-SenseWord Embeddings." *Proceedings of the 56th Annual Meeting of the Association for Computational Linguistics* (Melbourne, Australia): 1–11.

Amnesty International and Access Now. 2018. *The Toronto Declaration: Protecting the right to equality and non-discrimination in machine learning systems*. https://www.accessnow.org/cms/assets/uploads/2018/08/The-Toronto-Declaration_ENG_08-2018.pdf.

Atkinson, Robert. 2016. "'It's Going to Kill Us!' and Other Myths About the Future of Artificial Intelligence." *NCSSS Journal* 21(1): 8–11.

Boddington, Paula. 2017. *Toward a Code of Ethics for Artificial Intelligence*. Springer.

Bolukbasi, T. et al. 2016. "Quantifying and Reducing Stereotypes in Word Embeddings." *2016 ICML Workshop on #Data4Good: Machine Learning in Social Good Applications* (New York): 41–45.

Bolukbasi, Tolga, et al. 2016. "Man is to Computer Programmer as Woman is to Homemaker? Debiasing Word Embeddings." *NIPS'16 Proceedings of the 30th International Conference on Neural Information Processing Systems* (Curran Associates Inc.): 4356–4364.

Boden, Margaret A. 2006. *Mind as Machine: A History of Cognitive Science*. Oxford University Press.

Brooks, Rodney. 2008. "I, Rodney Brooks, Am a Robot." *IEEE Spectrum* 45(6): 68–74.

Brundage, Miles, et al. 2018. *The Malicious Use of Artificial Intelligence: Forecasting, Prevention, and Mitigation*. Future of Humanity Institute et al. https://arxiv.org/ftp/arxiv/papers/1802/1802.07228.pdf.

Chander, Anupam. 2017. "The Racist Algorithm?" *Mich. Law Review* 115(6): 1023–1045.

Clarke, Arthur C. 1962. "Hazards of Prophecy: The Failure of Imagination." In *Profiles of the Future*, edited by Arthur C. Clarke. London: Gollancs.

Cossins, Daniel. 2018. Discriminating algorithms: 5 times AI showed prejudice. https://www.newscientist.com/article/2166207-discriminating-algorithms-5-times-ai-showed-prejudice/#ixzz6ED4ETtoo.

Datta, Anupam, Shayak Sen, and Yair Zick. 2016. "Algorithmic Transparency via Quantitative Input Influence: Theory and Experiments with Learning Systems." *2016 IEEE Symposium on Security and Privacy* (San Jose, CA, USA).

Domingos, Pedro. 2018. *The Master Algorithm: How the Quest for the Ultimate Learning Machine Will Remake Our World*. Basic Books.

Doshi-Velez, Finale, and Mason Kortz. 2017. *Accountability of AI Under the Law: The Role of Explanation*. Berkman Klein Center for Internet and Society. http://nrs.harvard.edu/urn-3:HUL.InstRepos:34372584.

Eickhoff, Calsten. 2018. "Cognitive Biases in Crowdsourcing." *Proceedings of the Eleventh ACM International Conference on Web Search and Data Mining* (ACM): 162–170.

Ekbia, Hamid. 2008. *Artificial Dreams: The Quest for Non-Biological Intelligence*. Cambridge University Press.

Emmert-Streib, Frank, and Matthias Dehmer. 2019. "Evaluation of Regression Models: Model Assessment, Model Selection and Generalization Error." *Machine Learning & Knowledge Extraction* 1(1): 521–551.

Epstein , Robert. 2019. *Why Google Poses a Serious Threat to Democracy, and How to End That Threat*. America Institute for Behavioral Research and Technology. https://www.judiciary.senate.gov/imo/media/doc/Epstein%20Testimony.pdf.

Expert Committee on human rights dimensions of automated data processing and different forms of artificial intelligence (MSI-AUT). 2019. *A study of the implications of advanced digital technologies (including AI systems) for the concept of responsibility within a human rights framework* (*DGI*(*2019*)*05*). Council of Europe. https://rm.coe.int/responsability-and-ai-en/168097d9c5.

Eubanks, Virginia. 2018. *Automating Inequality: How High-Tech Tools Profile, Police, and Punish the Poor*. New York: St. Martin's Press.

European Commission Staff Working Paper. 2011. *Operational Guidelines on taking account of Fundamental Rights in Commission Impact Assessments*. Brussels. https://ec.europa.eu/info/sites/info/files/opperational-guidance-fundamental-rights-in-impact-assessments_en.pdf.

Europen Commision's High-Level Expert Group on Artificial Intelligence. 2018. *A Definition of AI: Main Capabilities and Scientific Disciplines*. Brussel. https://ec.europa.eu/digital-single-market/en/news/definition-artificial-intelligence-main-capabilities-and-scientific-disciplines.

Frankish, Keith, and William M. Ramsey, ed. 2014. *The Cambridge Handbook of Artificial Intelligence*. Cambridge University Press.

Giubilini, Alberto, and Julian Savulescu. 2018. "The Artificial Moral Advisor. The 'Ideal Observer' Meets Artificial Intelligence." *Philosophy and Technology* 31(2): 169–188.

Grimmelmann, James. 2011. *Some Skepticism About Search Neutrality*. The Next Digital Decade: Essays On The Future of The Internet (ed. B. Szoka & A. Marcus). Washington: TechFreedom: 435–459.

Gunkel, David J. 2012. *The Machine Question*. MIT Press.

Hajian, Sara, Francesco Bonchi, *and* Carlos Castiglo. 2016. "Algorithmic Bias: From Discrimination Discovery to Fairness-aware Data Mining." *Proceedings of the 22nd ACM SIGKDD International Conference on Knowledge Discovery and Data Mining* (ACM): 2125–2126.

Hardt, Moritz, Eric Price, and Nathan Srebro. 2016. "Equality of Opportunity in Supervised Learning." *NIPS'16 Proceedings of the 30th International Conference on Neural Information Processing Systems*: 3323–3331.

Haselton, Martie, Daniel Nettle, and Paul W. Andrews. 2015. *The Evolution of Cognitive Bias. The Handbook of Evolutionary Psychology*. Hoboken: John Wiley.

Haule, Romuald. 2006. "Some Reflections on the Foundation of Human Rights – Are Human Rights an Alternative to Moral Values?" *Max Planck Yearbook of United Nations Law* 10: 367–395.

Hazem, Amir, and Beatrice Daille. 2018. "Word Embedding Approach for Synonym Extraction of Multi-Word Terms." *LREC-2018 Proceedings of the Eleventh International Conference on Language Resources and Evaluation* (European Language Resource Association): 297–303.

Hofstadter, Douglas. 1999. *Godel, Escher, Bach: An Eternal Golden Braid.* New York: Basic Books.

Houston, James, ed. 2001. *Thesaurus of ERIC Descriptors.* Westport: Oryx Press.

Jha, Akshita, and Radhika Mamidi. 2017. "When does a Compliment become Sexist? Analysis and Classification of Ambivalent Sexism using Twitter Data." *Second Workshop on Natural Language Processing and Computational Social Science (NLP+CSS)* (ACL): 7–16.

Kliegr, Tomaš, Štepan Bahnik, and Johannes Furnkranz. 2018. *A Review of Possible Effects of Cognitive Biases on Interpretation of Rule-based Machine Learning Models.* https://arxiv.org/abs/1804.02969v3.

Latonero, Mark. 2018. "Governing Artificial Intelligence: Upholding Human Rights and Dignity." *Data&Society.* https://datasociety.net/wp-content/uploads/2018/10/DataSociety_Governing_Artificial_Intelligence_Upholding_Human_Rights.pdf.

Llorens, Marisa. 2018. "Text Analytics Techniques in the Digital World: Word Embeddings and Bias." *Irish Communication Review* 16(1): 76–89.

Marcus, Gary. 2015. "Is 'Deep Learning' a Revolution in Artificial Intelligence?" *The New Yorker*, November 25, 2012. https://www.newyorker.com/news/news-desk/is-deep-learning-a-revolution-in-artificial-intelligence.

Marcus, Gary. 2018. *Deep Learning: A Critical Appraisal.* https://arxiv.org/abs/1801.00631.

Marcus, Gary. 2018. *Innateness, AlphaZero, and Artificial Intelligence.* https://arxiv.org/abs/1801.05667.

Metz, Cade. 2019. *We Teach A.I. Systems Everything, Including Our Biases.* https://www.nytimes.com/2019/11/11/technology/artificial-intelligence-bias.html.

Moor, James. 2006. "The Nature, Importance, and Difficulty of Machine Ethics." *IEEE Intelligent Systems* (IEEE) 21(4): 18–21.

Nilson, Nils. 2009. *The Quest for Artificial Intelligence: A History of Ideas and Achievements.* Cambridge University Press.

Noble, Safiya Umoja. 2018. *Algorithms of Oppression: How Search Engine Reinforce Racism.* New York University Press.

Odlyzko, A., Network Neutrality, Search Neutrality, and the Never-Ending Conflict between Efficiency and Fairness in Markets. *Review of Network Economics*, Vol. 8, No. 1, pp. 40–60, March 2009.

O'Neil, Cathy. 2016. *Weapons of Math Destruction.* New York: Crown Publishers.

Pavaloiu, Alice, and Utku Kose. 2017. "Ethical Artificial Intelligence – An Open Question." *Journal of Multidisciplinary Developments* 2(2): 15–27.

Pereira, Luis Moninz, and Ari Saptawijaya. 2016. *Programming Machine Ethics*. Springer.

Pinker, Stewen. 1998. *How the Mind Works*. Penguin Press.

Raso, Filippo, et al. 2018. *Artificial Intelligence and Human Rights: Opportunities and Risks*. Berkman Klein Center for Internet and Society at Harvard University. https://cyber.harvard.edu/sites/default/files/2018-09/2018-09_AIHumanRightsSmall.pdf?.

Rich, Elaine, Kevin Knight, and Shivashankar B. Nair. 2009. *Artificial Intelligence*. New Delhi: Tata McGraw-Hill.

Russell, Stuart, and Peter Norvig. 2010. *Artificial Intelligence: A Modern Approach*. Pearson.

Russell, Stuart, Daniel Dewey, and Max Tegmark. 2015. "Research Priorities for Robust and Beneficial Artificial Intelligence." *AI Magazine* 36(4): 105–114.

Select Committee on Artificial Intelligence. 2018. *AI in the UK: Ready, Willing and Able?* London: Authorithy of the House of the Lords. https://publications.parliament.uk/pa/ld201719/ldselect/ldai/100/100.pdf.

Tegmark, Max. 2014. *Our Mathematical Universe*. New York: Random House LLC.

The European Commission's High-Level Expert Group on Artificial Intelligence. 2019. *Ethics Guidelines for Trustworthy AI*. Brussels: European Commission. https://ec.europa.eu/digital-single-market/en/news/draft-ethics-guidelines-trustworthy-ai.

Trappl, Robert, ed. 2015. *A Construction Manual for Robots' Ethical Systems: Requirements, Methods, Implementations*. Springer.

Turing, Alan. 1950. "Computing Machinery and Intelligence." *Mind* 59(236): 433–460.

Tzafestas, Spyros G. 2016. *Roboethic: A Navigation Overview*. Springer.

United Nations. 2011. *Guiding Principles on Business and Human rights*. New York, Geneva: United Nations. https://digitallibrary.un.org/record/720245/.

United Nations. 2012. *Guiding Principles on Business and Human Rights*. New York and Geneva: United Nations. https://www.ohchr.org/documents/publications/GuidingprinciplesBusinesshr_eN.pdf.

Verma, Sahil. 2018. "Fairness Definitions Explained." *2018 ACM/IEEE International Workshop on Software Fairness* (Sweden): 1–7.

Wachter-Boettcher, Sara. 2018. *Technically Wrong: Sexist Apps, Biased Algorithms and other Threats of Toxic Tech*. W.W. Norton Company.

Wallach, Wendell, and Colin Allen. 2009. *Moral Machines: Teaching Robots Right from Wrong*. New York: Oxford University Press.

World Economic Forum. 2018. *How to Prevent Discriminatory Outcomes in Machine Learning*. World Economic Forum. http://www3.weforum.org/docs/WEF_40065_White_Paper_How_to_Prevent_Discriminatory_Outcomes_in_Machine_Learning.pdf.

Zhao, Jhao, et al. 2018. "Gender Bias in Coreference Resolution: Evaluation and Debiasing Methods." *Conference: Proceedings of the 2018 Conference of the North American Chapter of the Association for Computational Linguistics: Human Language Technologies* 2, (ACL): 15–20.

Index

Page numbers in *italic* refer to figures. Page numbers followed by a *t* refer to tables.

www.ingramcontent.com/pod-product-compliance
Ingram Content Group UK Ltd.
Pitfield, Milton Keynes, MK11 3LW, UK
UKHW022208051026
15624UKWH00001B/63

9 789004 436732